Cancer Genomics and Proteomics

METHODS IN MOLECULAR BIOLOGY™

John M. Walker, SERIES EDITOR

METHODS IN MOLECULAR BIOLOGY™

Cancer Genomics and Proteomics

Methods and Protocols

Edited by

Paul B. Fisher

Departments of Pathology, Neurosurgery, and Urology
Columbia University Medical Center
College of Physicians and Surgeons, New York, NY

HUMANA PRESS ✴ TOTOWA, NEW JERSEY

Production Editor: Jennifer Hackworth

Cover design by Karen Schulz

Cover illustration: Panel D from Figure 3 in chapter 13, appearing on page 180.

For additional copies, pricing for bulk purchases, and/or information about other Humana titles, contact Humana at the above address or at any of the following numbers: Tel.: 973-256-1699; Fax: 973-256-8341; E-mail: orders@humanapr.com; or visit our Website: www.humanapress.com

Printed in the United States of America. 10 9 8 7 6 5 4 3 2 1

ISSN 1064-3745

E-ISBN 1-59745-335-8

Library of Congress Control Number:

Preface

Recent advances in molecular biology and the complete sequencing of important genomes, including that of human, are now providing the crucial tools for developing a working understanding of the genes that are relevant to normal physiology and disease. However, even with this information approaches for defining differentially expressed genes and characterizing these genes, the proteins they encode and their mechanisms of action remain primary research objectives. The present volume provides a compendium of techniques and applications that will be of wide utility to researchers interested in gene identification and function. Reviews are written by experts in specific aspects of gene identification and full-length gene cloning, gene profiling (microarrays), chromatin modification of gene regulation, bacterial artificial chromosomes, cancer cytogenetic analyses, gene methylation, phage display, yeast and mammalian two-hybrid systems, RNA silencing, monoclonal antibody production, kinases and signal transduction, RNA-dependent protein kinase (PKR), analyses of mouse embryo fibroblasts, protein microarrays, and protein crystallization.

The first two chapters describe unique and efficient subtraction hybridization protocols for cloning differentially expressed genes. The chapter by Sarkar et al. provides the detailed methodology for performing reciprocal subtraction differential RNA display (RSDD), an approach demonstrating wide applications for identifying novel genes involved in cancer progression. The chapter by Boukerche et al. outlines an efficient cloning strategy, called rapid subtraction hybridization (RaSH), which has been applied to identify genes involved in tumor metastasis, HIV-1 resistance, and gene expression changes in astrocytes infected/treated with neurtropic agents. Perhaps one of the most frequently used approaches for differential gene identification is differential RNA display. Ken et al. overview this approach and describe its application for identifying genes differentially expressed in different cellular contexts and following different treatments. A widely used scheme for defining global patterns of gene expression changes is serial analysis of gene expression. Van Ruissen and Baas provide a detailed protocol for performing serial analysis of gene expression (SAGE). The next three chapters deal with the application of microarrays as gene profiling tools. Basso and Dalla-Favera and Chen et al. provide applications for using cDNA microarrays for studying gene expression changes in tumors. Ma et al. describe applications of microarrays for identifying retinoid target genes. Many, if not all, of the approaches described in chapters 1–7 result in the initial identification of partial gene sequences. For known genes, this may not represent a problem, because the appropriate full-length sequence can frequently be obtained by informatic approaches. However, for novel genes, obtaining a full-length clone embodying the protein-coding region is essential for further characterization. Kang and Fisher provide a detailed description of an approach that can be used to efficiently obtain complete coding sequences for genes in which only partial sequence information is initially available, a technique called complete open reading frame (C-ORF). Transcription factors bind to specific consensus sequences in DNA, thereby promoting gene transcription and expression. Dasgupta and Chellappan provide the technical details for performing chromatin immoprecipitation assays to define differential recruitment of transcription factors to DNA consensus sequences. Phelps et al. provide details for using bacterial artificial chromosomes to study the organization of entire genes, replicons, and other large genomic loci.

Cancer cells frequently exhibit nonrandom and complex chromosomal abnormalities and Rao et al. describe detailed methodologies for the three most widely utilized cytogenetic methods, comparative genomic hybridization, spectral karyotype, and fluorescence *in situ* hybridization, which are used to identify and study these changes in cancer cells. Gene expression is regulated at many levels, one of which is the status of methylation of the promoter region of the gene. Carraway and Herman describe in detail methods for performing bisulfite modification of DNA and polymerase chain reaction and nested methylation specific PCR to evaluate the status of DNA methylation in target genes. Protein–protein interactions are critical regulators of gene expression.

Brissette and Goldstein describe phage display peptide library approaches for identifying short peptides or single chain antibodies capable of regulating target protein function. Matuszawa and Reed describe the yeast and mammalian two-hybrid system that is frequently employed to investigate protein–protein interactions in target cells. Baroni et al. describe unique technologies for defining the binding of specific mRNA subsets to RNA-binding proteins and the application of short hairpin RNA technology to investigate gene "knock-down" in the context of tumor dormancy.

Goldstein and Fisher describe an efficient approach for developing monoclonal antibodies targeting cell surface-expressed molecules, a technique called surface-epitope masking (SEM). The next four chapters deal with applications permitting investigation of signal transduction changes in cells, analysis of PKR functions and methods for studying gene regulation using mouse embryo fibroblasts. Protein kinases represent important components of cell signaling. Dent et al. describe methods for performing immune complex protein kinase assays and immunoblotting using phosphoro-specific antibodies against regulatory sites of phosphorylation in protein kinases. PKR is a pivotal antiviral protein found in most cells, which when activated potently inhibits both cellular and viral protein synthesis. In this context, PKR is important in innate immunity, cancer, and viral oncolysis. Balachandran and Barber describe methods for evaluating this important protein kinase.

Mouse embryo fibroblasts are significant cellular reagents for studying many important biological processes. By using mouse embryo fibroblasts that contain specific genes that are "knocked-out" by homologous recombination, it is possible to analyze the role of specific target genes in cellular physiology in comparison with wild type cells. Sun and Taneja and Sun et al. describe detailed methodologies for using these valuable resources to study transformation and tumorigenesis and cell cycle. Protein lysate arrays represent helpful tools for profiling the protein molecular pathways in a given cellular (patient) sample. Espina et al. describe the application of reverse phase protein microarrays for monitoring biologic responses. Defining protein structure offers significant potential for understanding biological function and for developing targeted therapeutic approaches. The final chapter in this volume by Deivanayagam et al. describes techniques used for protein crystallization.

The present volume will be of interest to a wide audience, including molecular biologists, geneticists, and biochemists involved in studying genes associated with and regulating important physiological processes, including cancer, growth control, senescence, response to chemotherapy, development, neurodegeneration, infectious diseases, autoimmunity, and apoptosis. The approaches that are described in detail are state-of-art and can be tailored to individual ongoing or planned research projects. In these contexts, this work will serve as a valuable laboratory resource for designing experiments to identify and analyze genes that are relevant to complex biological phenomena.

Paul B. Fisher, MPH, PhD

Contents

Contributors

JULIO A. AGUIRRE-GHISO • *Gen*NY*Sis Center for Excellence in Cancer Genomics and Department of Biomedical Sciences, School of Public Health, University of Albany, SUNY, Rensselaer, NY*

NICHOLE BAYER ARNOLD • *Department of Medicine, Dartmouth Medical School, Hanover, NH*

KATIA BASSO • *Institute of Cancer Genetics, Herbert Irving Comprehensive Cancer Center, Columbia University Medical Center, College of Physicians and Surgeons, New York, NY*

SIDDHARTH BALACHANDRAN • *University of Miami School of Medicine, Miami, FL*

GLEN N. BARBER • *Department of Microbiology and Immunology, Sylvester Comprehensive Cancer Center, University of Miami School of Medicine, Miami, FL*

TIMOTHY E. BARONI • *Gen*NY*Sis Center for Excellence in Cancer Genomics and Department of Biomedical Sciences, School of Public Health, University of Albany, SUNY, Rensselaer, NY*

FRANK BAAS • *Neurogenetics Laboratory, Academic Medical Center, Amsterdam, The Netherlands*

HABIB BOUKERCHE • *Department of Pathology, Columbia University Medical Center, College of Physicians and Surgeons, New York, NY, and INSERM, Lyon, France*

RENEE BRISSETTE • *Antyra, Inc., Edison, NJ*

VALERIE S. CALVERT • *Center for Applied Proteomics and Molecular Medicine, George Mason University, Manassas, VA*

HETTY CARRAWAY • *Department of Oncology, Sidney Kimmel Comprehensive Cancer Center at Johns Hopkins, Baltimore, MD*

KEN CHIEN-NENG CHANG • *Osteoporosis Research, Women's Health Research Institute, Wyeth Research, Collegeville, PA*

SRIKUMAR CHELLAPPAN • *Drug Discovery Program, Department of Interdisciplinary Oncology, H. Lee Moffitt Cancer Center and Research Institute, Tampa, FL*

ZHONG CHEN • *Head and Neck Surgery Branch, National Institute on Deafness and Other Communication Disorders, National Institutes of Health, Bethesda, MD*

DOUGLAS S. CONKLN • *Gen*NY*Sis Center for Excellence in Cancer Genomics and Department of Biomedical Sciences, School of Public Health, University of Albany, SUNY, Rensselaer, NY*

WILLIAM J. COOK • *Department of Pathology, University of Alabama at Birmingham, Birmingham, AL*

RICCARDO DALLA-FAVERA • *Institute for Cancer Genetics and Herbert Irving Comprehensive Cancer Center, Columbia University Medical Center, College of Physicians and Surgeons, New York, NY*

PIYALI DASGUPTA • *Department of Interdisciplinary Oncology, H. Lee Moffitt Cancer Center, Tampa, FL*

PAUL DENT • *Department of Biochemistry, Virginia Commonwealth University, Richmond, VA*

CHAMPION DEIVANAYAGAM • *Center for Biophysical Sciences and Engineering, Department of Vision Sciences, University of Alabama at Birmingham, Birmingham, AL*

ETHAN DMITROVSKY • *Department of Pharmacology and Toxicology, Department of Medicine, Norris Cotton Cancer Center, Dartmouth Medical School, Hanover NH, and Dartmouth-Hitchcock Medical Center, Lebanon, NH*

GANG DONG • *National Institute of Allergy and Infectious Disease, National Institutes of Health, Bethesda, MD*

VIRGINIA ESPINA • *Center for Applied Proteomics and Molecular Medicine, George Mason University, Manassas, VA*

QING FENG • *Department of Pharmacology and Toxicology, Dartmouth Medical School, Hanover NH, and Dartmouth-Hitchcock Medical Center, Lebanon, NH*

PAUL B. FISHER • *Herbert Irving Comprehensive Cancer Center, Departments of Pathology, Neurosurgery and Urology, College of Physicians and Surgeons, Columbia University Medical Center, New York, NY*

NEIL I. GOLDSTEIN • *Antyra, Inc., Edison, NJ*

STEVEN GRANT • *Department of Medicine, Virginia Commonwealth University, Richmond, VA*

NERIMAN TUBA GULBAGCI • *Brookdale Department of Molecular, Cell and Developmental Biology, Mount Sinai School of Medicine, New York, NY*

NICHOLAS H. HEINTZ • *Department of Pathology, University of Vermont, College of Medicine, Burlington, VT*

JAMES HERMAN • *Sidney Kimmel Comprehensive Cancer Center at Johns Hopkins, Baltimore, MD*

PHILIP B. HYLEMON • *Department of Microbiology and Immunology, Virginia Commonwealth University, Richmond, VA*

SHARON ILLENYE • *Department of Pathology, University of Vermont, College of Medicine, Burlington, VT*

DONG-CHUL KANG • *Hallym University, Ilsong Institute of Life Science, Anyang, Kyeonggi-do, Republic of Korea*

SUTISAK KITAREEWAN • *Department of Pharmacology and Toxicology, Dartmouth Medical School, Hanover NH, and Dartmouth-Hitchcock Medical Center, Lebanon, NH*

BARRY KOMM • *Osteoporosis Research, Women's Health Research Institute, Wyeth Research, Collegeville, PA*

MURRAY KORC • *Department of Medicine, Dartmouth Hitchcock Medical Center and Dartmouth Medical School, Hanover, NH*

MICHELE T. LASTRO • *Gen*NY*Sis Center for Excellence in Cancer Genomics and Department of Biomedical Sciences, School of Public Health, University of Albany, SUNY, Rensselaer, NY*

TIN-LAP LEE • *Head and Neck Surgery Branch, National Institute on Deafness and Other Communication Disorders, National Institutes of Health, Bethesda, MD*

LANCE A. LIOTTA • *Center for Applied Proteomics and Molecular Medicine, George Mason University, Manassas, VA*

AMY LOERCHER • *Head and Neck Surgery Branch, National Institute on Deafness and Other Communication Disorders, National Institutes of Health, Bethesda, MD*

YAN MA • *Department of Pharmacology and Toxicology, Dartmouth Medical School, Hanover NH, and Dartmouth-Hitchcock Medical Center, Lebanon, NH*

SHU-ICHI MATSUZAWA • *Burnham Institute for Medical Research, La Jolla, CA*

VUNDAVALLI V. MURTY • *Department of Pathology, Columbia University Medical Center, College of Physicians and Surgeons, New York, NY*

SUBHADRA V. NANDULA • *Department of Pathology, Columbia University Medical Center, College of Physicians and Surgeons, New York, NY*

EMANUEL F. PETRICOIN III • *Center for Applied Proteomics and Molecular Medicine, George Mason University, Manassas, VA*

STEPHANIE F. PHELPS • *Microbiology and Molecular Genetics, University of Vermont, College of Medicine, Burlington, VT*

APARNA C. RANGANATHAN • *Gen*NY*Sis Center for Excellence in Cancer Genomics and Department of Biomedical Sciences, School of Public Health, University of Albany, SUNY, Rensselaer, NY*

JOHN C. REED • *Burnham Institute for Medical Research, La Jolla, CA*

IAN PITHA-ROWE • *Department of Pharmacology and Toxicology, Dartmouth Medical School, Hanover NH, and Dartmouth-Hitchcock Medical Center, Lebanon, NH*

PULIVARTHI H. RAO • *Baylor College of Medicine, Texas Children's Cancer Center, Houston, TX*

FRED VAN RUISSEN • *Neurogenetics Laboratory, Academic Medical Center, Amsterdam, The Netherlands*

DEVANAND SARKAR • *Department of Urology, Columbia University Medical Center, College of Physicians and Surgeons, New York, NY*

ZAO-ZHONG SU • *Department of Urology, Columbia University Medical Center, College of Physicians and Surgeons, New, York, NY*

HONG SUN • *Brookdale Department of Molecular, Cell and Developmental Biology, Mount Sinai School of Medicine, New York, NY*

RESHMA TANEJA • *Brookdale Department of Molecular, Cell and Developmental Biology, Mount Sinai School of Medicine, New York, NY*

SCOTT A. TANENBAUM • *Gen*NY*Sis Center for Excellence in Cancer Genomics and Department of Biomedical Sciences, School of Public Health, University of Albany, SUNY, Rensselaer, NY*

CARTER VAN WAES • *Tumor Biology Section, Head and Neck Surgery Branch, National Institute on Deafness and Other Communication Disorders, National Institutes of Health, Bethesda, MD*

MARK R. WALTER • *University of Alabama in Birmingham-CBSE, Department of Microbiology, Birmingham, AL*

JULIA D. WULFKUHLE • *Center for Applied Proteomics and Molecular Medicine, George Mason University, Manassas, VA*

XIN-PING YANG • *Head and Neck Surgery Branch, National Institute on Deafness and Other Communication Disorders, National Institutes of Health, Bethesda, MD*

1

Reciprocal Subtraction Differential RNA Display (RSDD)

An Efficient Technology for Cloning Differentially Expressed Genes

Devanand Sarkar, Dong-chul Kang, and Paul B. Fisher

Summary

Identification of differentially expressed genes is an essential step in comprehending the molecular basis of complex physiological and pathological processes. Subtraction hybridization and differential RNA display (DDRT-PCR) are two methods that are widely and successfully employed to clone differentially expressed genes. Unfortunately, both methods have inherent problems and limitations requiring improvements in the technique. A combination of these two methods termed reciprocal subtraction differential RNA display is described here that considerably reduces the complexity of DDRT-PCR and facilitates the rapid and efficient identification and cloning of both abundant and rare differentially expressed genes.

Key Words: Cloning; differential RNA display; reciprocal subtraction differential RNA display; RSDD; subtraction hybridization; tumor progression; gene expression.

1. Introduction

An in-depth understanding of the regulation of gene expression, at transcriptional, processing (RNA), and translational (proteins) levels is mandatory to obtain a thorough picture of the multitude of independent, parallel, converging, and/or diverging gene expression changes underlying even simple biological processes. Accordingly, the identification, cloning, and characterization of differentially expressed genes holds promise for elucidating the molecular determinants of intricate processes such as growth, development, aging, differentiation, and maladies like cancer and autoimmune diseases, thereby resulting in the development of enhanced preventive and interventional medicines.

A number of different strategies have been developed to identify and clone differentially expressed mRNAs *(1)*, among which the differential RNA display (DDRT-PCR) approach developed by Liang and Pardee has gained wide

From: *Methods in Molecular Biology, vol. 383: Cancer Genomics and Proteomics: Methods and Protocols*
Edited by: P. B. Fisher © Humana Press Inc., Totowa, NJ

popularity in analyzing and cloning differentially expressed genes *(2–6)*. DDRT-PCR is a powerful methodology in which a vast number of mRNA species (>20,000, if no redundancy occurs) can be analyzed with only a small quantity of RNA (~5 μg) *(2)*. The general strategy of this method is to amplify partial cDNA sequences from subsets of mRNAs by reverse transcription and PCR. These short sequences are then displayed on a sequencing gel. Pairs of primers are designed so that each will amplify cDNA from approx 50 to 100 mRNAs, because this number is optimal for display on the gel.

The 3′-primers are designed to anchor at the poly(A) tail plus one additional 3′-base *(3,6)*. A primer such as $5′-T_{11}C$ would allow anchored annealing to mRNAs containing G located just upstream of their poly(A) tails. The primer allows initiation of reverse transcription of only this subpopulation and therefore three RT reactions for each sample need to be performed using the primers $5′-T_{11}C$, $5′-T_{11}G$, and $5′-T_{11}A$. Total RNAs or mRNAs are extracted from two experimental groups and reverse transcribed with the 3′-primer followed by PCR in the presence of different 5′-primers and $[\alpha-^{32}P]$-dATP. Because PCR is performed using short primers, a very low annealing temperature (40°C) is used and to achieve a high level of mismatched synthesis, a low dNTP concentration (2–5 μ*M*) is employed. The PCR products from each group are run on a DNA sequencing gel. The gel is exposed to an autoradiogram and the PCR products are displayed at approx 50 to 100 bands ranging from 100 to 500 bp. The intensities of the bands in the two lanes are compared and those showing differential levels are cut out of the gel and eluted. The recovered DNA is amplified by PCR, cloned, sequenced, and differential expression is confirmed by Northern blot analysis.

DDRT-PCR is often the method of choice when the RNA source is limiting, such as tissue biopsies. A direct advantage of DDRT-PCR is the ability to identify and isolate both up- and downregulated differentially expressed genes in the same reaction. Furthermore, the DDRT-PCR technique permits the display of multiple samples in the same gel, which is useful in defining specific diagnostic alterations in RNA species and for temporally analyzing gene expression changes. However, difficulties are encountered while performing standard DDRT-PCR that include a high incidence of false-positives and redundant gene identification, poor reproducibility, biased gene display, and lack of functional information about the cloned cDNA *(7,8)*. Furthermore, poor separation can mask differentially expressed genes of low abundance under the intense signals generated by highly expressed genes. The generation of false-positives and redundancy can be highly problematic, resulting in an inordinate expenditure of resources to confirm appropriate differential expression and uniqueness of the isolated cDNAs. The cDNAs must be isolated from the gels in pure form (contamination of bands with multiple sequences complicates clone identification), reamplified, placed in an appropriate cloning vector, analyzed for authentic differential expression and finally sequenced. These limitations of the standard DDRT-PCR approaches emphasize the need for improvements in this procedure to more efficiently and selectively identify differentially expressed genes.

Subtraction hybridization, in which hybridization between tester and driver is followed by selective removal of common gene products, enriches for unique gene products in the tester cDNA population and reduces the abundance of common cDNAs *(9)*. A subtracted cDNA library can be analyzed to identify and clone differentially expressed genes by randomly picking colonies or by differential screening *(10–12)*. Although subtraction hybridization has been used successfully to clone a number of differentially expressed genes, this approach is labor-intensive and does not result in isolation of the full spectrum of genes displaying altered expression *(9,13)*.

In principle, DDRT-PCR performed with subtracted RNA or cDNA samples should provide a powerful strategy to clone up- and down-regulated gene products. This approach should combine the benefits of both techniques, resulting in the enrichment of unique sequences, and a reduction or elimination of common sequences. This scheme also should result in a consistent reduction in band complexity on a display gel, thereby permitting a clearer separation of cDNAs, resulting in fewer false-positive reactions. Additionally, it should be possible to use fewer primer sets for reverse transcription and PCR reactions to analyze the complete spectrum of differentially expressed genes. Of particular importance for gene identification and isolation, rare gene products that are masked by strong common gene products should be displayed by using subtraction hybridization in combination with DDRT-PCR. In addition, the DDRT-PCR approach with subtraction libraries also could prove valuable for efficiently screening subtracted cDNA libraries for differentially expressed genes. Reciprocal subtraction differential RNA display (RSDD) approach combines the virtues of DDRT-PCR and subtraction hybridization and efficiently and consistently reduces the complexity of DDRT-PCR and results in the identification and cloning of genes displaying anticipated differential expression *(14)*.

2. Materials

2.1. Subtraction Hybridization

1. Qiagen RNeasy RNA extraction kit (Qiagen, Hilden, Germany): to be used according to the manufacturer's instruction.
2. Oligo(dT) cellulose for purification of poly(A)$^+$ RNA (Invitrogen, Carlsbad, CA): to be used according to the manufacturer's instruction.
3. λ-ZAP cDNA library kit (Stratagene, La Jolla, CA): to be used according to the manufacturer's instruction.
4. DNA purification column (Qiagen).

5. Adapter oligonucleotides obtained from Operon Technologies (Alameda, CA).
6. Restriction nucleases (*Eco*RI and *Xho*I; New England Biolabs, Beverly, MA): to be kept on ice.
7. Luria Bertani (LB) medium and plates.
8. Antibiotics (amplicillin and kanamycin, Sigma, St. Louis, MO).
9. Centricon 100 filters (Millipore, Billerica, MA).
10. Chemicals: Tris-HCl, ethylenediaminetetraacetic acid (EDTA), NaCl, Na-acetate, sodium dodecyl sulfate (SDS), HEPES, formamide, phenol, choloroform, HCl, butanol, ethanol, photoactivatable biotin, and streptavidin (Sigma).

2.2. Reciprocal Subtraction Differential RNA Display

1. Reagents for PCR including Taq DNA polymerase and PCR buffer (Invitrogen).
2. dNTP (Amersham, Piscataway, NJ).
3. Primers (Operon).
4. α-^{35}S-dATP (Amersham).
5. Reagents for making 5% sequencing gel: urea (40.4 g); 10X Tris-borate acetate (8 mL); long ranger gel solution (50%; 8 mL); ammonium persulfate (70 mg); distilled water (35.8 mL); TEMED 25 µL (obtained from Sigma except Long Ranger which is form Cambrex [San Diego, CA]).

2.3. Reverse Northern and Northern Blotting

1. Nylon membrane (Nytran, Schleicher & Schuell, Keene, NH).
2. α-^{32}P-dCTP (Amersham).
3. Superscript II (Invitrogen).
4. RNase H (Invitrogen).
5. Rediprime labeling kit (Amersham).

3. Methods

3.1. RNA Isolation and cDNA Library Construction

1. Total RNA is isolated from tester and driver using Qiagen RNeasy kit (Qiagen) and Poly(A)$^+$ RNA is purified with oligo(dT) cellulose chromatography *(15)*. Two λ-ZAP cDNA libraries are constructed from mRNAs from tester and driver using λ-ZAP cDNA library kit (Stratagene) *(16)*.
2. A primer-adapter consisting of oligo(dT) next to a unique restriction site (*Xho*I) is used for first strand synthesis. The double-stranded cDNAs are ligated to *Eco*RI adapters and then digested with the *Xho*I restriction endonuclease. The resultant *Eco*RI and *Xho*I cohesive ends allow the finished cDNAs to be inserted into λ ZAPII vector in a sense orientation with respect to the lacZ promoter. The λ ZAPII vector contains pBluescript plasmid sequences flanked by bacteriophage-derived f1 sequences that facilitate in vivo conversion of the recombinant phage clones into the phagemid *(17)*.
3. The tester and driver cDNA libraries thus created are packaged with Gigapack II Gold Packaging Extract (Stratagene) and amplified on PLK-F' bacterial cells.

3.2. Preparation of Double-Stranded DNA From Tester Library

1. The tester cDNA phagemid library is excised from λ ZAP using the mass excision procedure as described by Stratagene *(18)*. 1×10^7 pfu of tester cDNA library is mixed with 2×10^8 XL-1 blue strain of *Escherichia coli* and 2×10^8 pfu of ExAssist helper phage in 10 m*M* MgSO$_4$, followed by absorption at 37°C for 15 min *(19)*.

2. After the addition of 10 mL LB medium, the phage/bacteria mixture is incubated with shaking at 37°C for 2h, followed by incubation at 70°C for 20 min to heat inactivate the bacteria and the λ ZAP phage particles.

3. After centrifugation at 4000*g* for 15 min, the supernatant is transferred to a sterile polystyrene tube and stored at 4°C before use. To produce double-stranded DNA, 5×10^7 pfu of the phagemids is combined with 1×10^9 SOLR strain of *E. coli*, which are nonpermissive for the growth of helper phages and therefore prevent coinfection by the helper phages, in 10 m*M* MgSO$_4$, followed by absorption at 37°C for 15 min *(19)*.

4. The phagemids/bacteria are transferred to 250 mL LB medium containing 50 µg/mL ampicillin and incubated with shaking at 37°C overnight. The bacteria are harvested by centrifugation and the double-stranded phagemid DNA is isolated by the alkali-lysis method and purified through a Qiagen-tip 500 column (Qiagen).

3.3. Preparation of Single-Stranded DNA From Driver Library

1. The driver cDNA library is excised from λ ZAP using the mass excision procedure described in **Subheading 3.2.**, **step 1**. The phagemids (5×10^7) are combined with 1×10^9 XL-1 Blue in 10 m*M* MgSO$_4$, followed by absorption at 37°C for 15 min. The phagemids/ bacteria are transferred to 250 mL LB medium and incubated with shaking at 37°C for 2 h.

2. Helper phage VCS M13 (Stratagene) is added to 2×10^7 pfu/mL and after incubation for 1 h, kanamycin sulfate is added to 70 µg/mL. The bacteria are grown overnight, the phagemids are harvested and single-stranded DNAs are prepared using standard protocols *(20)*.

3.4. Pretreatment of Double- and Single-Stranded DNA Before Hybridization

1. To excise the inserts from the vector, double-stranded DNA from the tester cDNA library is digested with *Eco*RI and *Xho*I and extracted with phenol and chloroform, followed by ethanol precipitation. After centrifugation, the pellet is resuspended in distilled water. Single-stranded DNA from driver cDNA library is biotinylated using photoactivatable biotin (Photobiotin, Sigma) *(21)*.

2. In a 650-µL microcentrifuge tube, 50 µL of 1 µg/µL single-stranded DNA is mixed with 50 µL of 1 µg/µL photoactivatable biotin in water. The solution is irradiated with the tube slanted on crushed ice at a distance of 10 cm from a 300-watt sunlamp for 15 min.

3. The DNA is further biotinylated by adding 25 µL of photoactivatable biotin to the solution, which is then exposed to an additional 15 min of irradiation as described

earlier. To remove unlinked biotin, the reaction is diluted to 200 µL with 100 m*M* Tris-HCl, 1 m*M* EDTA, pH 9.0, and extracted 3X with 2-butanol. Sodium acetate, pH 6.5 is added to a concentration of 0.3 *M* and the biotinylated DNA is precipitated with two volumes of ethanol.

3.5. Subtraction Hybridization and Construction of Subtracted cDNA Libraries

1. In a 650-µL siliconized microcentrifuge tube, 400 ng of *Eco*RI- and *Xho*I-digested tester cDNA library and 12 µg of biotinylated driver cDNA library is mixed in 20 µL of 0.5 *M* NaCl, 0.05 *M* HEPES, pH 7.6, 0.2% SDS and 40% deionized formamide *(22)*. The mixture is boiled for 5 min and incubated at 42°C for 48 h. The hybridization mixture is diluted to 400 µL with 0.5 *M* NaCl, 10 m*M* Tris-HCl, pH 8.0, 1 m*M* EDTA and then 15 µg of streptavidin in water is added, followed by incubation at room temperature for 5 min.
2. The sample is extracted twice with phenol/chloroform (1:1), followed by back-extraction of the organic phase with 50 µL of 0.5 *M* NaCl in TE buffer, pH 8.0. An additional 10 µg of streptavidin is added and phenol/chloroform extraction is repeated. After removal of excess chloroform by brief lyophilization, the final solution is diluted to 2 mL with TE, pH 8.0 and passed through a Centricon 100 filter (Amicon, Beverly, MA) twice.
3. The concentrated DNA solution (~50 µL) is then lyophilized. The subtracted cDNAs are ligated to *Eco*RI- and *Xho*I-digested and calf intestinal alkaline phosphatase-treated arms of the λ ZAPII vector and packaged with Gigapack II Gold packaging extract.
4. The library is then amplified using the PLK-F' bacterial cells. Based on the methods described in **Subheading 3.5., step 1**, reciprocal subtraction between driver and tester is performed and two subtracted cDNA libraries (tester minus driver and driver minus tester) are constructed. Plasmid cDNA libraries from the subtracted λ-ZAP cDNA libraries are obtained by in vivo excision as previously described and the plasmids are isolated using Qiagen columns.

3.6. RSDD Methodology

1. The purified plasmids of reciprocally subtracted cDNA libraries are subjected directly to differential display with minor modifications *(6,14)*. The plasmids of reciprocally subtracted cDNA libraries are PCR-amplified with the combination of three single-anchor 3′ primers ($T_{13}A$, $T_{13}C$, or $T_{13}G$) and 18 arbitrary 5′ 10-mer primers obtained from Operon Technologies (Alameda, CA; OPA1-20 except OPA1 and 3). The 20-µL PCR reaction consists of 10 m*M* Tris-HCl (pH 8.4), 50 m*M* KCl, 1.5 m*M* MgCl$_2$, 2 µ*M* dNTP, 0.2 µ*M* 5′ arbitrary primer, 1 µ*M* 3′ anchor primer, 50 ng of plasmid of a subtracted library, 10 µCi α-^{35}S-dATP (3000 Ci/mmol from Amersham Biosciences) and 1 U of *Taq* DNA polymerase (Invitrogen). The parameters of PCR are 30 s at 95°C, 40 cycles of 30 s at 95°C, 2 min at 40°C and 30 s at 72°C and an additional 5 min at 72°C. After the cycling, 10 µL of 95% formamide, 0.05% bromophenol blue and 0.05% xylene cyanol are added to each PCR reaction (*see* **Notes 1–6**).

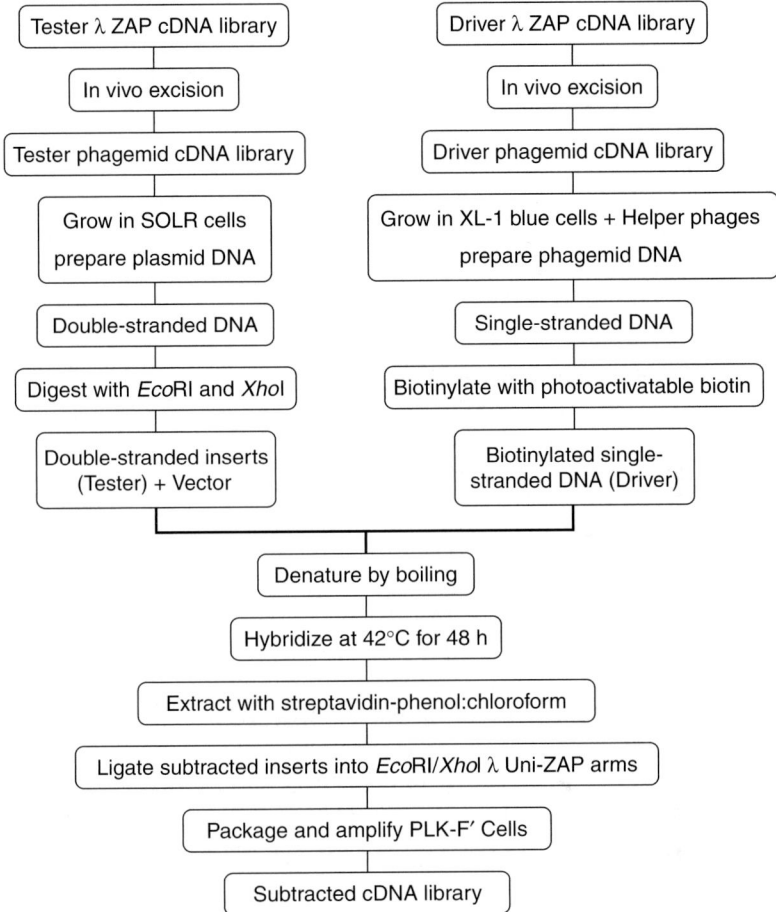

Fig. 1. Flowchart for constructing a subtracted cDNA library. Please refer to the text for the details of the methods.

2. The mixture is heated at 95°C for 2 min and is separated in a 5% denaturing DNA sequencing gel maintained at 50°C. PCR reactions of plasmids from each subtracted library in a primer set are run side-by-side. Differentially amplified bands from plasmids of each subtracted library are marked with an 18-G needle through the film and are cut out with a razor.

3. The gel slice is put in 100 μL of 10 m*M* Tris and 1 m*M* EDTA (TE, pH 8.0) and is incubated at 4°C overnight. After the incubation, the mixture is boiled for 5 min and microcentrifuged for 2 min. The supernatant is collected and stored at −20°C until reamplification. The band extract is reamplified with the same cycling parameters in a 50-μL reaction consisting of 10 m*M* Tris-HCl (pH 8.4), 50 m*M* KCl, 1.5 m*M* MgCl$_2$, 20 μ*M* dNTP, 0.2 μ*M* 5′ arbitrary primer, 1 μ*M* 3′ anchor primer, 5 μL of band extract and 2.5 U of *Taq* DNA polymerase (**Figs. 1** and **2**).

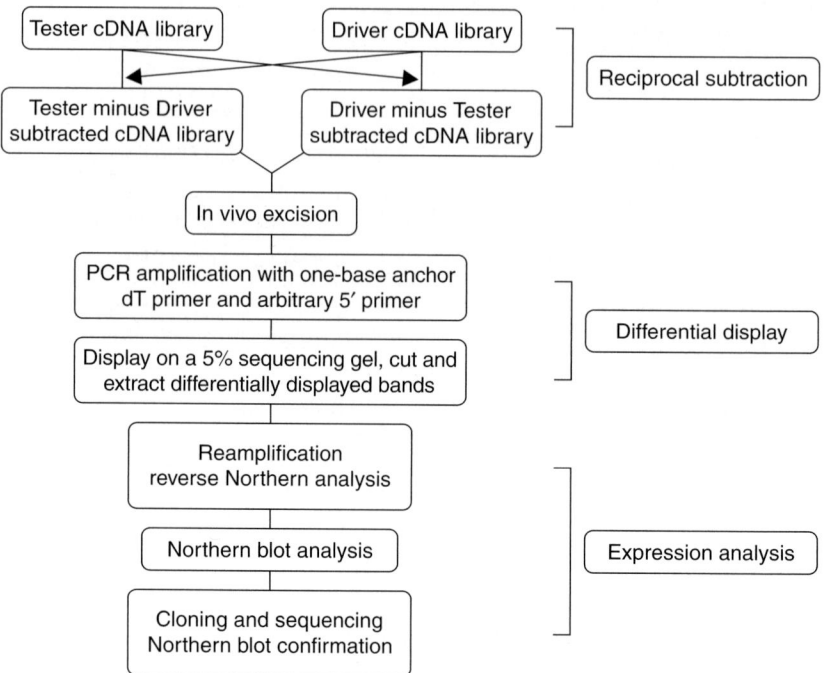

Fig. 2. Schematic outline of RSDD protocol. Please refer to the text for the details of the methods.

3.7. Reverse Northern Blotting

1. Differential expression of the reamplified DNA fragment is scrutinized by reverse Northern analysis and Northern blot analysis *(14)*. In reverse Northern analysis, after confirmation in a 1% agarose gel, the reamplified DNA fragment (10 μL of PCR reaction) is mixed with 90 μL of TE and is spotted on a positively charged nylon membrane with a 96-well vacuum manifold. The membrane is soaked with denaturing and neutralizing solution successively and the spotted DNA is crosslinked to the membrane with a ultraviolet crosslinker (Stratagene).

2. ^{32}P-labeled first strand cDNA is prepared by reverse transcription of total RNA. After heating at 70°C for 10 min and quenching on ice for 2 min, 0.4 μ*M* each $T_{13}A$, $T_{13}C$, and $T_{13}G$, and 10 μg total RNA mixture is added with 50 m*M* Tris-HCl (pH 8.3), 75 m*M* KCl, 3 m*M* $MgCl_2$, 10 m*M* DTT, 0.5 m*M* dATP, 0.5 m*M* dGTP, 0.5 m*M* dTTP, 0.02 m*M* dCTP, 0.5 μL of RNase inhibitor (Invitrogen), 100 μCi α-^{32}P-dCTP (3000 Ci/mmol from Amersham Biosciences) and 200 U of Superscript RT II (Invitrogen) in a final 25-μL reaction.

3. The reaction mixture is incubated at 42°C for 1 h and at 37°C for 30 min after addition of 2 μL of RNase H (10 U, Invitrogen). The membrane is hybridized at 42°C overnight in a 50% formamide hybridization solution. The hybridized membrane is washed at room temperature for 15 min with 2X standard saline citrate

(SSC) containing 0.1% SDS twice and at 55°C for at least 1 h with 0.1X SSC containing 0.1% SDS, successively. The membrane is probed with the ^{32}P-labeled cDNA of driver, stripped off and is probed with the ^{32}P-labeled cDNA of tester.

4. The signal intensity of each spot is normalized against that of glyceraldehydes-3-phosphate dehydrogenase and is compared between driver and tester. Reamplified DNA fragments displaying differential expression levels ≥1.8-fold higher between the two groups are selected and analyzed by Northern blot analysis.

3.8. Northern Blotting Analysis

1. In Northern blot analysis, 10 µg of total RNA from driver and tester are run side-by-side in a 1% agarose gel with formaldehyde and are transferred to a positively charged Nylon membrane *(14)*.
2. Reamplification reaction (5 µL) is ^{32}P-labeled with a Rediprime labeling kit (Amersham) and is used to probe the membrane as described in **Subheading 3.7., step 3**. DNA fragments expressed differentially between tester and driver in Northern blot analyses are cloned into the *Eco*RV site of the pZero-3.1 cloning vector (Invitrogen) and sequenced.

4. Notes

1. A representation of the differential RNA display pattern obtained from standard DDRT-PCR and RSDD is shown in **Fig. 3** *(14)*. The differential RNA display pattern of RSDD is significantly less complex than that of DDRT-PCR. These experiments demonstrate that subtraction hybridization before differential RNA display is effective in simplifying display patterns, permitting the efficient identification of differentially expressed cDNAs. Because RSDD substantially reduced the number of bands displayed, single anchor oligo dT primers, which can increase band numbers, were used successfully in subsequent amplifications of the RSDD approach (**Fig. 3**, right panel).
2. Using RSDD between an adenovirus transformed rat embryo cell line E11 and its more aggressive oncogenic progression variant E11-NMT *(23,24)*, 234 differentially displayed cDNAs in this tumor progression model system was identified *(14)*. Gel-extracted cDNA fragments were reamplified, dot-blotted on nylon membrane and successively probed with reverse transcribed ^{32}P-cDNA from E11 or E11-NMT RNAs. A representation of the reverse Northern blot is shown in **Fig. 4** that clearly demonstrates certain transcripts being upregulated in either E11 or E11-NMT cells. The genes that were upregulated in the progressed E11-NMT cells were termed progression elevated genes (PEGen) and those that were downregulated in E11-NMT cells were termed progression suppressed genes (PSGen) *(14)*. **Figure 5** shows a representative Northern blot analysis of a number of differentially identified genes that clearly demonstrate the successful application of RSDD approach to identify differentially expressed genes. With three single anchored oligo dT primers and 18 arbitrary 5′ primers, 72 bands were identified that displayed differential expression by using reverse Northern blot analysis. Thirty eight cDNA species were analyzed by Northern blot analysis and 31 (~82%) displayed differential expression

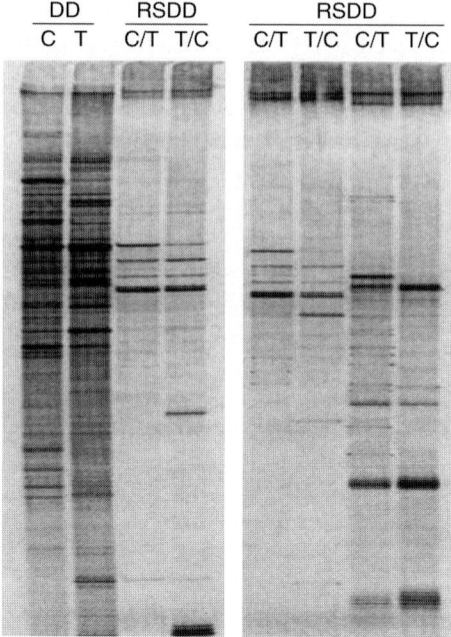

Fig. 3. Identification of differentially expressed sequence tags by RSDD. (Left) Differential RNA display pattern of conventional DDRT-PCR with RNA from Driver (C) and Tester (T) and an RSDD analysis of reciprocally subtracted Driver minus Tester (C/T) and Tester minus Driver (T/C) cDNA libraries. (Right) Representative RSDD patterns using different sets of primers. (From Kang et al. *[14]*.)

in E11 against E11-NMT cells. Sequence analysis of the cloned cDNA fragments revealed 16 different genes, including 11 novel genes not reported in recent DNA databases. All these data point to the efficient and specific aspects of differential gene identification by RSDD.

3. The combination of subtraction hybridization and DDRT-PCR in RSDD clearly provides improved results over both methods performed alone. RSDD significantly reduces band complexity of DDRT-PCR, which often obscures the identification of differentially expressed genes and generates false-positive signals. Another important aspect of successful application of RSDD is to use an appropriate subtraction hybridization protocol. The subtraction hybridization protocol described here involves a single round of reciprocal subtraction without intermediate amplification. Other subtraction hybridization procedures that employ multiple rounds of PCR *(25)* might not be efficient for this method because of PCR-bias during amplicon generation resulting in reduced or inefficient subtraction and also the generation of PCR artifacts.

4. It should be noted that RSDD and DDRT-PCR do not resolve the same differentially expressed bands. Unique bands identified in DDRT-PCR that are expressed

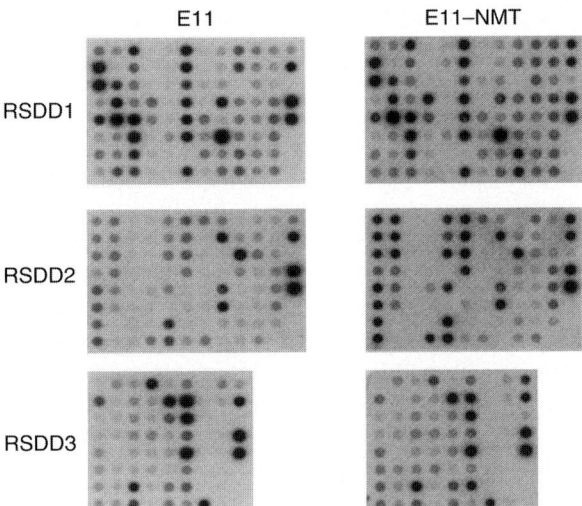

Fig. 4. Reverse Northern blot analysis of differentially expressed sequence tags identified by RSDD. Differentially expressed sequence tags obtained from RSDD were dot-blotted onto Nylon membrane and were probed with ^{32}P-cDNA reverse transcribed from RNA samples of E11 and E11-NMT cells. (From Kang et al. *[14].*)

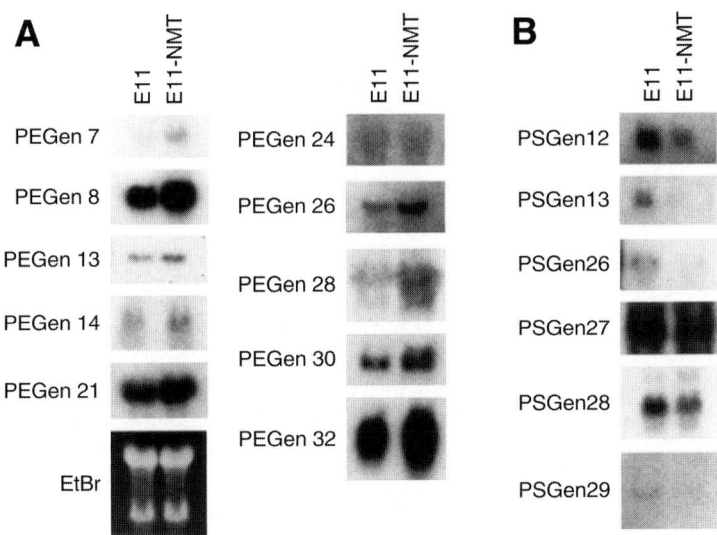

Fig. 5. Differential expression of representative PEGen and PSGen genes identified by RSDD and reverse Northern blotting. Northern blots of E11 and E11-NMT RNA samples were probed with radiolabeled expressed sequence tags identified by RSDD and reverse Northern blotting. Equal loading of E11 and E11-NMT RNA is demonstrated by ethidium bromide staining. (From Kang et al. *[14].*)

differentially when analyzed by Northern blotting are not the same as those found by using RSDD and vice versa *(14)*. Moreover, random isolation and analysis of differentially expressed cDNAs derived from subtraction hybridization also do not overlap with sequences obtained by using DDRT-PCR or RSDD. Apparently, specific differentially expressed genes are lost during subtraction hybridization and differential RNA display of subtracted cDNAs. As such, it might be necessary to use multiple gene discovery approaches to identify and clone the complete spectrum of differentially expressed genes in any experimental context.

5. A problem frequently encountered in DDRT-PCR, which is reduced but still can occur in RSDD, is the isolation of multiple cDNA species from what appears to be a single amplified band. When this occurs, these multiple species can produce spurious results when analyzed by reverse Northern blot analysis. For example, if two distinct species are isolated, one displaying modified expression and a second not displaying modified expression, an accurate estimate of differential expression will not be obtained by reverse Northern blot analysis. In this case, a number of potential false-positives generated by using Northern blot analysis might, in reality, not be false-positives but instead might represent multiple cDNAs. This problem can be ameliorated by performing single-strand conformational polymorphism or Northern blot analysis using cloned cDNA populations *(26,27)*.

6. RSDD represents a method of choice either as a more efficient and less time-consuming modification of the differential RNA display strategy or as a screening methodology for identifying differentially expressed genes in reciprocally subtracted cDNA libraries. Moreover, the ability of RSDD to identify differentially expressed genes that are dissimilar to those recognized by using standard DDRT-PCR or subtraction hybridization indicates that this approach should be a valuable adjunct to these approaches in identifying and cloning differentially expressed genes occurring between complex genomes and resulting from changes in cellular physiology.

Acknowledgments

The present study was supported in part by National Institutes of Health Grants CA035675, CA097318, CA098712, GM068448, NS31492, and P01 CA104177; the Samuel Waxman Cancer Research Foundation; and the Chernow Endowment. P.B.F. is the Michael and Stella Chernow Urological Cancer Research Scientist in the departments of pathology and urology, College of Physicians and Surgeons of Columbia University, and a SWCRF Investigator.

References

1. Sarkar, D., Kang, D. C., Goldstein, N. I., and Fisher, P. B. (2004) Approaches for gene discovery and defining novel protein interactions and networks. *Curr. Genomics* **5,** 231–244.
2. Liang, P. and Pardee, A. B. (1992) Differential display of eukaryotic messenger RNA by means of the polymerase chain reaction. *Science* **257,** 967–971.

3. Liang, P., Zhu, W., Zhang, X., et al. (1994) Differential display using one-base anchored oligo-dT primers. *Nucleic Acids Res.* **22,** 5763–5764.

4. Liang, P., Bauer, D., Averboukh, L., et al. (1995) Analysis of altered gene expression by differential display. *Methods Enzymol.* **254,** 304–321.

5. Liang, P. and Pardee, A. B. (1995) Recent advances in differential display. *Curr. Opin. Immunol.* **7,** 274–280.

6. Liang, P. and Pardee, A. B. (1997) Differential display. A general protocol. *Methods Mol. Biol.* **85,** 3–11.

7. Debouck, C. (1995) Differential display or differential dismay? *Curr. Opin. Biotechnol.* **6,** 597–599.

8. Sompayrac, L., Jane, S., Burn, T. C., Tenen, D. G., and Danna, K. J. (1995) Overcoming limitations of the mRNA differential display technique. *Nucleic Acids Res.* **23,** 4738–4739.

9. Sagerstrom, C. G., Sun, B. I., and Sive, H. L. (1997) Subtractive cloning: past, present, and future. *Ann. Rev. Biochem.* **66,** 751–783.

10. Rangnekar, V. V., Waheed, S., and Rangnekar, V. M. (1992) Interleukin-1-inducible tumor growth arrest is characterized by activation of cell type-specific "early" gene expression programs. *J. Biol. Chem.* **267,** 6240–6248.

11. Wong, B., Park, C. G., and Choi, Y. (1997) Identifying the molecular control of T-cell death; on the hunt for killer genes. *Semin. Immunol.* **9,** 7–16.

12. Maser, R. L. and Calvet, J. P. (1995) Analysis of differential gene expression in the kidney by differential cDNA screening, subtractive cloning, and mRNA differential display. *Semin. Nephrol.* **15,** 29–42.

13. Wan, J. S., Sharp, S. J., Poirier, G. M., et al. (1996) Cloning differentially expressed mRNAs. *Nat. Biotechnol.* **14,** 1685–1691.

14. Kang, D. C., LaFrance, R., Su, Z. Z., and Fisher, P. B. (1998) Reciprocal subtraction differential RNA display: an efficient and rapid procedure for isolating differentially expressed gene sequences. *Proc. Natl. Acad. Sci. USA* **95,** 13,788–13,793.

15. Jiang, H. and Fisher, P. B. (1993) Use of a sensitive and efficient subtraction hybridization protocol for the identification of genes differentially regulated during the induction of differentiation in human melanoma cells. *Mol. Cell Differ.* **1,** 285–299.

16. Gubler, U. and Hoffman, B. J. (1983) A simple and very efficient method for generating cDNA libraries. *Gene* **25,** 263–269.

17. Short, J. M., Fernandez, J. M., Sorge, J. A., and Huse, W. D. (1988) Lambda ZAP: a bacteriophage lambda expression vector with in vivo excision properties. *Nucleic Acids Res.* **16,** 7583–7600.

18. Short, J. M. and Sorge, J. A. (1992) In vivo excision properties of bacteriophage lambda ZAP expression vectors. *Methods Enzymol.* **216,** 495–508.

19. Hay, B. and Short, J. M. (1992) ExAssist™ helper phage and SOLR™ for lambda ZAP II excisions. *Strategies* **5,** 16–18.

20. Sambrook, J. and Russell, D. W. (ed.) (2001) *Molecular Cloning: A Laboratory Manual.* Cold Spring Harbor Lab. Press, Cold Spring Harbor, NY.

21. Sive, H. L. and St. John, T. (1988) A simple subtractive hybridization technique employing photoactivatable biotin and phenol extraction. *Nucleic Acids Res.* **16,** 10,937.
22. Herfort, M. R. and Garber, A. T. (1991) Simple and efficient subtractive hybridization screening. *Biotechniques* **11,** 598–604.
23. Babiss, L. E., Zimmer, S. G., and Fisher, P. B. (1985) Reversibility of progression of the transformed phenotype in Ad5-transformed rat embryo cells. *Science* **228,** 1099–1101.
24. Su, Z. Z., Shi, Y., and Fisher, P. B. (1997) Subtraction hybridization identifies a transformation progression-associated gene PEG-3 with sequence homology to a growth arrest and DNA damage-inducible gene. *Proc. Natl. Acad. Sci. USA* **94,** 9125–9130.
25. Hakvoort, T. B., Leegwater, A. C., Michiels, F. A., Chamuleau, R. A., and Lamers, W. H. (1994) Identification of enriched sequences from a cDNA subtraction-hybridization procedure. *Nucleic Acids Res.* **22,** 878–879.
26. Mathieu-Daude, F., Cheng, R., Welsh, J., and McClelland, M. (1996) Screening of differentially amplified cDNA products from RNA arbitrarily primed PCR fingerprints using single strand conformation polymorphism (SSCP) gels. *Nucleic Acids Res.* **24,** 1504–1507.
27. Zhang, H., Zhang, R., and Liang, P. (1996) Differential screening of gene expression difference enriched by differential display. *Nucleic Acids Res.* **24,** 2454–2455.

2

Cloning Differentially Expressed Genes Using Rapid Subtraction Hybridization (RaSH)

Habib Boukerche, Zao-zhong Su, Dong-chul Kang, and Paul B. Fisher

Summary

Differential gene expression represents the entry point for comprehending complex biological processes. In this context, identification and cloning of differentially expressed genes represent critical elements in this process. Many techniques have been developed to facilitate achieving these objectives. Although effective in many situations, most currently described approaches are not trouble-free and have limitations, including complexity of performance, redundancy of gene identification (reflecting cloning biases) and false-positive gene identification. A detailed methodology to perform a rapid and efficient cloning approach, called rapid subtraction hybridization is described in this chapter. This strategy has been applied successfully to a number of cell culture systems and biological processes, including terminal differentiation and cancer progression in human melanoma cells, resistance or sensitivity to HIV-1 in human T cells and gene expression changes following infection of normal human fetal astrocytes with HIV-1 or treatment with neutrotoxic agents. Based on its simplicity of performance and high frequency of genuine differential gene identification, the rapid subtraction hybridization (RaSH) approach will allow wide applications in diverse systems and biological contexts.

Key Words: RaSH; gene cloning; terminal differentiation; metastasis; reverse Northern; Northern blotting.

1. Introduction

Differentially expressed genes are potentially of great interest, as they might be associated with a phenotypic difference, thereby providing insight not only into gene function but also the fundamental molecular mechanism mediating variations in closely related biological systems *(1)*. Notwithstanding the fact that the number of genes estimated to be expressed in the human genome has been decreased to approx 30,000 from the initial estimate of approx 100,000, monitoring global gene expression levels still remains a challenging task for

From: *Methods in Molecular Biology, vol. 383: Cancer Genomics and Proteomics: Methods and Protocols*
Edited by: P. B. Fisher © Humana Press Inc., Totowa, NJ

most laboratories. The past decade has witnessed an explosion in the number of techniques available to identify networks of genes aberrantly expressed in various disease states, such as those manifest in normal vs cancer cells *(1)*.

Cloning of mRNAs for genes whose expression varies between two physiological states is an important entry point for defining the role of theses genes in regulating important cellular processes. Subtractive hybridization has been employed to clone differentially expressed genes between two systems for the last 30 yr *(2)*. The essential principle underlying subtractive hybridization is that cDNAs common to both control (i.e., driver) and experimental samples (i.e., the "tester") are removed or suppressed to enrich differentially expressed mRNA species. With wide use of the PCR, various methods based on PCR were developed to clone differentially expressed genes by direct side-by-side comparison or by subtractive hybridization. Among these approaches are mRNA differential display, RNA fingerprinting by arbitrary primed PCR, representational differential analysis (RDA), reciprocal subtraction differential RNA display (RSDD), and suppression subtractive hybridization (SSH) *(3–8)*. Differential screening employed in oligonucleotide or cDNA microarray analysis is gaining significant popularity as a means of analyzing global alterations in gene expression patterns and to clone differentially expressed genes with the support of bioinfomatics *(9,10)*.

mRNA differential display and RNA fingerprinting by arbitrary primed PCR have been widely used to detect and isolate genes whose expression correlates with specific disease states *(11,12)*. However, both of these methods are not trouble-free, generating high levels of false-positive signals and they are inefficient for experiments where the expression levels of relatively few genes are expected to vary. In addition, sensitivity of these approaches is highly dependent on the primer sequences, the concentration of the template, and its potential binding site. Several modifications in the original differential mRNA display approach, including reciprocal subtraction differential RNA display and RFLP-based differential display methods, have been developed to overcome these limitations *(4,13)*. However, these methods still do not completely resolve the problem of false-positive signals. Other widely used techniques, such as RDA or SSH, which combine normalization and suppression steps in a single reaction allowed up to 10^6-fold enrichment of mRNAs leading to the isolation of qualitative and quantitative differences between mRNA populations in a single round of subtraction *(6,8)*. Despite the fact that numerous genes have been successfully identified with these methods, they require multiple rounds of hybridization, amplification, and digestion, resulting in amplification bias and generation of spurious clones that do not correspond to authentic differentially expressed genes.

In this chapter, a simple and very effective method, called rapid subtraction hybridization (RaSH) is described. RaSH has wide applications for the rapid identification and cloning of genes displaying both minor and significant changes

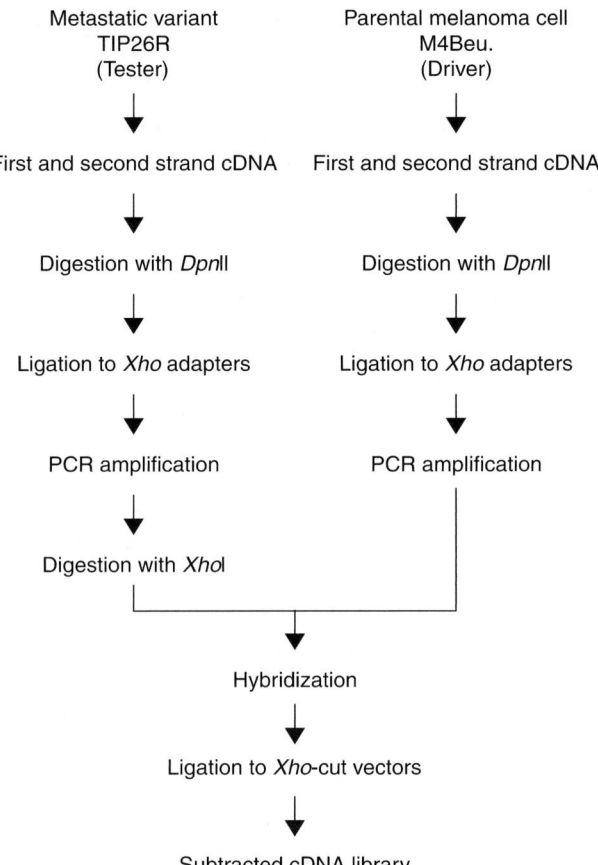

Metastatic variant TIP26R (Tester)	Parental melanoma cell M4Beu. (Driver)

Fig. 1. Schematic representation of the RaSH approach as applied to a highly metastatic variant T1P26R selected in vivo after orthotopic inoculation in immuno-suppressed newborn rats of the parental M4Beu. melanoma cell line. This scheme involves construction of tester (T1P26R) and driver (M4Beu.) melanoma cDNA libraries, followed by digestion of only the tester library with *Xho*I. After hybridization, differentialy expressed sequences are cloned into *Xho*I-digested vectors, resulting in a subtracted cDNA library enriched for metastasis elevated genes (MEGs) displaying elevated expression in melanoma cells displaying an enhanced metastatic phenotype. (From Boukerche et al. *[22]*).

in message expression *(14–19)*. A schematic flow diagram of the RaSH approach is shown in **Fig. 1**. The entry material includes two populations of mRNAs that are to be subtracted. The "driver" is defined as the population of mRNAs that will be eliminated during the subtraction procedure, whereas, the "tester" population of mRNAs contains both commonly and differentially expressed genes. Both mRNA populations are reverse transcribed and digested with a frequent cutting restriction endonuclease *Dpn*II that generates small fragments on average of about 256 bp. Fragments are then ligated to adaptor oligonucleotides containing an additional restriction enzyme site, *Xho*I, allowing subsequent cloning into vector. Primers complementary to the adaptors are then used to generate representations of the cDNA populations by PCR and only the tester cDNAs are digested with the cloning restriction enzyme *Xho*I. Subtractive hybridization is performed by incubation of an aliquot of restriction enzyme digested tester cDNA with a large excess of the control sample (i.e., driver), and the duplexes are melted and allowed to reanneal. In this manner, unique cDNAs in the tester containing exposed *Xho*I sites are cloned into plasmid and propagated in bacteria without further manipulation. Randomly selected clones from the RaSH library are PCR-amplified and evaluated by reverse Northern blot hybridization, in which equal amounts of PCR fragments of cloned cDNAs immobilized on a membrane are hybridized with first-strand ^{32}P-labeled-cDNA derived from "driver" and "tester" mRNA, respectively. Validation of the differential expression of positive clones is achieved using conventional Northern blotting analyses.

The RaSH method combines many aspects of previously described PCR-based subtractive hybridization techniques, including cDNA fragmentation with *Dpn*II, ligation to suitable linkers, PCR, and selection of subtracted cDNA species by matching restriction sites in the cDNA fragments and in the cloning vector, thereby reducing cost and decreasing drastically the workload on the experimenter involved in constructing conventional subtractive cDNA libraries *(4,6,8,20,21)*. However, in contrast to other subtractive hybridization techniques, such as RDA or SSH, RaSH employs a different strategy *(6,8)*. Specifically, multiple rounds of subtraction and amplification are avoided after the posthybridization step, thus rendering the RaSH approach robust, less time-consuming and highly effective in cloning truly differentially expressed genes, without potentially inducing artificial bias by multiple rounds of PCR. Cloning subtracted products into plasmids by restriction site matching and application of differential screening by reverse Northern blot analyses are also unique combinations in RaSH, which reduces the false-positive incidence significantly *(4,16,20,21)*.

This RaSH strategy has previously documented high sensitivity and effectiveness in identifying genes that are differentially expressed as a function of induction of terminal differentiation in human melanoma cells, resistance vs sensitivity to HIV-1 infection of human T cells, perturbation in gene expression in normal human fetal astrocytes infected with HIV-1 or treated with HIV-1 gp120 viral envelope glycoprotein, or tumor necrosis factor-α and searching for genes potentially associated with melanoma tumor progression *(14–19)*. Presently, we describe the successful application of the RaSH technique to search for genes potentially associated with melanoma tumor progression, resulting in eight known differentially expressed genes, including two novel genes associated with elevated metastatic potential in human melanoma cells *(22)*. This procedure can be readily adapted to other model systems for isolating regulatory genes and in this context will have wide application for efficiently and rapidly identifying differentially expressed genes between two genomic samples.

2. Materials

1. Nuclease-free microcentrifuge tubes.
2. DEPC-treated water.
3. Phenol:chloroform:isoamylic alcohol (25:24:1) and phenol:chloroform (1:1).
4. TE (Tris-EDTA) buffer: 10 mM Tris-HCl, pH 8.0, 1 mM ethlenediaminetetraacetic acid. Autoclave.
5. 5 M NaCl. Autoclave.
6. 100% Ethanol.
7. 70% Ethanol. Store at −20°C.
8. RNAsy Maxi kit (Qiagen).
9. MessageMaker mRNA (Gibco-Brl).
10. SuperScript II RNase H$^-$ Reverse Transcriptase amplification System (InVitrogen, cat. no. 18064-014).
11. RNasin (10 U/uL) (InVitrogen).
12. Oligo(dT)$_{12-18}$ (500 µg/mL) (InVitrogen).
13. Retriction enzymes (*Dpn*II, *Xho*I) (New England BioLabs), *Escherichia coli* DNA polymerase I (10 U/µL) (InVitrogen), T4 DNA ligase *E. coli* DNA ligase (400 U/µL) (New England BioLabs), *E. coli* RNA H (2 U/µL) (InVitrogen), Taq DNA polymerase (5 U/µL) (InVitrogen).
14. 10X Second strand buffer (InVitrogen).
15. dNTP 10 mM.
16. RaSH adaptors and PCR oligos:
    a. XDPN18: CTG ATC ACT CGA GAG ATC. Dissolve to 2 mg/mL in nuclease-free water. Store at −20°C.
    b. XDPN14: CTG ATC ACT CGA G. Dissolve to 100 mM in nuclease-free water. Store at −20°C.
    c. XDPN12: GAT CTC TCG AGT. Dissolve to 100 mM in nuclease-free water. Store at −20°C.
17. T7 promotor primer.
18. T3 promotor primer.
19. PCR II vector (TA Cloning Kit, InVitrogen).
20. Competent *E. coli* DH5-α.
21. Mixture 1:
    a. 1 µg Poly(A) RNA template.
    b. 2.5 µL Oligo(dT)$_{12-18}$ (500 µg/mL).

22. Mixture 2:
    a. 5 µL 5X First strand buffer.
    b. 1.25 µL 10 mM dNTP mix.
    c. 2.5 µL 0.1 M DTT.
    d. 0.5 µL RNaseOUT recombinant ribonuclease inhibitor (40 U/µL).
    e. 1.25 µL SuperScript II for RT-PCR (200 U/µL).
23. Mixture 3:
    a. 145 µL Cold diethyl pyrocarbonate (DEPC)-treated water.
    b. 10.5 µL 10 mM dNTP mix.
    c. 20 µL 10X Second strand buffer.
    d. 5 µL *E. coli* DNA polymerase I (10 U/µL).
    e. 0.85 µL RNase H (2 U/µL).
    f. 0.65 µL *E. coli* DNA ligase (10 U/µL).
24. Mixture 4:
    a. 30 µL Nuclease-free water.
    b. 5 µL 10X *Dpn*II buffer.
    c. 5 µL *Dpn*II (10 U/µL).
    d. 10 µL cDNA tester or driver.
25. Mixture 5:
    a. 16 µL *Dpn*II-digested cDNA.
    b. 4 µL 100 mM XDPN-12.
    c. 4 µL 100 mM XDPN-14.
    d. 3 µL 10X ligase buffer.
26. Mixture 6:
    a. 80 µL Distilled water.
    b. 10 µL 10X PCR buffer.
    c. 4 µL 25 mM MgCl$_2$.
    d. 2 µL 20 mM dNTP.
    e. 1 µL 2 mg/mL XDPN-18.
    f. 2 µL cDNA ligation-mixt.
    g. 1 µL 5 U/µL Taq DNA polymerase.
27. Mixture 7:
    a. 80 µL Nuclease-free water.
    b. 10 µL 10X *Dpn*II buffer.
    c. 10 µL cDNA tester.
    d. 10 µL *Xho*I (20 U/µL).
28. Mixture 8:
    a. 2.5 µL Subtracted cDNA.
    b. 1 µg *Xho*I-digested pCRII plasmid.
    c. 1 µL 10X Ligation buffer.
    d. 1 µL T4 DNA ligase (400 U/µL).
    e. 4.5 µL Nuclease-free water.
29. Mixture 9:
    a. 1 µL Culture.

b. 5 µL 10X Colony PCR buffer.
c. 1.5 µL 50 mM MgCl$_2$.
d. 1 µL 10 mM dNTP.
e. 1 µL 10 µM T7 promotor primer.
f. 1 µL 10 µM SP6 promotor primer.
g. 0.2 µL Taq DNA polymerase.
h. 39.3 µL Nuclease-free water.

3. Methods

3.1. Total RNA and Polyadenylated RNA Purification From Cells

1. Total RNA is isolated from cells using the Quiagen Rneasy Maxi kit following the manufacturer's recommendation (*see* **Notes 1** and **2**).
2. Purification of mRNA is necessary for RaSH and is performed from total RNA using the InVitrogen MessageMaker mRNA Isolation System as described in the product profile.

3.2. cDNA Synthesis of Tester and Driver

3.2.1. First-Strand cDNA Synthesis

1. SuperScript II® RNase H⁻ Reverse Transcriptase amplification System can be used for this purpose following the manufacturer's suggestions. The entire procedure is briefly described before (*see* **Notes 3** and **4**).
2. Thaw the following components and place on ice. Set up two reverse transcription reactions for each mRNA sample in two nuclease-free microcentrifuge tubes (0.5 mL size). Each tube contains either the mRNA tester or the driver. A 20-µL reaction volume can be used for 1 µg of mRNA. Add the components from mixture 1 listed in the materials to a nuclease-free microcentrifuge tube (*see* **Note 1**).
3. Heat the mixture at 70°C for 5 min and quick chill on ice.
4. Collect the contents of the tube by brief centrifugation and then add the components of mixture 2 listed in the materials.
5. Incubate the mixture at 37°C for 1.5 h then chill on ice.

3.2.2. Second-Strand cDNA Synthesis

1. Add the reagents from mixture 3 listed in the materials section to the large-scale first-strand reaction mixture.
2. Mix the reagents gently and centrifuge the reactions mixture briefly in a microfuge tube to eliminate any bubbles. Incubate the mixture for 3 h at 16°C. At the end of the incubation, extract the mixture once with phenol:chloroform and one with chloroform.
3. Recover the DNA by precipitation with 2.5 vol of 100% ethanol in the presence of 0.05 vol of 5 M NaCl.

4. Spin tubes for 10 min at 4°C.
5. Wash the pellet once with 0.5 mL of 70% ethanol.
6. Spin tubes again for 10 min at 4°C.
7. Remove ethanol and air-dry the pellet for 10 min.
8. Dissolve the DNA in 10 μL TE buffer (*see* **Note 2**).
9. Store the DNA at −20°C if needed, before proceeding with digestion.

3.3. Digestion of Both Tester and Driver cDNAs With DpnII

1. For *Dpn*II digestion in a 0.5-mL tube, add mixture 4 (**Subheading 2.**, **item 24**).
2. Mix well and incubate for 4 h at 37°C (longer incubations are not advisable).
3. Dilute the mixture to 100 μL and extract once with phenol/chloroform followed by precipitation with ethanol as described in **Subheadings 3.2.1.** and **3.2.2.**
4. Spin tubes for 10 min at 10,000*g* at 4°C, then remove supernatants. A white pellet is clearly visible.
5. Wash the cDNA pellet with 0.5 mL 70% ethanol. Spin tubes for 10 min, remove ethanol and air-dry the pellet for 10 min or dry in a Speedvac.
6. Resuspend each cDNA tester and driver in 16 μL of TE buffer.

3.4. Ligation of Adaptors to Both Tester and Driver cDNAs

1. The resulting cDNA fragments are each ligated to XDPN-12/XDPN-14 adaptors with mixture 5.
2. Incubate adaptor mix at 55°C for 1 min, and allow annealing to occur during a slow cooling to room temperature over approx 1 h. This slow cooling can be performed by incubating the reaction in a water bath that cools from 55°C to room temperature.
3. Add 3 μL T4 DNA ligase (400 U/μL).
4. Incubate the mixture overnight at 14°C.
5. At the end of incubation, dilute the reaction mixture to 100 μL with TE buffer.

3.5. PCR Amplification of Both Tester and Driver cDNA Libraries and Purification of Amplified Product (see Note 5)

1. For each cDNA library, 20 PCR reactions are performed. In addition, a control reaction (without Taq DNA polymerase) is also used. Thus, a total of 42 reactions are performed (*see* **Note 3**). Amplify each cDNA library in a total volume of 100 μL by adding mixture 6.
2. Mix well by pipetting up and down, add 25 μL of mineral oil if needed.
3. Program the PCR machine as follows: 94°C for 1 min, 25 cycles at 94°C for 1 min, 55°C for 1 min, 72°C for 2 min and a final extension at 72°C for 5 min. Then store product at 4°C until used.
4. Separate 5 μL of the PCR products on a 1% agarose gel containing ethidium bromide (the expected result is a smear with an average size of approx 256 bp).
5. Combine each PCR-amplified cDNA library and purify the cDNAs using the Quiaquick PCR purification kit protocol (Quiagen, cat. no. 28304).

6. Measure the DNA concentration at OD_{260} with a spectrophotometer by diluting 5 μL of the DNA sample in 100 μL of TE buffer, pH 7.6.
7. Add TE buffer, pH 7.6 to 1 μg/μL.

3.6. Xho-I Digestion of Tester cDNA

1. Digest about 10 μg of tester cDNA with the restriction enzyme *Xho*I with mixture 7.
2. Mix well and incubate the mixture at 37°C for 4 h.
3. Extract twice with phenol/chloroform and precipitate the DNA at –70°C with 2.5 vol of 100% ethanol in the presence of 0.05 vol of NaCl. Wash the subsequent pellet with 70% ethanol.
4. Air dry the DNA pellet for 10 min or dry in a Speedvac.
5. Resuspend the DNA pellet in TE buffer at 1 μg/μL.

3.7. Construction of Minisubtracted cDNA Libraries (see Note 6)

1. Mix 0.1 μg of *Xho*I-digested tester cDNA with 3 μg of driver cDNA (*see* **Note 4**).
2. Precipitate the DNA with 2.5 vol of 100% ethanol in the presence of 0.05 vol of 5 *M* NaCl at –70°C for 1 h.
3. Spin and dry the DNA in a Speedvac.
4. Resuspend the DNA in 10 μL of hybridization buffer (50 m*M* HEPES, pH 7.5/0.2%, SDS/40%, formamide, pH 8.0).
5. Denature the DNA by boiling 5 min at 98°C and then allow annealing to occur during a slow cooling to 42°C.
6. Briefly, microcentrifuge the Eppendorf tube and overlay with mineral oil to prevent evaporation.
7. Incubate at 42°C for 48 h.
8. Extract the hybridization mixture with phenol/chloroform as in **Subheading 3.2.**
9. Add 2.5 vol of 100% ethanol (do not add NaCl), precipitate at –70°C for 10 min and wash the subsequent pellet with 70% ethanol.
10. Resuspend the DNA in 20-μL of TE buffer.

3.8. Cloning and Analysis of the Subtracted cDNA (see Notes 6 and 7)

1. In this ligation reaction, only the restriction site of the unique tester cDNA remains available for ligation into the digested vector, while the restriction site for common tester cDNAs remain masked following hybridization with driver cDNAs. The resulting subtracted cDNAs are cloned into the pCR II vector, digested with *Xho*I and then treated with calf intestinal alkaline phosphatase. Set up the ligation reaction as follows:
2. In a 0.5-mL tube, add the reagents from mixture 8 for a 10-μL ligation reaction.
3. Ligate the vector and the subtracted cDNA overnight at 16°C
4. Transform high-efficiency, chemically competent *E. coli* DH5-α cells by standard methods (10-μL ligation mix with 100 μL cells). Grow cells on X-gal/ampicillin plates. Plate the cells onto 10 ampicilin plates.
5. Incubate the plates overnight at 37°C until small colonies are visible. Usually, 50–100 colonies are typically obtained. Continue incubation at 4°C until blue/white staining can be clearly distinguished.

6. Randomly isolate white colonies from each subtracted cDNA library plate with a sterile toothpick, place each colony into 2 mL of LB medium containing 100 µg/mL ampicilin and grow overnight at 37°C.
7. The next day, perform the colony-PCR assay using mixture 9 (in "master" mix format).
8. Heat the mixture to 95°C for 3 min, then incubate in a thermal cycler for 35 cycles at 95°C for 1 min, 55°C for 1 min, and 72°C for 1 min 30 s, and heat to 72°C for 10 min.
9. Remove 5 µL of the PCR reaction and analyze on a 1.2% agarose gel.
10. Analyze the product visually, and record insert sizes. Exclude clones that generate more than one insert.

3.9. Reverse Northern Blot

1. Primary selection of potentially differentially expressed clones is achieved by reverse Northern blot analyses. In this procedure, equal amounts of PCR fragments of cloned cDNAs are electrophoresed in an agarose gel, denatured and alkaline blotted onto nylon membranes (*see* **Note 1**).
2. The filters are then hybridized under stringent conditions with equivalent amounts of ^{32}P-labeled double-stranded cDNA of approximately equal amounts of specific activity of driver and tester mRNA, respectively. Filters are then washed under stringent conditions and exposed at −80°C.

3.10. Northern Blot

1. Appropriate expression of the differentially expressed genes that were identified by reverse Northern blotting is confirmed by Northern blotting using standard procedures *(4,14–19)* (*see* **Note 1**).

4. Notes

1. The RaSH strategy has been used to identify genes displaying elevated expression as a consequence of metastatic progression in human melanoma *(22)*. Eight metastasis elevated genes (MEGs) were detected by reverse Northern blotting as overexpressed in the metastatic variant with respect to the weakly metastatic cells based on a criteria requiring a minimum of twofold difference in hybridization signal **(Fig. 2)**. The eight MEG cDNA clones were then hybridized to membrane-blotted total RNA isolated from normal SV-40 immortalized melanocytes (FM516-SV) or the poorly metastatic melanoma cell line (M4Beu.) and three metastatic melanoma cell lines (TIP26, 7GP122, and C8161). As shown in **Figs. 3** and **4**, most of these MEGs displayed true differential expression in the metastatic melanoma cells vs normal immortalized melanocytes and poorly nonmetastatic melanoma cell lines. Although the basal expression levels differ slightly in non-metastatic cells, expressions of the eight MEGs were up-regulated in all of the metastatic cells. These results support the robustness of RaSH protocol in cloning differentially expressed genes.

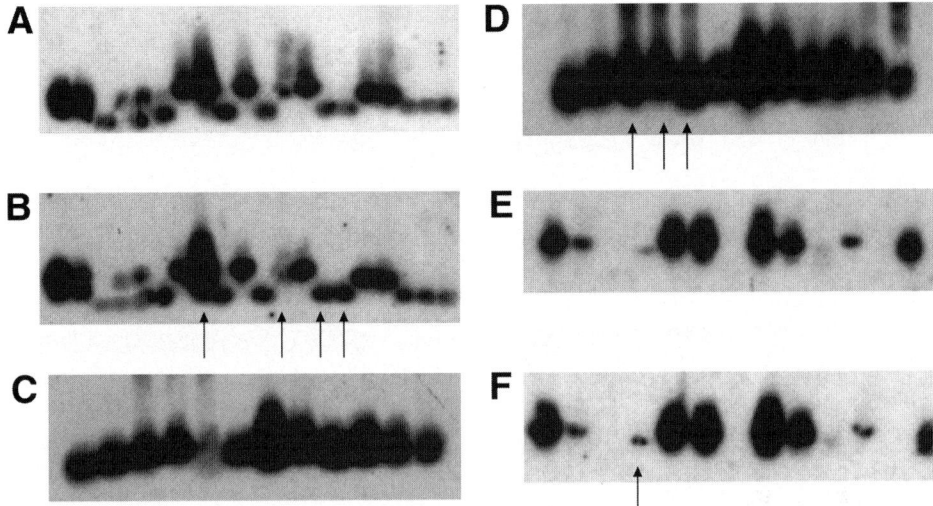

Fig. 2. Reverse Northern blot analysis of differentially expressed sequences identified by RaSH. Equal amounts of PCR amplified products (5 µL) from random bacterial clones of RaSH-derived libraries were loaded onto 1.2% agarose gels. Samples were electrophoresed and transferred onto nylon membranes. The membranes were then hybridized with ^{32}P-labeled cDNA reverse transcribed RNA samples from poorly metastatic parental M4Beu cell line (**A,C,E**) or highly metastatic variant T1P26R (**B,D,F**). Blots were exposed to autoradiography. Arrows indicate differentially expressed cDNA fragments in the T1P26R variant melanoma cell line compared with M4Beu. parental melanoma cells. (From Boukerche et al. *[22]*.)

2. Previous applications of RaSH in cloning genes associated with terminal differentiation of human melanoma cells provide important quantitative parameters regarding size of DNA fragments in subtractive hybridization, enrichment factor and superior performance compared with conventional subtracted cDNA library screening *(14–16)*. The effect of cDNA size on subtraction efficiency and redundancy was compared by digesting cDNAs with two restriction enzymes of differing average restriction fragment size, *Dpn*II (~256 bp) and *Eco*RII (~512 bp). Restriction by *Dpn*II was expected to perform better in subtraction, but to generate more frequent cuts and thereby more redundancy than that by *Eco*RII. However, unexpectedly, the *Dpn*II library manifested higher enrichment and less redundancy (higher representativeness) than the *Eco*RII library did, suggesting that hybridization efficiency is a more critical factor in this cloning procedure. The enrichment factor of *Dpn*II-RaSH for fibronectin and leukemia inhibitory factor was calculated as 75- and 544-fold, respectively. The hit rate of RaSH, representing number of positive clones in randomly picked colonies (~45%) was about two-fold higher in comparison with that of previously reported subtracted cDNA library screening under the same experimental conditions (~20%). Thus, RaSH,

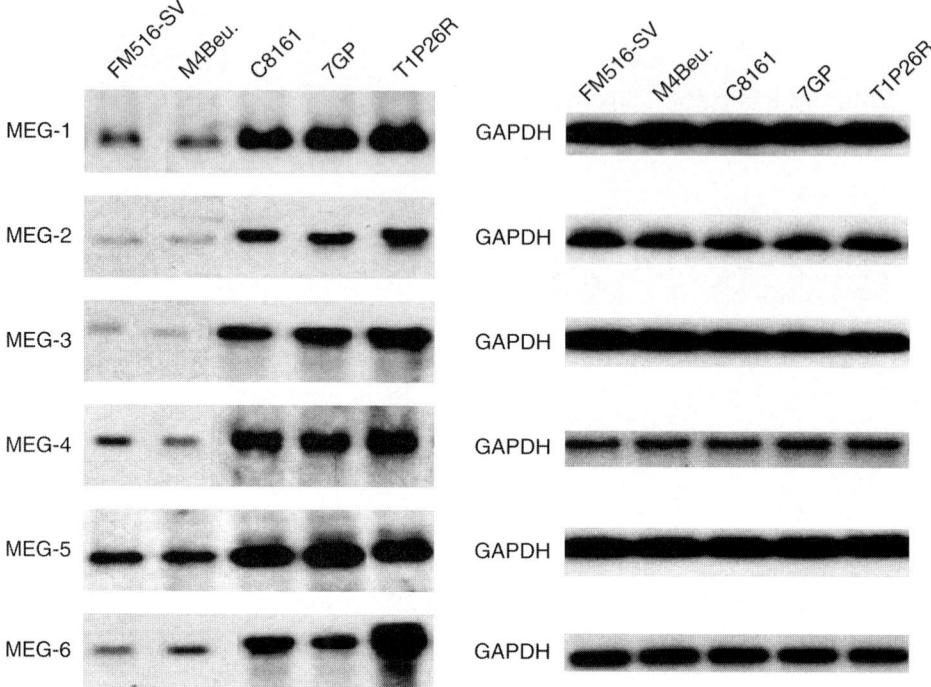

Fig. 3. Confirmation of RaSH-selected cDNA clones (MEG-1 to MEG-6) representing authentic differentially expressed genes upregulated in highly metastatic cell lines by
Northern blot analysis. Total RNA from the indicated immortal melanocyte (FM516-SV)
or melanoma cell line were transferred to positively charged nylon membranes and
probed with [α-^{32}P]-dCTP differentially expressed MEG fragments identified by RaSH.
Blots were stripped and subsequently probed with GAPDH. (From Boukerche et al. *[22]*.)

which significantly simplifies the procedure of conventional subtractive hybridization is an extremely efficient method for cloning differentially expressed genes
between two divergent or closely related biological systems.

3. Poly(A) RNA can also be reverse transcribed using anchored oligo(dT) primers
   (i.e., with either G, C, or A at its 3′ end) (*see* **Subheading 3.2.**).
4. If the pellet is too big, then dissolve it in 100 µL TE buffer, then precipitate the
   DNA for 1 h to overnight at –80°C with 0.5 vol of 7.5 *M* ammonium acetate and
   2.5 vol of 100% ethanol. In our experience, an appropriate differentially expressed
   clone is invariably indentified by reverse Northern blotting, thereby permiting a
   more economical use of experimental RNA *(4,15,16)*. This is one of the most
   important steps in the RaSH approach (*see* **Subheading 3.2.**).
5. Prepare stock mixtures when ever possible to avoid pipetting errors (e.g., aliquot
   cDNA ligation mix and primers individually). Otherwise, it will be difficult to
   pipet 0.2 µL of Taq DNA polymerase. Mix well by pipetting up and down (*see*
   Subheading 3.5.).

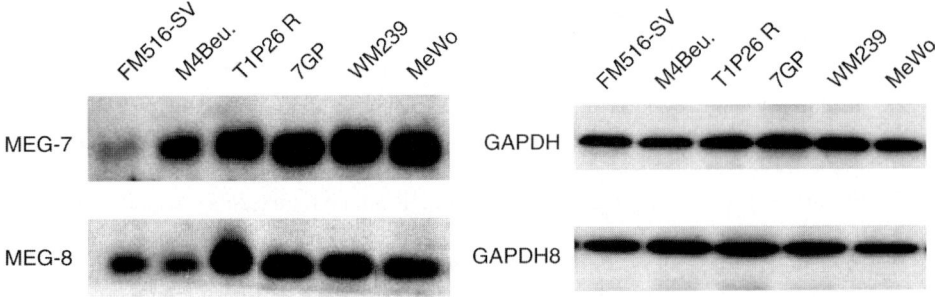

Fig. 4. Confirmation of RaSH-selected cDNA clones (MEG-7 and MEG-8) representing authentic differentially expressed genes upregulated in highly metastatic cell lines by Northern blot analysis. Total RNA from immortal melanocytes (FM516-SV) or the indicated melanoma cell line were transferred to positively charged nylon membranes and probed with (α-^{32}P)-dCTP differentially expressed MEG fragments identified by RaSH. Blots were stripped and subsequently probed with GAPDH. (From Boukerche et al. *[22]*.)

6. Ratios of tester to driver can be varied empirically to optimize RaSH for different applications (*see* **Subheading 3.7.**).
7. In a few cases, RaSH clones contain two concatenated gene fragments, which can affect reverse Northern blotting results *(14–16)*. However, the existence of an additional fragment can be easily detected by examining internal *Dpn*II restriction sites.

Acknowledgments

The present study was supported in part by National Institutes of Health Grants CA035675, CA097318, CA098712, GM068448, P01 CA104177, and P01 NS31492; the Samuel Waxman Cancer Research Foundation; and the Chernow Endowment. P.B.F. is the Michael and Stella Chernow Urological Cancer Research Scientist in the Departments of Pathology and Urology, College of Physicians and Surgeons of Columbia University, and a SWCRF Investigator.

References

1. Sarkar, D., Kang, D. -c., Goldstein, N. I., and Fisher, P. B. (2004) Approaches for gene discovery and defining novel protein interactions and networks. *Curr. Genomics* **5,** 231–244.
2. Sagerstrom, C. G., Sun, B. I., and Sive, H. L. (1997) Subtractive cloning: past, present, and future. *Ann. Rev. Biochem.* **66,** 751–783.
3. Liang, P. and Pardee, A. B. (1992) Differential display of eukaryotic messenger RNA by means of the polymerase chain reaction. *Science* **257,** 967–971.
4. Kang, D. -c., LaFrance, R., Su, Z. -z., and Fisher, P. B. (1998) Reciprocal subtraction differential RNA display: an efficient and rapid procedure for isolating differentially expressed gene sequences. *Proc. Natl. Acad. Sci. USA* **95,** 13,788–13,793.

5. Welsh, J., Chada, K., Dalal, S. S., Cheng, R., Ralph, D., and McClelland, M. (1992) Arbitrarily primed PCR fingerprinting of RNA. *Nucleic Acids Res.* **20,** 4965–4970.

6. Hubank, M. and Schatz, D. G. (1994) Identifying differences in mRNA expression by representational difference analysis of cDNA. *Nucleic Acids Res.* **22,** 5640–5648.

7. Velculescu, V. E., Zhang, L., Vogelstein, B., and Kinzler, K. W. (1995) Serial analysis of gene expression. *Science* **270,** 484–487.

8. Diatchenko, L., Lau, Y. F., Campbell, A. P., et al. (1996) Suppression subtractive hybridization: a method for generating differentially regulated or tissue-specific cDNA probes and libraries. *Proc. Natl. Acad. Sci. USA* **93,** 6025–6030.

9. Chee, M., Yang, R., Hubbell, E., et al. (1996) Accessing genetic information with high-density DNA arrays. *Science* **274,** 610–614.

10. Schena, M., Shalon, D., Davis, R. W., and Brown, P. O. (1995) Quantitative monitoring of gene expression patterns with a complementary DNA microarray. *Science* **270,** 467–470.

11. Liang, P., Bauer, D., Averboukh, L., et al. (1995) Analysis of altered gene expression by differential display. *Methods Enzymol.* **254,** 304–321.

12. McClelland, M. and Welsh, J. (1994) DNA fingerprinting by arbitrarily primed PCR. *PCR Methods Appl.* **4,** S59–S65.

13. Ivanova, I. B., Fesenko, I. V., and Beliavskii, A. V. (1994) A new method of comparative analysis of gene expression and identification of differentially expressed mRNA. *Mol. Biol. (Mosk)* **28,** 1367–1375.

14. Jiang, H., Kang, D. -c., Alexandre, D., and Fisher, P. B. (2000) RaSH, a rapid subtraction hybridization approach for identifying and cloning differentially expressed genes. *Proc. Natl. Acad. Sci. USA* **97,** 12,684–12,689.

15. Kang, D. -c., Jiang, H., Su, Z. -z., Volsky, D. J., and Fisher, P. B. (2002) RaSH-Rapid subtraction hybridization, in *Analysing Gene Expression* (Lorkowski, S., and Cullen, P., eds.), Germany: Wiley-VCH Verlag GmbH, pp. 206–214.

16. Kang, D. -c., Su, Z. -z., Boukerche, H., and Fisher, P. B. (2006) Identification of differentially expressed genes using rapid subtraction hybridization (RaSH): detailed methodology for performing RaSH, in *Immunohistochemistry and In Situ Hybridization of Human Carcinomas: Molecular Genetics, Liver Carcinoma, and Pancreatic Carcinoma,* (Hyat, M. A., ed.), Elsevier/Academic Press, CA, Vol. 3, in press.

17. Simm, M., Su, Z. -z., Huang, E. Y., Chen, Y., Jiang, H., Volsky, D. J., and Fisher, P. B. (2001) Cloning of differentially expressed genes in an HIV-1 resistant T cell clone by rapid subtraction hybridization, RaSH. *Gene* **269,** 93–101.

18. Su, Z. -z., Kang, D. -c., Chen, Y., et al. (2003) Identification of gene products suppressed by human immunodeficiency virus type 1 infection or gp 120 exposure of primary human astrocytes by rapid subtraction hybridization. *J. Neurovirol.* **9,** 372–389.

19. Su, Z. -z., Kang, D. -c., Chen, Y., et al. (2002) Identification and cloning of human astrocyte genes displaying elevated expression after infection with HIV-1 or

exposure to HIV-1 envelope glycoprotein by rapid subtraction hybridization, RaSH. *Oncogene* **21,** 3592–3602.

20. Corton, J. C. and Gustafsson, J. A. (1997) Increased efficiency in screening large numbers of cDNA fragments generated by differential display. *Biotechniques* **22,** 802–804, 806, 808.
21. Jiang, H. and Fisher, P. B. (1993) Use a sensitive and efficient subtraction hybridization protocol for the identification of genes differentially regulated during the induction of differentiation in human melanoma cells. *Mol. Cell Differ.* **1,** 285–299.
22. Boukerche, H., Su, Z. -z., Kang, D. -c., and Fisher, P. B. (2004) Identification and cloning of genes displaying elevated expression as a consequence of metastatic progression in human melanoma cells by rapid subtraction hybridization. *Gene* **343,** 191–201.

3

The Application of Differential Display as a Gene Profiling Tool

Ken Chien-Neng Chang, Barry Komm, Nichole Bayer Arnold, and Murray Korc

Summary

Differential display is an effective expression profiling tool which was first introduced in 1992. The original technique is discussed along with modifications that have been described over the last several years. A highly reproducible, semihigh-throughput differential display protocol used in our laboratories is described along with an example of its successful application using pancreatic cancer cells. In addition to the work performed in our laboratories, several examples of successful applications of differential display under a number of scenarios are reviewed. Differential display is one of several expression profiling technologies available and is compared with some of them. The future of differential display remains bright and is as applicable today as it was in 1992.

Key Words: Differential display; gene expression; RT-PCR; pancreatic cancer; thioredoxin; apoptosis.

1. Introduction

1.1. The Evolution and Accomplishments of Differential Display

Expression/transcriptional profiling has become the center stage of the cutting edge sciences in the field of biotechnology and pharmaceutical research in the last 10 yr, akin to the terms that first grasped the attention of the general public such as "high-throughput screening" in the late 1980s and "automated DNA sequencing" in the mid 1990s. Among the many different transcriptional profiling technologies, differential display is probably one of the most important break-through technologies that produced a large number of successful examples over the timeframe of its short existence. Differential display, a tool for identifying and cloning differentially expressed genes, was first introduced

From: *Methods in Molecular Biology, vol. 383: Cancer Genomics and Proteomics: Methods and Protocols*
Edited by: P. B. Fisher © Humana Press Inc., Totowa, NJ

in 1992 by Liang and Pardee *(1)*. The technology took advantage of reverse transcriptase (RT-PCR) with both anchored and random priming to generate labeled cDNA followed by polyacrylamide gel electrophoresis separation and analysis to characterize differentially regulated bands. Hypothetically, all regulated mRNAs would be represented (up- or downregulated) without a major concern for a particular mRNA's abundance. Unlike gene chip profiling, the bands putatively representing mRNA can be excised, reamplified, cloned, and sequenced to identify corresponding genes, if known. In general, the number of primer sets and the design of the primer sequences will determine the degree of coverage. Greater than 90% coverage of the expressed mRNA population of a cell using a reasonable number of primer sets has been achievable. For the first 5 yr after the introduction of differential display technique, the technology evolved as it went through a period of refinement and validation. Shortly after the original design of differential display was disclosed many publications showed successful applications of the technique, yet a 70–80% false-positive rate plagued investigators *(2–4)*. A new report introduced fluorescent differential display using arbitrarily primed PT-PCR fingerprinting on an automated DNA sequencer *(5)*. However, the fluorescent approach was limited to the analysis and viewing of differential expression profiles such as screening for target genes *(5)*, as there is no easy way of excising the differentially regulated gel bands for cloning and sequencing.

1.2. Technology Modifications and Improvements

Two major obstacles identified shortly after the original differential display design was released were redundancy and underrepresentation of certain mRNA species as well as the high false-positive rate. Liang and Pardee *(6)* published an improved design to overcome these difficulties using one-base anchored oligo-dT primers and a rationally designed 13-mer primer with a HindIII restriction site at the 5' ends of both primers. This modification minimized the redundancy and under-representation of certain RNA species owing to the degeneracy of the primers. The introduction of a restriction enzyme site also allowed for easier, more efficient cDNA cloning. Most recent use of this technology has focused on primer design to improve the reproducibility and coverage. Many successful designs have been commercialized or have become proprietary. Although most specially designed proprietary primer sets claim to provide more than 90% coverage, it might still be species or tissue dependent. Frost et al. *(7)* suggested that in many cases using identical set of primers, some tissues yield simpler banding patterns than others. They also provided evidence that mispriming of one or more bases in the anchor region of the primer and priming at an internal site on the mRNA rather than to the poly (A) tail are common in differential display reverse transcription, and therefore, repetitive sampling occurs extensively in differential display. One very recent publication used commercially available primer sets for their differential display experiment, yet it excluded one of the most important genes supposed to be regulated *(8)*. The gene was located 185 bases away from the other differentially regulated gene that they had identified. By sequencing and database mining as well as literature searching, they were able to identify the gene, but this example points out an inherent problem that can be encountered using this technology.

Commercially available kits such as the delta differential display kit from Clontech have their own primer design, and were developed based on the publication by Zhao et al. in 1995 *(9)*. In general, the sequences of arbitrary primers correspond to the most common sequence motifs found in mammalian mRNA. These sequences were obtained by computer analysis of the coding region of a large number of mammalian mRNAs. This kit includes 10 arbitrary (25-nt) P primers and nine T primers (29-nt), to generate 90 possible primer combinations, and it is claimed that the confirmation rate is about 85%. It has also been shown that increasing the A/T content of the random primers by up to 60–80% with a concomitant slight increase in primer length (12- to 14-bp), leads to increased coverage of differential display *(10)*, because higher A/T content is typically seen in the 3′-UTR. If cDNA and primers are in equilibrium, sequences with favorable primer on/off rates will compete out the amplification of other bands. It is believed that DD-PCR must be viewed as a competitive equilibrium type of PCR.

A mathematical model has predicted that 80–160 arbitrary 13-mers in combination with three one-base anchored oligo-dT primers, would allow any given mRNA within a eukaryotic cell to have 74–93% detection probability *(11)*. In general, improvements such as increasing the length of the primers, using optimized annealing temperatures, increasing the dNTP concentration in the reaction, and halting the reaction within the exponential phase of PCR amplification are modifications that enhance the generation of very reproducible banding patterns. Other modifications, such as preventing depurination during the elution step *(12)* could also potentially improve the success of confirmation, cloning, and sequencing.

1.3. Successful Examples From the First Decade of Differential Display's Application

The second 5 yr after differential display's appearance on the technological landscape, revealed a number of examples where the application was successfully applied. Soon after differential display was described, Terry Maratos-Flier used this technology to identify a neuropeptide called melanin-concentrating hormone (MCH), a gene that is two to three times more active in the brains of ob/ob mice (an obese leptin-deficient strain) than in normal mice *(13)*. Another group used differential display to identify an mRNA that increases in sensory neurons after exposure to serotonin *(14)*. Based on the amino acid sequence they linked this mRNA to a developmentally regulated family of genes like *Drosophila* tolloid gene and the human bone morphogenetic protein (BMP)-1 gene. They suggested that this gene product might regulate the morphology and efficacy of synaptic connections between sensory and motor neurons. Another study has shown that an alternative form of human lactoferrin (a multifunctional protein involved in many aspects of the host defense against infection) mRNA that is expressed differentially in normal tissues and tumor-derived cell lines was identified using the differential display technique *(15)*. Yet another study revealed novel targets for the mood-stabilizing drug Valproate using differential display PCR technique *(16)*. Other successful examples in the early stage of differential display can be found in several review articles *(17–20)*.

1.4. More Recent Applications

After more than 10 yr, differential display has matured and still has its value in the field of expression profiling sciences. Over the last few years, many publications have taken advantage of the more evolved differential display technology to further identify new targets for cancers and women's health-related disorders *(21–25)*, study gene expression profiles in human bone marrow stromal cells *(26)* and rat white adipose tissue *(27)*, uncover the secret of drug resistance *(8)*, exploit DNA damage and mutation-related gene expression responses *(28)*, and discover the additional transcriptional factors implicated in the pathobiology of the immunodeficiency virus *(29)*. The breadth and depth of these publications reveal that the differential display technology is firmly ensconced in the biomedical and drug discovery research processes.

Among these publications, several significant accomplishments should be noted. An extremely specific marker called DD3[PCA3] for prostate cancer was identified by differential display *(21)*. This marker is not expressed in any other normal human tissues and is highly overexpressed in the prostate tumors by comparison with normal prostatic tissue. Other technologies might have been able to identify this gene, but the lack of dependence on abundance of a gene's transcript in differential display often will make it more likely to find a regulated gene than in the transcript level dependent gene chip profiling. Another paper reported that *in situ* hybridization was used to confirm the differential display results with RNA isolated from human mammary carcinomas *(22)*. They used the differentially regulated cDNA bands as the probes for *in situ* hybridization to further characterize other breast cancer cases. This study demonstrated that *in situ* hybridization could be used in conjunction with differential display as a screening test for studying biopsy material.

Using a slight modification of differential display called restriction fragment differential display, a group from Italy compared the gene expression of bone marrow stromal cells (BMSC) and BMSC derived osteoblasts *(26)*. Their results showed that 34 out of 47 candidate regulated genes were confirmed by RT-PCR analysis to be preferentially expressed in BMSC, in at least one out of three cell clones examined. According to their analysis, different individuals appeared to have substantial discrepancy in gene expression profiles among primary BMSC cultures. Even with the same donor, this type of discrepancy exists. They suggested that BMSC cultures are a heterogeneous population of different, and possibly uncharacterized cell types, whereas cell populations derived from a single cell type could represent different stages of differentiation. In this experimental paradigm, the focus was on a restricted set of genes downregulated when BMSCs differentiate to osteoblasts. With this goal in mind, restriction fragment differential display might appear to be a more appropriate method compared with microarray/gene chip approach. Because without concentrating on a subset of genes the heterogeneous nature of these cells might cause difficulty in sorting out a focused set of meaningfully regulated transcripts. Furthermore, several of the genes listed in the publication appear to be low abundance genes, and since the focus was on downregulated species, the use of certain microarray systems might encounter difficulties.

A group of investigators at University of Navarra, Spain, studied gene expression changes in rat white adipose tissue after a high-fat diet using differential display *(27)* resulted in the identification of eight genes that were upregulated, and four genes that were downregulated. Among the four that were downregulated, two had been previously shown to be related to obesity (adiponectin and caveolin-2), whereas most of the other up- or downregulated genes were involved in metabolic signaling or transcriptional regulation pathways that might be related to energy metabolism. The authors indicated that another group using microarray analyses performed a similar study, and although not totally consistent, some genes were upregulated in both studies *(30)*.

1.5. Application in Our Laboratory

Using the technique of differential display, thioredoxin-1 (TRX) was recently identified as a gene whose basal expression is increased in COLO-357 pancreatic cancer cells engineered to overexpress Smad7, a protein implicated in the negative feedback regulation of transforming growth factor (TGF)-β signaling pathways *(31)*. To delineate the biological consequences of TRX overexpression, TRX mRNA levels were assessed in pancreatic ductal adenocarcinoma, an aggressive human malignancy in which Smad7 is commonly overexpressed. Analysis of laser-captured pancreatic cancer cells revealed parallel increases in Smad7 and TRX mRNA levels. Furthermore, retroviral infection of an antisense TRX cDNA suppressed TRX protein levels and blunted the increased capacity of Smad7 overexpressing cells to form colonies in soft agar. High levels of *cis*-diamminedichloroplatinum (II) (CDDP), a chemotherapeutic agent, induced apoptosis in Sham-transfected COLO-357 cells, as evidenced by induction of DNA laddering, PARP cleavage, and increased caspase-3/9 activities. However, these proapoptotic actions were greatly attenuated in Smad7 overexpressing cells. Thus, the use of differential display identified TRX as a downstream mediator of Smad7 in a pathway that confers a growth advantage to pancreatic cancer cells and that increases their resistance to CDDP-mediated apoptosis. It was also determined that Smad7 overexpressing exhibited a more prolonged association of TRX with the apoptosis inducer apoptosis signal-regulating kinase-1, and enhanced NF-kB activation in response to CDDP. We were thus able to extend the observations made by differential display to advance our understanding of the mechanisms that modulate chemoresistance in pancreatic cancer, underscoring the many avenues for studies that can be pursued following observational findings made through the use of the differential display technique.

1.6. What a Reproducible Differential Display Protocol Looks Like

To provide readers with a comprehensive scope of what is involved, a very successful modified semihigh throughput differential display protocol, based on the 1994 publication of Liang et al. *(6)*, and timelines used in our laboratories is summarized in **Subheading 3.** This protocol took advantage of three anchored primers, AAGCT(11)N, where N equals A, G, or C, and a set of 80 specially designed/selected 13-mer random primers (AAGCTT-N7) *(6)* to obtain broad coverage with minimized redundancy. A robotic liquid handling station is set up to generate 3×80 reactions for each sample. This provides a high level of reproducibility, but sacrifices sensitivity to some degree.

2. Materials

1. For the initial screen, 20 µg of DNase1-treated total RNA (batch 1) should be targeted.
2. For the confirmation screen, 10 µg DNase1-treated RNA (batch 2) can be used.
3. RNA purification protocols can be found in Chapters 1 (Sarkar et al.) and 2 (Boukerche et al.) of this volume.

3. Methods

1. Approximately 1000–2000 µg of total RNA for Northern analysis, isolated from either batch 1 or 2, for a yield of around 25–50 µg polyA RNA, should be planned. RNA from a tissue or cell line is prepared using the Trizol method or any dependable RNA isolation procedure (*see* Chapters 1 [Sarkar et al.] and

2 [Boukerche et al.] this volume). To assure reproducibility, the confirming run is done using a different batch of similarly treated samples (such as two different pools of animals). To maximize efficiency, the two independent RNA isolations should be included in the experimental design. One isolation can be as little as 100 µg of RNA; the second should aim for 1–2 mg of RNA. This will serve as the source for Northern blotting.

2. It is important to perform polyA blotting if possible, because in general less than 20% of the genes can be detected by total-RNA hybridization, thus making quantification difficult. For quantitation by more sensitive methods, such as RT-PCR, or *in situ*, less material is acceptable.

3. A typical timeline for a two-sample comparison should require two full days beginning with the reverse transcription (RT) reaction through all the PCR reactions, followed by about 2 d of gel runs (15 gels in total). RT and PCR conditions will depend on the primers used and might need to be optimized. For commercial primer sets, these conditions are available through the vendors. DNase1-treated total RNA is used for three RT reactions using three anchored primers. The resulting three RT reactions are used to set up 80 PCR reactions each using 80 random primers in the presence of [^{33}P]-dATP to generate labeled cDNA.

4. The fifth day should be set aside for making the confirming batch of RNA because the last set of gels will not have the film developed until the sixth day. On the sixth day, all of the gels should be analyzed for candidate bands. The primer pairs should be identified, allowing for set up of the confirming PCR reactions. It is likely that the number of candidates chosen will allow the repeat gels to be finished in 1 d (the seventh day). The final film will be available the morning of the ninth day.

5. At this point, it should be possible to make the final candidate selections, isolate and elute the bands followed, and precipitate the RNA overnight to maximize recovery. The next day, an aliquot (one-third) is used to perform reamplification reactions and set up ligations for the standard pCR2.1 TA cloning vector to clone the differentially regulated candidates. It will then be possible to perform the transformation, pick colonies, and screen six to ten colonies from each reaction by PCR for comparison with the originally reamplified products. One clone chosen from each candidate band is subjected to overnight culture for miniDNA preparation. After checking on gel, the samples are submitted for DNA sequencing.

4. Notes

1. The probes used in Northern blotting should be the reamplified products and not the final clones. This avoids the possibility of choosing the wrong clone and thus missing a truly regulated gene. After a clone is confirmed, a cleaner signal is generally obtained using the cloned fragment as a probe. If regulation could not be confirmed for a candidate that had very obvious differential regulation, then use the reamplified, gel purified fragment as a probe to prove that the candidate is indeed represented (**Fig. 1**).

Differential display

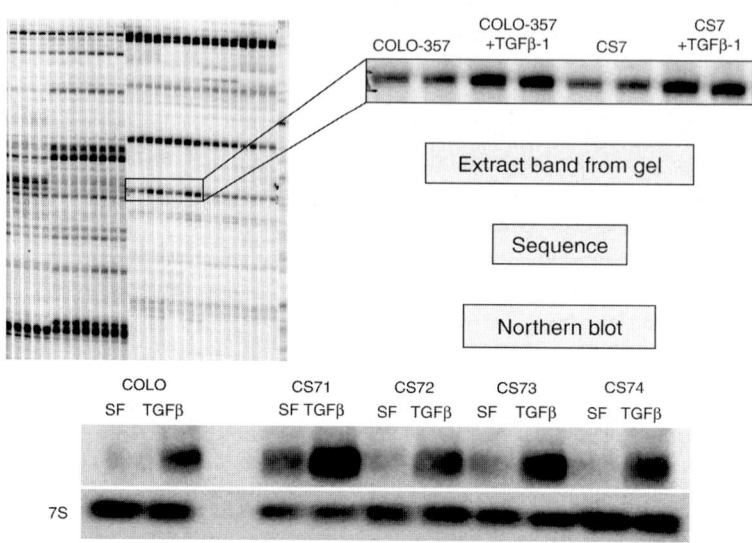

Fig. 1. RNA was isolated from Sham-transfected COLO-357 human pancreatic cancer cells and Smad7 overexpressing clones (CS7) that were incubated for 24 h in the absence or presence of 500 pM TGF-β1. RNA was then treated with RNase-free DNase in the presence of RNasin RNase inhibitor, and reverse transcribed to cDNA using an anchored (HT$_{11}$A) primer. The cDNA was PCR amplified in the presence of [^{33}P] dCTP in duplicate using the anchored primer and multiple arbitrary primers. PCR products were separated on an 8% denaturing polyacrylamide gel, exposed to Kodak Biomax, and excised from the gel. The desired PCR products were then amplified using the same set of primers from the initial differential display, subcloned, and sequenced. Total RNA was then extracted, size fractioned, transferred onto nylon membranes and ultraviolet crosslinked. In the example shown at the bottom of the figure, RNA was isolated from Sham-transfected COLO-357 cells and from four different Smad7 overexpressing clones (SS71, CS72, CS73, and CS74) that were incubated for 24 h in the absence or presence of 500 pM TGF-β1. Blots were hybridized overnight with the [^{33}P] dCTP-labeled cDNA probe, targeting the differentially expressed gene. A 190 bp 7S cDNA probe (indicated with the notation 7S) was used to confirm equal RNA loading and transfer.

References

1. Liang, P. and Pardee, A. B. (1992) Differential display of eukaryotic messenger RNA by means of the polymerase chain reaction. *Science* **257**, 967–971.

2. Mou, L. J., Miller, H., Li, J., Wang, E., and Chalifour, L. (1994) Improvements to the differential display method for gene analysis. *Biochem. and Biophy. Res. Comm.* **199,** 564–569.

3. Sun, Y., Hegamyer, G., and Colburn, N. H. (1994) Molecular cloning of five messenger RNAs differentially expressed in preneoplastic or neoplastic JB6 mouse epidermal cells. *Cnacer Research* **54,** 1139–1144.

4. Nishio, Y., Aiello, L. P., and King, G. L. (1994) Glucose induced genes in bovine aortic smooth muscle cells identified by mRNA differential display. *FASEB* **8,** 103–106.

5. Ito, T., Kito, K., Adati, N., Mitsui, Y., Hagiwara, H., and Sakaki, Y. (1994) Fluorescent differential display: arbitrarily primed RT-PCR fingerprinting on an automated DNA sequencer. *FEBS Lett.* **351,** 231–236.

6. Liang, P., Zhu, W., Zhang, X., et al. (1994) Differential display using one-base anchored oligo-dT primers. *Nucleic Acids Res.* **22,** 5763–5764.

7. Frost, M. R. and Guggenheim, J. A. (1999) Mammalian polyadenylation sites: implications for differential display. *Nucleic Acids Res.* **27,** 1386–1391.

8. Francia, G., Man, S., Teicher, B., Grasso, L., and Kerbel, R. S. (2004) Gene expression analysis of tumor spheroids reveals a role for suppressed DNAmismatch repair in multicellular resistance to alkylating agents. *Mol. and Cell Biol.* **24,** 6837–6849.

9. Zhao, S., Ooi, S. L., and Pardee, A. B. (1995) New primer strategy improves precision of differential display. *Biotechniques* **18,** 842–846.

10. Graf, D., Fisher, A. G., and Merkenschlager, M. (1997) Rational primer design greatly improves differential display-PCR (DD-PCR). *Nucleic Acids Res.* **25,** 2239–2240.

11. Yang, S. and Liang, P. (2004) Global analysis of gene expression by differential display: a mathematical model. *Mol. Biotech.* **27,** 197–208.

12. Frost, M. R. and Guggenheim, J. A. (1999) Prevention of depurination during elution facilitates the reamplification of DNA from differential display gels. *Nucleic Acids Res.* **27,** E6.

13. Gura, T. (2000) Tracing leptin's partners in regulating body weight. *Science* **287,** 1738–1741.

14. Liu, Q., Hattar, S., Endo, S., et al. (1997) A developmental gene (Tolloid/BMP-1) is regulated in *Aplysia* neurons by treatments that induce long-term sensitization. *J. Neurosci.* **17,** 755–764.

15. Siebert, P. D. and Huang, B. C. B. (1997) Identification of an alternative form of human lactoferrin mRNA that is expressed differentially in normal tissues and tumor-derived cell lines. *Proc. Natl. Acad. Sci. USA* **94,** 2198–2203.

16. Wang, J., Bown, C., and Young, L. T. (1999) Differential display PCR reveals novel targets for the mood-stabilizing drug Valproate inducing the molecular chaperone GRP78. *Mol. Pharmacol.* **55,** 521–527.

17. Liang, P. and Pardee, A. B. (1995) Recent advances in differential display. *Curr. Opin. Immunol.* **7,** 274–280.

18. Matz, M. V. and Lukyanov, S. A. (1998) Different strategies of differential display: areas of application. *Nucleic Acids Res.* **26,** 5537–5543.
19. Ali, M., Markham, A. F., and Isaacs, J. D. (2001) Application of differential display to immunological research. *J. Immu. Methods* **205,** 29–43.
20. Liang, P. (2002) A decade of differential display. *BioTechniques* **33,** 338–346.
21. Schalken, J. A., Hessels, D., and Verhaegh, G. (2003) New targets for therapy in prostate cancer: differential display CODE 3 (DD3^{PAC3}), a highly prostate cancer-specific gene. *Urology* **62,** 34–43.
22. Kao, R. H., Francia, G., Poulsom, R., Hanby, A. M., and Hart, I. R. (2003) Application of differential display, with *in situ* hybridization verification, to microscopic samples of breast cancer tissue. *Int. J. Exp. Pathol.* **84,** 207–212.
23. Greenland, K. J., Rekaris, G., MacLean, H. E., Warne, G. L., and Zajac, J. D. (2004) Application of differential display in the identification of androgen-regulated genes. *Endocr. Res.* **30,** 69–82.
24. Gladney, C. D., Bertani, G. R., Johnson, R. K., and Pomp, D. (2004) Evaluation of gene expression in pigs selected for enhanced reproduction using differential display PCR and human microarrays. *J. Anim. Sci.* **82,** 17–31.
25. Zhang, W. M., Liu, W. T., Xu, Y., Xuan, Q., Zheng, J., and Li, Y. Y. (2004) Study of genes related to gastric cancer and its premalignant lesions with fluorescent differential display. *Ai Zheng.* **23,** 264–268.
26. Monticone, M., Liu, Y., Tonachini, L., et al. (2004) Gene expression profile of human bone marrow stromal cells determined by restriction fragment differential display analysis. *J. Cell Biochem.* **92,** 733–744.
27. Lopez, I. P., Milagro, F. I., Marti, A., Moreno-Aliaga, M. J., Martinez, J. A., and De Miguel, C. (2004) Gene expression changes in rat white adipose tissue after a high-fat diet determined by differential display. *Biochem. and Biophy. Res. Commu.* **318,** 234–239.
28. Kote-Jarai, Z., Williams, R. D., Cattini, N., et al. (2004) Gene expression profiling after radiation-induced DNA damage is strongly predictive of BRCA1 mutation carrier status. *Clin. Cancer Res.* **10,** 958–963.
29. Troung, M., Delsart, V., and Bahr, G. M. (2004) Differentially expressed genes in HIV-1-infected macrophages following treatment with the virus-suppressive immunomodulator murabutide. *Virus Res.* **99,** 25–33.
30. Moraes, R. C., Blondet, A., Birkenkamp-Demtroeder, K., et al. (2003) Study of the alteration of gene expression in adipose tissue of diet-induced obese mice by microarray and reverse transcription-polymerase chain reaction. *Endocrinology* **144,** 4773–4782.
31. Arnold, N. B., Ketterer, K., Kleeff, J., Friess, H., Buchler, M. W., and Korc, M. (2004) Thioredoxin is downstream of Smad7 in a pathway that promotes growth and suppresses cisplatin-induced apoptosis in pancreatic cancer. *Cancer Res.* **64,** 3599–3606.

4

Serial Analysis of Gene Expression (SAGE)

Fred van Ruissen and Frank Baas

Summary

In 1995, serial analysis of gene expression (SAGE) was developed as a versatile tool for gene expression studies. SAGE technology does not require pre-existing knowledge of the genome that is being examined and therefore SAGE can be applied to many different model systems. In this chapter, the SAGE procedure was described, which cannot only be used as a tool to generate gene expression profiles but also to identify new transcripts. SAGE data can easily be exchanged between laboratories and a large database with SAGE data is now available on the internet.

Key Words: Gene discovery; gene expression; transcriptome; serial analysis of gene expression; SAGE; gene expression.

1. Introduction
1.1. Gene Expression Profiling Techniques

Over the past 10 yr several techniques for high throughput analysis of gene expression have been developed. These high throughput techniques have been used to obtain information about changes in the transcriptome. Initially the analysis of gene expression was limited by the knowledge of the genome. The use of expressed sequence tags (EST) clones on either macro (nylon) or micro (glass) arrays allowed the simultaneous analysis of thousands of gene transcripts. However, genes for which no sequence information was available cannot be analyzed in this way.

In 1995 the group of Vogelstein and Kinzler (1) described a method, which allowed analysis of overall gene expression without pre-existing knowledge of the entire genome of the species that is analyzed. This is the most important difference between serial analysis of gene expression (SAGE) and other technologies for high throughput gene expression analysis. SAGE can be used to identify and quantitate new genes, therefore, it is also a method for gene discovery.

From: *Methods in Molecular Biology, vol. 383: Cancer Genomics and Proteomics: Methods and Protocols*
Edited by: P. B. Fisher © Humana Press Inc., Totowa, NJ

SAGE is based on the following principles:

1. A short sequence tag (10–14 bp) contains sufficient information to identify a transcript provided that the tag is obtained from a unique position within each transcript.
2. Concatemers of sequence tags form long serial molecules that can be cloned and sequenced.
3. Quantification of the number of times a particular tag is observed provides the expression level of the corresponding transcript. *See* **Fig. 1** for a schematic representation of the SAGE technology.

These simple assumptions and the fact that SAGE can be performed without the need for expensive specialized equipment is probably one of the reasons why SAGE was initially applied for gene identification in many different laboratories. However, SAGE is a technically demanding procedure and only relatively small number of laboratories managed to exploit this technique.

The other option for large gene expression studies was to use high-density expression arrays. During its initial phase of development high quality expression arrays were very scarce and most were generated by homemade equipment. In the subsequent years, owing to the enormous progress that had been made by sequencing consortia the genome of many species became known. Also, the quality of microarrays increased and with the full genome sequence information becoming available, synthetic microarrays, which can incorporate all known and predicted genes in a given species can be produced.

One might think that the development of whole genome microarrays will make SAGE obsolete. However, the contrary is true. SAGE is still a method of choice to be used when whole genome microarrays are available. A primary reason is that SAGE generates high quality expression data, which can easily be stored for future use and exchange between researchers is achievable.

1.2. SAGE in Cancer Research

The cancer genome anatomy project includes a network of cancer researchers who collaborate to obtain the common goal of cataloguing genetic changes related to cancer. In the cancer genome anatomy project, tools necessary for easy dissemination and analysis of SAGE expression data have been developed *(2)*. The SAGE map website at NCBI (http://www.ncbi.nlm.nih.gov/SAGE) holds 277 human libraries, 143 mouse libraries, and a variety of libraries of other species (December 2004). This unique resource allows investigators to query the expression of a gene in a large sample of human tissues in cancer. The SAGE map website also hosts programs to analyze SAGE data. Digital Northerns can be applied to any gene for which a SAGE tag is known. In this way, the expression of a gene of interest can be determined in a large collection of tissues and cancers. In addition statistical comparisons of SAGE libraries are available through SAGEmap. It was believed that rapid and public availability

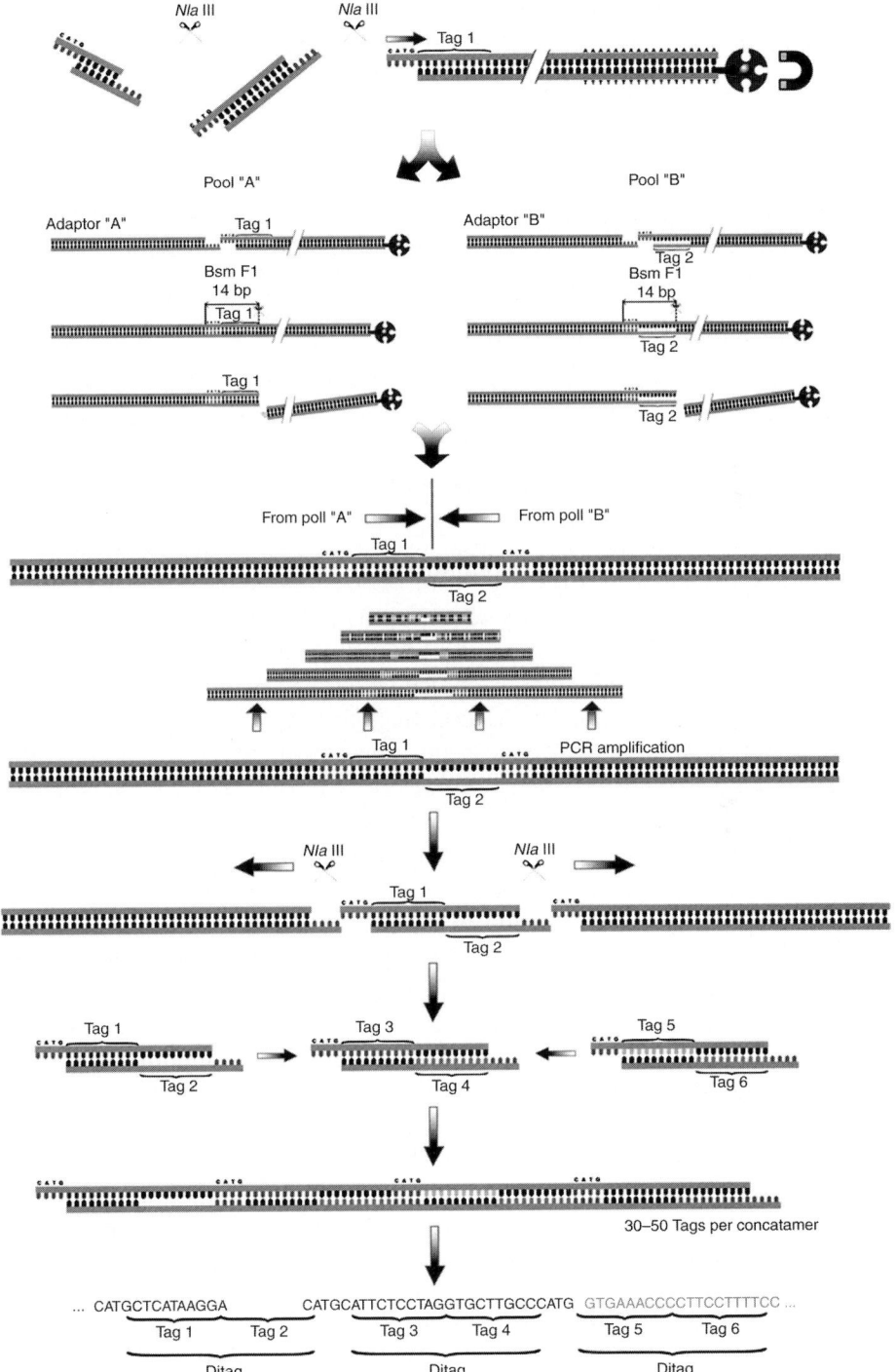

Fig. 1. Schematic representation of the SAGE technology.

of large data sets, easy access, and good informatics support have contributed significantly to the wide use of SAGE in cancer research.

1.3. SAGE Data Analysis

SAGE 2000 is the commonly used analysis program for SAGE and can be obtained from the Kinzler group (*see* www.sagenet.org). Other groups, including ours, have developed programs for SAGE data analysis *(3–8)* and even planning of SAGE experiments *(4)*. Like SAGE data, most of these programs are freely available for academic groups.

1.4. Developments in SAGE Technology

The initial SAGE method, described by Velculescu et al. (1995), uses the type-2 restriction enzyme *BsmF*I to generate SAGE tags of approx 14 nucleotides (nt). These SAGE tags were derived from the 3′-end of a messenger RNA. They were located 3′ of the last *Nla*III restriction enzyme site (**Fig. 1**). Velculescu and colleagues *(1)* argued that the combination of knowing the position of a SAGE-tag and the 14-base sequence information was sufficient to map SAGE tags to unique mRNA sequences. However, it turned out that many SAGE tags could initially not be mapped to mRNA sequences *(1)*. These so called "no-match" tags have been the subject of intensive research and resulted in the discovery of many novel genes. Several methods, which convert an unknown SAGE tag to a cDNA, have been described. The high throughput GLGI approach *(10)* and the reverse SAGE technique *(11)* have been used frequently. The reverse SAGE protocol is also available from the SAGENET website. However, not all the "no-match" tags represent novel genes. Tag to transcript assignment is not a trivial task. Especially in view of the high frequency of alternative 3′-ends of mRNA molecules. Current estimates are that a given mRNA has on average three different 3′-ends. In addition, single nucleotide polymorphisms can also affect tag mapping. If only full matches of the SAGE tag to the reference genome sequence are allowed, a large series of tags would be scored as "no-match," whereas, this is just owing to natural sequence variation.

In order to improve SAGE tag mapping, long-SAGE, was introduced *(12)*. This modification of the SAGE technique uses the restriction enzyme *Mme*I, which cleaves 20-nt from its recognition site and thus generates SAGE tags of 21-nt. This size is sufficient to allow mapping to the human genome. Therefore, long-SAGE is a powerful tool for gene discovery. However, longer SAGE tags mean that less SAGE tags can be read in a sequence run and thus more sequence runs have to be done to generate a library of a certain size. In a comparison of short and long-SAGE the authors and others found that the long tags do not contribute substantially to the SAGE tag mapping procedure. Still many SAGE tags remain "no-matches." Why long-SAGE does not live up to its expectations, might be a consequence of alternative 3′-end processing and single nucleotide polymorphisms in the mRNA sequences.

1.5. Performing a SAGE Experiment

As mentioned earlier, SAGE is a technically complicated procedure, but thanks to the continuous development of the SAGE protocol one with appropriate skills in molecular biology can generate a SAGE library. Protocols for SAGE, long-SAGE and micro-SAGE are available on the SAGENET website (section protocols). A SAGE kit has also been developed (Invitrogen, Carlsbad, CA). In our hands the SAGE kit, either normal or long-SAGE, performs well. It contains all the necessary reagents and every lot is tested for efficiency. The kit also comes with a detailed protocol. In the following section, the general SAGE procedures, note the steps which requires specific attention and provide some examples of the expected results.

2. Materials

All the reagents are provided with the I-LongSAGE kit, but several additional solutions for RNA extraction, gel loading, colony PCR and so on are described here.

1. I-SAGE Long kit (Invitrogen, Breda, The Netherlands).
2. Trizol reagent (Invitrogen BRL, cat. no. 15596-018).
3. Enzymes (*Sph*I [10 U/l] Roche Woerden, The Netherlands; cat. no. 606120; T4 Ligase regular concentration [1 U/μL] Roche, cat. no. 716359; Amplitaq [5 U/μL] Perkin Elmer, Groningen, The Netherlands, cat. no. N808-0167; Big Dye Terminator mix Perkin Elmer, cat. no. 4303150).
4. Zeozin (100 mg/mL) (Invitrogen, cat. no. R250-01).
5. 100 mM dNTPs (Amersham, Buckinghamshire, UK; cat. no. 27-20350).
6. Dimethyl sulfoxide (DMSO) (Merck, Amsterdam, The Netherlands; cat. no. 802912).
7. 10-bp ladder (1 μg/μL) (Gibco, cat. no. 10821-015) and 1-kb ladder.
8. DH10B electromax (Invitrogen, cat. no. 18290-015).
9. Accugel 40% 19:1 (Biozym, Landgraff, The Netherlands; cat. no. EC850).
10. Accugel 40% 29:1 (Biozym, cat. no. EC852).
11. T7a primer: 5′ TAA TAC GAC TCA CTA TAG GGC G 3′.
12. SP6a primer: 5′ GCC AAG CTA TTT AGG TGA CA 3′.
13. LoTE: Create a solution of Tris-HCl (pH 7.5) and ethylenediaminetetraacetic acid (EDTA) pH 7.5, diluted in water to produce a final concentration of 3 mM Tris-HCl (pH 7.5) and 0.2 mM EDTA (pH 7.5), and store at 4°C.
14. PC8 (phenol/cloroform/isoamylalcohol): obtained from Invitrogen.
15. 10X PCR buffer colony PCR: prepare a mixture of KCl, Tris-HCl and gelatine to obtain a final concentration of 500 mM KCl; 100 mM Tris-HCl pH 8.0 and 0.1% gelatine (store at –20°C).
16. 5X Tris-boric-EDTA (TBE): dissolve 54 g Tris base, 27.5 g boric acid, 2.9 Na$_2$EDTA· 2H$_2$O in distilled water and adjust volume to 1 L; sterilize by autoclaving.

17. 50X Tris-acetate-EDTA (TAE): dissolve 242 g Tris base, 57.1 mL glacial acetic acid, 37.2 g Na$_2$EDTA·2H$_2$O and adjust volume to dH$_2$O to 1 L, sterilize by autoclaving.
18. Super optimal broth (SOB) medium: dissolve 20 g tryptone, 5 g yeast extract, 0.5 g NaCl in 950 mL dH$_2$O; add 10 mL of a 250 mM KCl stock; adjust pH to 7.5 with 5 M NaOH; add dH$_2$O to 1 L; sterilize by autoclaving. And cool to 55°C; finally add 10 mL sterile 1 M MgCl$_2$.
19. Super optimal catabolite (SOC) medium: take 1 L SOB and add 7.2 mL of 50% glucose.
20. Low salt Luria Bertanii (LB) agar: dissolve 10 g tryptone, 5 g yeast extract, 5 g NaCl and 15 g agar in 1 L of dH$_2$O, sterilize by autoclaving, let cool to 55°C and add Zeozin to a final concentration of 100 µg/mL, mix well and pour plates.
21. Sequence dilution buffer: 200 mM Tris-HCl pH 9.0, 5 mM MgCl$_2$.
22. Water baths set at 16, 37 or 42, 50, 65, and 75°C, respectively.
23. 0.5- and 1.5-mL Sterile microcentrifuge tubes (Eppendorf tubes).
24. 1.5-mL Sterile siliconized microcentrifuge tubes.
25. 10 mg/mL Ethidium bromide solution for staining of the gels.
26. 18-Gauge needles.
27. 70 and 100% Ethanol.
28. 96-Well plates.
29. Bio-Rad Genepulser electroporator (Bio-Rad, Veenendaal, The Netherlands).
30. Electroporation cuvets (Bio-Rad).
31. DH10B cells.
32. Ice.
33. Mineral oil.
34. Razor blades.
35. Sterile toothpicks.

3. Methods

The following protocol is a combination of the original Velculescu and Kinzler protocol (**1**), the protocol of the Invitrogen (LongSAGE kit) and the protocols employed at the department of Neurogenetics of the AMC. This is the protocol were used for the SAGE courses. For the first time, starters are recommended to use commercially available I-SAGE kit, this contains all reagents needed and everything is tested. For additional information, troubleshooting, and performance check you can use the manual of the I-LongSAGE kit, of course you can also follow the complete I-LongSAGE protocol when you are first doing SAGE at your lab (*see* **Notes 1–4**).

3.1. Isolating RNA

Isolating high-quality RNA using a method of choice, before using the I-LongSAGE kit. Use cells (5×10^5–2×10^6), total RNA (5–50 µg), or mRNA (50–100 ng) as starting material. Prepare total RNA from tissue or cells of choice using standard methods, trizol reagent were used (*see* **Note 5**).

3.1.1. Binding of mRNA to Magnetic Beads

1. Mixing and washing of the magnetic beads must be done carefully to prevent sample loss owing to drying or clumping of the beads. Either mix the beads by flicking with your finger or using a slow speed vortex (use a setting of five to six).
2. Do not mix by pipetting up and down as this may result in the loss of beads.
3. During all washing steps of the beads, add buffer to the tube containing the beads. Do not allow the beads to dry out as that will reduce the efficiency the beads.

3.1.2. Washing the Beads

1. Place the tube on a magnetic stand for 1–2 min. Place the pipet tip at the opposite side of the tube, away from the beads, and slowly slide the pipet tip to the bottom of the tube. Carefully remove supernatant and discard. Do not disturb or remove any beads.
2. Add the appropriate volume of buffer.
3. Remove the tube from the stand and mix the contents of the tube by gentle vortexing or flicking the tube with a finger.
4. Centrifuge briefly to collect any beads that may stick to the cap of the tube.
5. Return the tube to the magnetic stand for 1–2 min and carefully remove the supernatant.
6. Repeat **steps 2–5** until all the washing steps are complete.
7. After the last wash, resuspend the beads in an appropriate buffer.

3.1.3. Preparing Oligo (dT) Beads

1. Thoroughly resuspend the oligo (dT) beads and transfer 100 µL to an RNase-free 1.5-mL siliconized microcentrifuge tube.
2. Place the tube on a magnetic stand for 1–2 min and carefully remove the supernatant and discard it.
3. Wash beads in 500 µL of Lysis/binding buffer as described (*see* **Note 6**).
4. Place the tube on a magnetic stand for 1–2 min.
5. Prepare your RNA sample for binding to the beads (*see* **steps 1** and **2**, mRNA binding **Subheading 3.1.4.**).
6. Carefully remove the supernatant and immediately add your RNA sample to the beads as described in mRNA binding **(Subheading 3.1.4.)**.

3.1.4. mRNA Binding

1. Adjust the volume of the 25 µg total RNA to 1 mL with lysis/binding buffer.
2. Load the entire 1 mL of RNA sample or cell lysate to oligo (dT) magnetic beads equilibrated with lysis/binding buffer **(step 4, Subheading 3.1.3.)**.
3. Mix the beads and the RNA sample by slowly rocking the tube on a rocking platform or vortexing the tube intermittently on a slow vortex for 10–30 min at room temperature (*see* **Note 7**).
4. Place the tube on a magnetic stand for 1–2 min and carefully remove the supernatant. (You might save the supernatant here, in case binding did not occur or was insufficient.)

5. Wash the beads twice with 1 mL of wash buffer A as described earlier (*see* washing the beads).
6. Wash with 1 mL of wash buffer B as described earlier (*see* washing the beads).
7. Wash the beads four times with 100 µL of 1X first strand buffer by placing the tube on magnetic stand for 1–2 min and removing the supernatant between the washes. On the fourth wash, do not remove the supernatant. Proceed to **step 8**.
8. Before removing the supernatant after the fourth wash, prepare the first strand cDNA synthesis reaction as described in **Subheading 3.2., step 1**.
9. Remove the supernatant after the fourth wash and proceed immediately to first strand cDNA synthesis (**Subheading 3.2.1.**). Do not store the beads at this point.

3.2. Synthesizing cDNA

3.2.1. First Strand cDNA Synthesis

1. Mix the following reagents for the first strand synthesis on ice: 18 µL 5X first strand buffer, 1 µL RNaseOUT, 54.5 µL DEPC water, 9 µL 0.1 *M* dithiothreitol (DTT) (*see* **Note 8**), 4.5 µL dNTP mix (10 m*M* each).
2. Resuspend the beads containing the mRNA sample (from **Subheading 3.1.4., step 9**) in the first strand mix (87 µL).
3. Mix gently by flicking the tube with a finger or by slow vortexing.
4. Place the tube at 37°C for 2 min to equilibrate the reagents.
5. Add 3 µL Superscript II RT. Mix gently and incubate at 37°C for 1 h. Mix gently every 10 min on a slow vortex. Note: while performing cDNA synthesis at 42°C, if the RNA is known to contain more secondary structure and to prevent oligodT mispriming at internal sites (*see* **Note 9**).
6. Meanwhile equilibrate another water bath to 16°C for the second strand synthesis.
7. Chill the first strand reaction on ice for 2 min and proceed to second strand cDNA synthesis (**Subheading 3.2.2.**).

3.2.2. Second Strand cDNA Synthesis

1. Add the following second strand reagents to the tube containing 90 µL of the first strand reaction (**step 7**): 465 µL diethylpyrocarbonate (DEPC) water, 150 µL 5X second strand buffer, 15 µL dNTP Mix (10 m*M* each), 5 µL *Escherichia coli* DNA Ligase, 20 µL *E. coli* DNA polymerase, 5 µL *E. coli* RNase H (*see* **Note 10**).
2. Mix the contents by vortexing and centrifuge the tube briefly in a microcentrifuge. (If beads are sticking on the plastic surface, you can scrabe them off using the tip of the pipet.)
3. Incubate the reaction mixture at 16°C for 2 h. Always be sure to resuspend the beads by mixing gently for every 10–15 min. During incubation preheat wash buffer C to 75°C.
4. Place the tube on ice and add 45 µL 0.5 *M* EDTA to inhibit the reaction.
5. Place the tube on a magnetic stand for 1–2 min and carefully remove the supernatant. Add 750 µL of warm wash buffer C to inactivate *E. coli* DNA polymerase (*see* **Note 11**).

6. Mix well and heat the sample to 75°C for 10–12 min with intermittent mixing to completely inactivate the polymerase. Place the tube on a magnetic stand for 1–2 min. Remove the supernatant and wash again with 750 μL of wash buffer C. Perform the wash steps quickly to prevent precipitation of sodium dodecyl sulfate, which may trap the beads.
7. Wash three to four times with 750 μL of wash buffer D (*see* **Note 12**).
8. Place the tube on a magnetic stand for 1–2 min.
9. Add 200 μL of 1X Buffer 4 to the tube and gently resuspend the beads. Transfer the contents of the tube to a new microcentrifuge tube to avoid any traces of exonuclease activity from *E. coli* DNA polymerase (*see* **Note 11**). If the beads are sticking to the sides of the tube, gently scrape off the beads from the tube using a pipet tip. Wash the old tube once with 200 μL of the same buffer and transfer the contents to the new tube containing the reaction mix.
10. Place the tube on a magnetic stand for 1–2 min.
11. Remove the supernatant and discard it. Wash the tube once with 200 μL of 1X Buffer 4 (*see* **Note 13**).
12. Remove the supernatant and proceed to digest the cDNA with *Nla*III (**Subheading 3.3.**).

3.3. Digesting the cDNA With Nla***III***

1. After cDNA synthesis we recommend putting a small volume on gel to analyze the integrity of the cDNA before cleavage of the cDNA with anchoring enzyme *Nla*III. Check integrity of the cDNA by gel electrophoresis (run 1 μL of the cDNA on a 0.8% agarose gel, put 10 min at 65°C before loading.) The cDNA should yield a uniform intense smear ranging from several hundred-bp to over 10 kb.
2. Resuspend the beads from **step 12** in the following mix: 172 μL LoTE, 2 μL 100X bovine serum albumin, 20 μL 10X buffer 4, 6 μL *Nla*III.
3. Incubate at 37°C for 1 h. Mix occasionally by flicking the tube with a finger or slow vortexing.
4. Place the bottle containing wash buffer C in a water bath set at 37°C to prevent precipitation of sodium dodecyl sulfate. (First preheat the solution at 65°C until the solution is clear, and then place the bottle at 37°C.)
5. After the reaction is complete, place the tube containing the beads on a magnetic stand for 1–2 min and carefully discard the supernatant. Inactivate *Nla*III by washing the tube twice with 750 μL of warm wash buffer C.
6. Wash the beads three to four times with 750 μL of wash buffer D. After the last wash, resuspend the beads in 750 μL of wash buffer D. Note: Wash the beads five to six times with wash buffer D, if clumping of the beads occurs.
7. Store samples at 4°C overnight. The next day, proceed to **Subheading 3.4.**

3.4. Ligating Adapters to the cDNA

The adapters contain a Type IIS restriction endonuclease site at the 3′ end, cohesive overhangs complementary to the *Nla*III recognition site, and priming sites for PCR amplification. The 3′-end of the adapters is modified with an amino group to prevent self-ligation. Two different adapters (A and B) are used to allow efficient PCR in the amplification reactions.

1. Place the tube on a magnetic stand for 1–2 min and carefully remove the supernatant.
2. Wash the beads twice with 150 µL of 1X ligase buffer. After the final wash, resuspend the beads in 100 µL of 1X ligase buffer and divide the sample equally into two new tubes, A and B. Be careful to divide the beads, while they are suspended, as the beads may stick to the original tube or pipet tips.
3. Wash tubes A and B containing the beads once with 100 µL 1X ligase buffer. Resuspend the beads in 1X ligase buffer.
4. Place the tubes from **step 8**, on a magnetic stand for 1–2 min and carefully remove the supernatant.
5. Add the following reagents to the beads on ice:

Contents	Tube A (µL)	Tube B (µL)
cDNA beads	Beads	Beads
LS-Adapter A (40 ng/µL)	1.5	–
LS-Adapter B (40 ng/µL)	–	1.5
LoTE	14	14
10X Ligase buffer	2	2
Total volume	17.5	17.5

6. Resuspend the beads by flicking the tube with a finger. Heat the tubes for 2 min at 50°C.
7. Cool the tubes to room temperature for 15 min and chill the samples on ice.
8. Add 2.5 µL T4 DNA ligase and mix well.
9. Incubate for 2 h at 16°C. Mix every 10–15 min by flicking the tube with a finger.
10. Wash each tube four times with 500 µL of wash buffer D, resuspend in 500 µL of wash buffer D.

3.5. Cleaving With Tagging Enzyme

Cleavage of the ligation products with the tagging enzyme, *Mme*I, results in release of the adapter with a short piece of the cDNA (tag) from the beads.

3.5.1. Preparing 10X SAM

Prepare a fresh dilution of 10X SAM (400 µM) from the kit-supplied 32-mM SAM for use in the *Mme*I digestion. About 20 µL of 10X SAM per library is required for the following dilution prepares 80 µL to enable easy pipetting. Discard excess dilution, which is not use immediately. The 10X SAM should be prepared fresh shortly before use to ensure stability.

3.5.2. Preparing 1X Buffer 4/1X SAM

Prepare a 1X buffer 4/1X SAM solution from the kit-supplied 32-m*M* SAM and 1X buffer 4 for use in the *Mme*I digestion. About 800 μL of this solution per library is required. The solution should be prepared fresh shortly before use to ensure stability.

3.5.3. Mme*I* Digestion

Place the tube from **Subheading 3.5.1., step 10**, on a magnetic stand for 1–2 min and remove the supernatant.

1. Wash each tube twice with 200 μL of 1X buffer 4/1X SAM (*see* preparation instructions listed previously). Carefully remove and discard the supernatant, and place the tubes on ice.
2. Add the following reagents to each tube: 70 μL LoTE, 10 μL 10X buffer 4, 10 μL 100X SAM (400 μ*M*) (prepare as previously described), and 10 μL *Mme*I.
3. Incubate the tubes at 37°C for 2.5 h with occasional gentle mixing.
4. Place on a magnetic stand for 1–2 min. Do not discard the supernatant. Carefully remove the supernatant and transfer to a new microcentrifuge tube. The tags are present in the supernatant after *Mme*I digestion. Proceed directly to **Subheading 3.6.**

3.6. Ligating the Tags to Create Ditags

3.6.1. Ethanol Precipitation

1. Pool the ditags by transferring 100 μL from tube A to B (200 μL total).
2. Wash tube A with 100 μL LoTE and then transfer it to tube B to yield a total of 300 μL.
3. Add 300 μL of phenol/chloroform to the tube B and vortex thoroughly. Centrifuge for 5 min at maximum speed in a microcentrifuge at room temperature.
4. Transfer 300 μL of the aqueous (top) phase to a new microcentrifuge tube, and mix thoroughly.
5. Remove 200 μL of the aqueous (top) phase from the previous step into another tube and set aside. This is the sample.
6. Add 100 μL of DEPC water to the remaining 100 μL to yield a final volume of 200 μL. This will be used as a negative control (no ligase).
7. Add the following to 200 μL of the sample and the negative control (no ligase) and mix vigorously.
8. Precipitate as described in the **(Subheading 3.16.4.).**
9. Carefully remove the supernatant and discard it. Be careful not to disturb the pellet.
10. Wash the pellet twice with 1 mL of cold 70% ethanol.
11. After the final wash, centrifuge again to collect any residual ethanol.
12. Carefully remove the ethanol by pipet and air-dry the pellet for 5–10 min.
13. Resuspend the sample pellet in 4 μL of LoTE and the negative control in 2 μL of

LoTE and incubate it at 37°C for 10–15 min to aid in solubilization. Proceed directly to forming ditag.

3.6.2. Forming Ditags

1. Prepare the following reaction mix on ice in two sterile microcentrifuge tubes:

	2X Ditag reaction mix (µL)	2X Negative control mix (µL)
3 m*M* Tris-HCl, pH 7.5	1.5	2.25
10X Ligase buffer	0.9	0.75
DEPC water	0.9	0.75
T4 DNA ligase	1.2	–

2. Add 4 µL of 2X ditag reaction mix to the sample tags resuspended in 4 µL LoTE (**Subheading 3.6.1., step 13**), and add 2 µL of 2X negative control mix to the negative control (no ligase **Subheading 3.6.1., step 13**).
3. Incubate overnight at 16°C.
4. After the overnight incubation from **step 3**, add 6 µL LoTE to the template ligation mixture and 10 µL LoTE to the negative control ligation mixture. Store at –20°C or proceed to performing PCR (**Subheading 3.7.**).

3.7. Performing PCR

In this step you will select the optimal conditions for the amplification of ditags using different dilutions of the ligated product (template) and different cycle numbers. Once you have determined the optimal PCR conditions, you will perform a scale-up PCR to generate sufficient amount of ditags for forming concatemers. This part is very important. Do not perform more PCR cycles then necessary.

1. Dilute the PCR templates with sterile water (*see* **Notes 15** and **16**).
2. Negative control (no ligase) to 1/20 and 1/50.
3. Template (ligated product) to 1/10, 1/50, and 1/100.
4. The sixth sample of each series is the blank.
5. Those six PCRs are pipeted together three times; to perform the same six PCR reactions at three different cycle numbers (26, 28, and 30 cycles).
6. Setup on ice the following mix for 20 PCR reactions (50 µL each).

	1 Reaction (µL)	20 Reactions (µL)
10X BV buffer	5	100
DMSO	3	60
dNTP mix (10 m*M* each)	7.5	150
LS-DTP-1	2	40
LS-DTP-2	2	40
DEPC water	28.5	570
Amplitaq	1	20

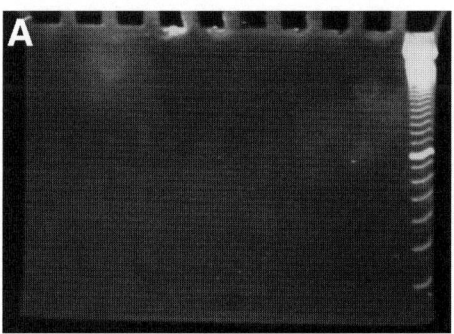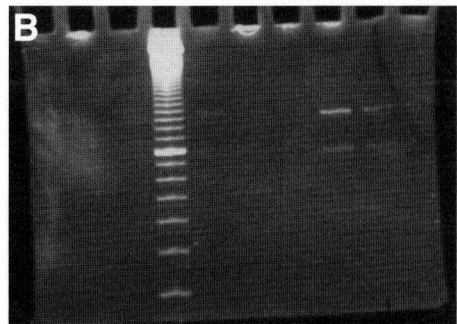

Fig. 2. Test LongSAGE PCR. The following samples were amplified as described in the text. **Figure 3** Lane 10: 10-bp ladder *26 cycles.* Lanes 1 and 2: Negative control (no ligase) 1/10 and 1/50. Lane 3: Negative control (no template) *28 cycles.* Lanes 4 and 5: Negative control (no ligase) 1/10 and 1/50. Lane 6: Negative control (no template) *30 cycles.* Lanes 7 and 8: Negative control (no ligase) 1/10 and 1/50. Lane 9: Negative control (no template). Right gel *26 cycles.* Lanes 1–3: 1/10, 1/50, and 1/100 dilution of the template, respectively. Lane 4: 10-bp ladder *28 cycles.* Lanes 5–7: 1/10, 1/50, and 1/100 dilution of the template, respectively, *30 cycles.* Lanes 8–10: 1/10, 1/500, and 1/100 dilution of the template, respectively.

7. Pipet 49 µL mix in 18 seperate PCR tubes, and if necessary, add a drop of oil to tubes 6, 12, and 18 (blancs) and close the tubes.
8. Add 1 µL template to the rest of the tubes in the following order: tubes 1, 7, 13: negative control 1/20; tubes 2, 8, 14: negative control 1/50 and if necessary, add a drop of oil and close the tubes; tubes 3, 9, 15: template 1/10; tubes 4, 10, 16: template 1/50; tubes 5, 11, 17: template 1/100.
9. PCR amplify tubes 1–6 for 26 cycles, tubes 7–12 for 28 cycles, and tubes 13–18 for 30 cycles.
10. Perform PCR at the following temperatures: 95°C 5 min; 26, 28 or 30 cycles: 95°C 30 s; 55°C 1 min; 70°C 1 min followed by 70°C 5 min.
11. Remove 10 µL from each reaction, add 2 µL of loading, place 10 min. at 65°C and run on a 12% polyacrylamide gel (**Subheading 3.16.1.**) (Bio-Rad small vertical electrophoresis system, thin spacers; 10 well comb), using a 10-bp ladder as a marker.
12. Run gel for 2 h at 120 V.
13. Stain the gel using ethidium bromide (50 µL 10 mg/mL in 100 mL 1X TBE buffer); let soak for 10 min (amplified ditags should be 130-bp in size). A background band of equal or lower intensity occurs around 100-bp (and in the case a 120-bp fragment). All other background bands should be of substantially lower intensity. The ligase samples should not contain any amplified product of the size of the ditags. **Figure 2** gives an indication what the gel should look like.
14. After PCR conditions have been optimized, large scale PCR (two to three 96-well plates containing 50 µL reactions/well) can be performed.
15. One hundred reaction PCR premix were usually used, which is aliquot into two 96-well plates (50 µL/well).

3.8. Gel-Purification of the 130-bp Ditag

This section describes the purification of the 130-bp ditag produced after scale-up PCR. The PCR reaction is electrophoreses on a 12% polyacrylamide gel to separate the 130-bp ditag from the 100-bp linker and other contaminants. The 130-bp ditags are excised, eluted from the gel, and purified using spin columns.

1. Collect 50 µL PCR from each well in 2-mL Eppendorf tube (use 10 tubes in total for 200 PCRs).
2. Extract with equal volume of PC8.
3. Precipitate 330 µL PCR product.
4. Wash two to three times with 70% ethanol, resuspend each tube in 10 µL LoTE (300 µL LoTE total).
5. Add 60 µL loading buffer to this sample (360 µL total volume) and incubate 10 min at 65°C.
6. Load 10 µL of the sample in 9 wells of each of four 12% polyacrylamide 10 well gels (samples are loaded in 36 total lanes; 10 µL sample per lane). Ten microliters of a 10-bp ladder is used as a marker on each gel.
7. Run gel for 2 h at 120V.
8. Stain gels with ethidium bromide (50 µL 10 mg/mL in 100 mL 1X TBE buffer, 10 min). Amplified ditags should run at 130-bp, while a background band runs at about 90-bp. Visualize bands under ultraviolet (UV) light (**Fig. 3**).
9. Cut out only amplified ditags from the gel, and place three cut-out-bands in a 0.5-mL microcentrifuge tube whose bottom has been pierced with an 18-gauge needle to form a small hole of about 0.5 mm diameter.
10. Place 0.5-mL microcentrifuge tubes in 2-mL microcentrifuge tubes (remove lid) and spin in microcentrifuge at full speed for 2 min (this serves to break up the cut-out-bands into small fragments at the bottom of the 2-mL microcentrifuge tubes).
11. Discard 0.5-mL tubes, add 250 µL LoTE and 50 µL 7.5 *M* ammonium acetate (125:25) to each 2-mL tube (close with the cut off lid). Tubes can remain at this point at 4°C overnight.
12. Then, vortex each tube, and place at 65°C for 2 h.
13. Transfer the contents of the tube to either the provided S.N.A.P-column or to a Spin-X column (Costar, Zwijndrecht, The Netherlands) fitted to a collection tube. Centrifuge at maximum speed in a microcentrifuge for 2 min.
14. PC8 extract and ethanol precipitate 300 µL fractions of eluate as described in **Subheadings 3.16.3.** and **3.16.4.**
15. Wash twice with 70% ethanol and dry the pellets briefly at room temperature.
16. Resuspend DNA in 126 LoTE in total, and store at −20°C overnight.

3.9. Digesting the 130-bp Ditag With NlaIII

1. Aliquot the 100-bp product (126 µL) equally into three separate sterile microcentrifuge tubes (*see* **Note 17**).

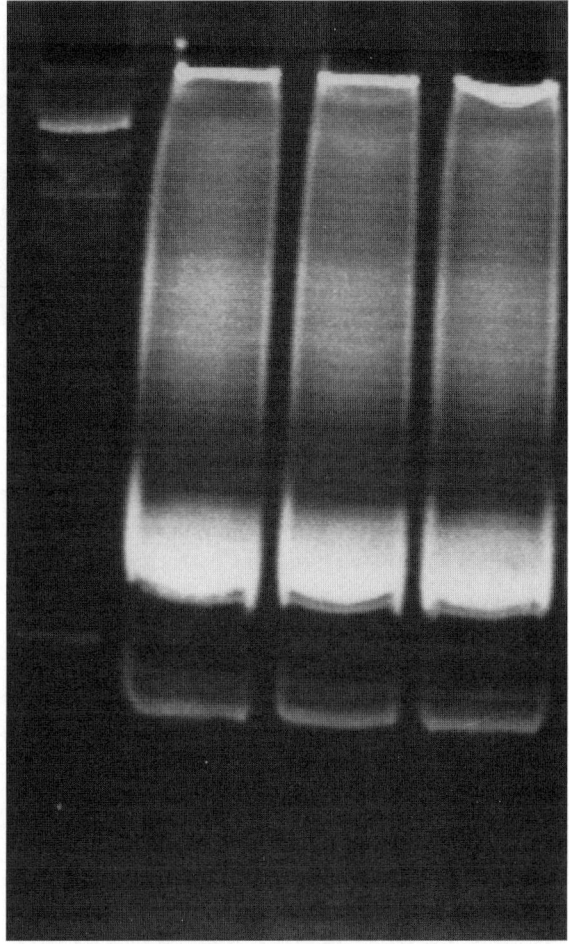

Fig. 3. LongSAGE bulk PCR. Amplified ditags should run at 130-bp while a background band runs at about 90-bp. On this gel too much DNA was loaded, which results in a thick 130-bp fragment not running on the exact 130-bp length. Try to prevent this overloading.

2. Add the following reagents to each of the three tubes: 42 μL 100-bp ditag, 15 μL 10X Buffer 4, 2 μL 100X bovine serum albumin, 12 μL *Nla*III, and 79 μL DEPC water.
3. Mix the contents well and incubate at 37°C for 1–2 h.

3.10. Purification of Ditags

1. Important: Perform everything on ice (*see* **Note 18**).
2. Adjust the volume in each tube to 200 μL with LoTE.
3. Add an equal volume of phenol/chloroform to each tube and centrifuge at maximum speed in a microcentrifuge for 5 min at 4°C.

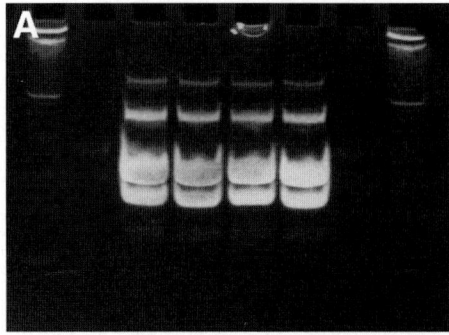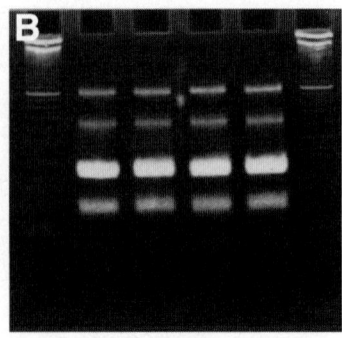

Fig. 4. Digestion of 130-bp fragment with *Nla*III. (**A**) On the gel, approx 34-bp fragment should be observed for the ditags and approx 43-bp fragment, which contains the
adapters. The other fragments are generated owing to incomplete *Nla*III digestion. If
separation was not optimal, check to see whether the isolated 34-bp fragment is pure
(*see* **Fig. 5**). A good separation should look like the one demonstrated for normal SAGE
in **Fig. 4B**. The adapter band of about 43-bp should be about two times as intense as the
ditag band. The extra linker purification was not performed.

4. Transfer aqueous phase (top) to a new tube.
5. Ethanol precipitate 200 µL sample as described in **Subheading 3.16.4.**
6. Wash twice with 1 µL of 70% ethanol; and air-dry the pellets for 5–10 min.
7. Combine the pellets and resuspend in 32 µL LoTE total. Make sure the pellets are
   completely dissolved.
8. Add 3.2 µL 0.5 *M* NaCl and 8 µL 5X loading buffer.
9. Load this sample into four lanes of a 12% polyacrylamide gel (10-well) (*see*
   Subheading 3.16.1.) use a 10-bp ladder as a marker.
10. Run at 80 V for 1–2 h (until Orange G is about 1-cm from the bottom of the gel)
    stain gel using ethidium bromide (**Figs. 4** and **5**).
11. Cut out the 34-bp band from the four lanes, (determine which band to cut: start to
    count down from the 100-bp band from the marker; the 10-bp fragment might
    have run off the gel).
12. Place four cut-out-bands in a 0.5-mL microcentrifuge tube, which is pierced with
    an 18-gauge needle.
13. Place the tube in 2-mL microcentrifuge tube (with cut off lid) and spin in
    microfuge at full speed for 2 min.
14. Discard 0.5-mL tube, add 475 µL TE and 25 µL 10 *M* ammonium acetate to 2-mL
    tube.
15. Vortex the tubes, and place at 37°C (not 65°C) for 2 h.
16. Incubate at 4°C overnight, if there is not enough time to complete the elution.
17. Use S.N.A.P. column as above to isolate eluate.
18. PC8 extract.
19. Ethanol precipitate 500-µL sample in a 2-mL tube as described in **Subheading
    3.16.4.**

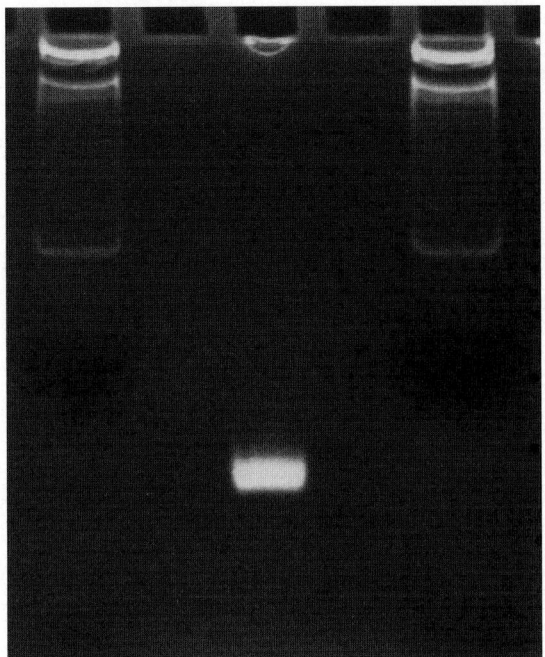

Fig. 5. Confirmation of ditag purity. If the separation of the ditags from the adapters is not efficient then check on gel if the isolated ditags are pure.

20. Wash twice with 1 mL of cold 75% ethanol. Carefully remove supernatants.
21. Resuspend in 7.75 μL cold TE (*see* **Note 19**).

3.11. Ligating the 34-bp Ditag to Yield Concatemers

Once the gel-purified 34-bp ditags are ready to ligate to form concatemers. The protocols for this step differ. Some protocols use a short ligation time. In the hands, this was not very reliable. A long ligation followed by digestion with *Sph*I were performed, in this way to generate concatemers with good adhesive, and thus easy to clone, ends. The concatemer bands of various sizes are separated on a polyacrylamide gel. The high molecular weight bands are excised and purified from the gel.

3.11.1. Ligation Reaction and Digestion

1. Set up a ligation reaction on ice using the gel-purified 34-bp DNA: 7.75 μL 34-bp DNA, 1 μL 10X ligase buffer, 1.25 μL T4 DNA ligase.
2. Incubate for 5 h to overnight at 16°C, to obtain long concatemers.
3. Add LoTE to 200 μL and PC8 extract.
4. Ethanol precipitate 200 μL sample as described in **Subheading 3.16.4.**
5. Mix well and place on ice for 10 min.

6. Spin 15 min at full speed.
7. Wash with 70% ethanol.
8. Dissolve pellet in 8 μL LoTE (note this step is different from the Long-SAGE manual): add: 2 μL buffer M, 2 μL *Sph*I (Roche 10 U/μL) and 8 μL H$_2$O.
9. Incubate 30 min (exact) at 37°C.
10. Add LoTE to 200 μL and PC8 extract.
11. Ethanol precipitate 200 μL sample as described in **Subheading 3.16.4.**
12. Mix well and place on ice for 10 min.
13. Spin 15 min at full speed.
14. Wash twice with 70% ethanol.
15. Air-dry the pellet and resuspend in 8 μL of LoTE.

3.11.2. Gel Purification of Concatemers

1. Add 2 μL 5X loading buffer to the concatemers (8 μL).
2. Incubate at 65°C for 10 min, then place on ice.
3. In the first lane of an 8% polyacrylamide gel (*see* **Subheading 3.16.2.**, using a small vertical gel system, thin spacers, 10 well comb, Bio-Rad), load 10 μL 1 kb ladder (25 ng/μL) as a marker.
4. Load entire concatenated sample into the third well (i.e., one lane).
5. Samples are run at 100 V for 2 h.
6. Stain gel with ethidium bromide and cut out the desired fraction. Concatemers will form a smear on gel with a range from about 100 bp to several kilobytes. The following regions were usually isolated: fraction G: 1200 bp—top of the gel; fraction M: 500–1200-bp and fraction S: bottom of the gel—500-bp **(Fig. 6)**.
7. Place each of these gel pieces into a 0.5-mL microcentrifuge tube of which the bottom is pierced with an 18-gauge needle, and place the tubes in a 2-mL microcentrifuge tube (with cut off lid) and spin in microfuge at full speed for 2 min.
8. Discard 0.5-mL tubes, add 300 μL LoTE to 2-mL tubes.
9. Vortex the tubes, and place at 65°C for 2 h.
10. Transfer contents of each tube to an S.N.A.P. column fitted to a collection tube.
11. Spin S.N.A.P. column in microcentrifuge for 5 min at full speed.
12. Transfer 300 μL eluate to a 1.5-mL tube and ethanol precipitate as described in **Subheading 3.16.4.**
13. Wash twice with 70% ethanol, and centrifuge and remove ethanol.
14. Resuspend purified concatemer DNA in 5 μL of LoTE.

3.12. Cloning Concatemers Into pZero-1

In this section, you will clone the gel-purified concatemers into the pZero-1 vector. Prior to the cloning you need to linearize pZero with *Sph*I. This step can be performed while gel-purifying the concatemers. Once you have linearized the vector should test it using a test insert. *See* **Subheading 3.16.5.**, for the protocols. We routinely store the digested vector at −20°C and test the linearized pZero-1 before use (*see* **Note 20**).

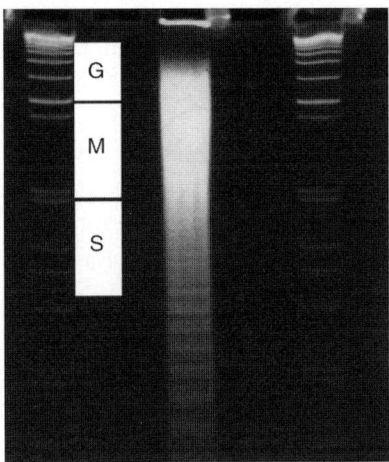

Fig. 6. Visualization of concatemers. The created concatemers will form a smear on gel in a range from about 100 bp to several kilobytes. The following regions were usually isolated; fraction G: 1200-bp—top of the gel; fraction M: 500–1200 bp; fraction S: bottom of the gel—500 bp. Only fraction M will be used for the generation of the library. If this will not result in an appropriate library one can use the other two fractions, but in our hands the fraction M is the most efficient one.

1. Prepare the following mix:

Contents	Sample (µL)	Vector alone control (µL)
Purified concatemer (fraction 1, 2 or 3)	5	–
dH$_2$O	2	7
pZero cut with *Sph*I (25ng/µL)	1	1
10X Ligase buffer	1	1
T4 ligase (1 U/µL)	1	1

2. Incubate 2 h at 16°C.
3. Bring sample volume to 200 µL with LoTE.
4. Extract 200 µL sample with equal volume PC8, ethanol precipitate as described in **Subheadings 3.16.3.** and **3.16.4.**
5. Wash four times with 70% ethanol, air-dry the pellet.
6. Resuspend in 6 µL LoTE.
7. Transform 2 µL ligation product into 25–40 µL ElectroMAX DH10Bs by electroporation (*see* **Subheading 3.16.** for protocol). Note this is different from the SAGE manual.
8. Plate each transformation on 10 big plates (low salt LB containing 100 µg/mL Zeozin), 200 µL per plate. For the vector control one plate containing 200 µL transformed bacteria is sufficient.

9. Incubate overnight at 37°C. Insert containing plates should have hundreds to thousands of colonies whereas control plates should have 0 to tens of colonies (at least 10 times less).

3.13. Screening Transformants

The following steps of the SAGE protocol are different from the SAGE manual. If enough colonies are obtained you are going to check insert sizes of the transformants by colony PCR. For this PCR you use the vector primers SP6a and T7a.

3.13.1. Test Colony PCR

1. Pick colonies and inoculate in 96-well plates containing 150 μL of LSLB + 100 μg/mL Zeozin per well.
2. Grow overnight at 37°C.
3. Set up 96 PCR reactions per ligated fraction of concatemers using the following conditions per reaction: 2.5 μL 10X PCR buffer, 2 μL MgCl$_2$ (25 mM), 4 μL dNTPs (1.25 mM), 1 μL T7a primer (10 ng/μL), 1 μL SP6a primer (10 ng/μL), 12.25 μL dH$_2$O and 0.25 μL Taq polymerase.
4. Aliquot 23 μL of PCR mix in 96-well plate.
5. Add 2 μL of the cultures as template in the PCR reactions.
6. Perform PCR at the following temperatures: 1 cycle 95°C 2 min at start; 30 cycles: 95°C 30 s, 50°C 1.30 min, 70°C 1.30 min; 1 cycle 70°C 5 min at end, and finally 4°C.
7. Run 5 μL of these PCR products on a 3% agarose gel containing ethidium bromide.
8. Visualize on a UV tray **(Fig. 7)**.

3.13.2. Scale Up Colony PCR

In this last part of the SAGE protocol we perform the pipetting steps using a Beckman Biomek Robot, which has a MJ Research Tetrad PCR machine attached to it.

1. Pick colonies and inoculate in 96-well plates containing 150 μL of LSLB + 100 μg/mL Zeozin per well.
2. Grow overnight at 37°C.
3. Make a bacterial lysate by adding 95 μL water to 5 μL bacterial suspension, heat 5 min at 95°C. The lysates are stored at −20°C (*see* **Note 21**).
4. Perform colony PCRs as above, making the mix for 22 plates (20 plates + 20% extra mix) use as template 2 μL of bacterial lysate.
5. Check PCR products of several plates (e.g., plate 1, 10, and 20) by running 5 μL on 3% agarose gel.
6. If the PCR products are of good quality, and give the expected insert size and percent of insert, proceed to sequencing.

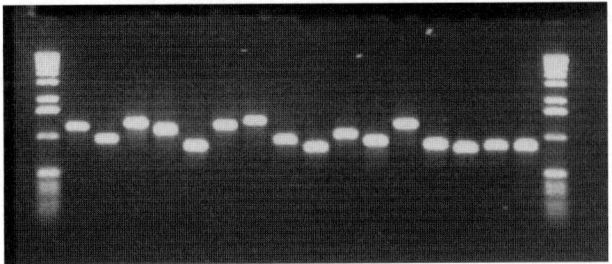

Fig. 7. Test colony PCR. The percentage of good inserts for each ligation and determine, which ligation to use for sequencing depending on insert size (>500 bp) and percentage of insert (>80%) were determined. Calculate how many clones require to PCR, to obtain the desired amount of tags after sequencing. Most of the time ligation M was used (containing concatemers fraction M: 500–1000 bp inserts) and inoculate 20X 96-well plates with colonies for the colony PCR.

3.14. Sequencing

Mix for one sequence reaction: 1 µL T7a primer (10 ng/µL), 0.7 µL Big Dye Terminator mix V1.1, 3.3 µL 2.5X Sequence dilution buffer, and 3 µL dH$_2$O (*see* **Note 22**).

1. Make the right amount of mix.
2. Add 100 µL H$_2$O to the PCR reactions.
3. Combine: 2 µL diluted PCR product with 8 µL sequence reaction mix.
4. Do not add oil. So use a PCR machine with a heated lid and PCR tubes with lids.
5. Perform sequence reaction: 1 cycle 96°C for 15 s, 30 cycles: 96°C for 30 s, 50°C for 10 s, 60°C for 4 min, and forever 4°C.
6. Raise the volume to 20 µL with dH$_2$O, purify the sequence samples and run the sequence on a capillary sequencer.

3.15. Analysis of the Data

For data analysis many different programs are available. The SAGE2000 software (*see* **Subheading 1.**) is very easy to use and freely available for academic institutions (*see* www.sagenet.org). Use this software for tag extraction and the data can be analyzed with the appropriate statistical programs (*see* **ref. 4**).

3.16. Gel Electrophoresis and Miscellaneous Methods

3.16.1. Gelelectrophoresis 12% PAGE

1. Prepare the following mixture for 1 gel: 1.5 mL 40% polyacrylamide (19:1 acrylamide: bis), 2.4 mL dH$_2$O, 1 mL 5X TBE buffer, 50 µL 10% APS, and 4.3 µL TEMED.

2. Mix above and add to vertical gel apparatus (Bio-Rad small system, thin spacers and comb, 10 wells). Add comb and let gel sit at least 30 min to polymerize.
3. Run in 1X TBE buffer.

3.16.2. Gel Electrophoresis 8% PAGE (for Separating Concatemers)

1. Prepare the following mixture for 1 gel: 1 mL 40% polyacrylamide (29:1 acrylamide: bis), 3.9 mL dH$_2$O, 100 µL 50X TAE buffer, 50 µL 10% APS, and 4.3 µL TEMED.
2. Mix above and add to vertical gel apparatus (Bio-Rad small system, thin spacers and comb, 10 wells). Add comb and let gel sit at least 30 min to polymerize.
3. Run in 1X TAE buffer.

3.16.3. PC8 Extraction

1. Add equal volume PC8 to sample.
2. Vortex for several second.
3. Spin for 2 min at full speed in microcentrifuge.
4. Transfer aqueous (top) layer to a new microcentrifuge tube.

3.16.4. Ethanol Precipitation

1. Precipitate the DNA with one-third of 10 M ammonium acetate (final conc. 2.5 M).
2. Add 3 µL glycogen (20 mg/mL).
3. Mix well and finally add 2.5 vol of ethanol 100%.
4. Incubate for 30 min at −80°C and centrifuge 30 min at 4°C full speed, or store at −20°C overnight and centrifuge 30 min at 4°C full speed.

3.16.5. PZero Preparation and Testing

3.16.5.1. DIGESTION

1. Digest 2 µL pZero (=2 µg) in 2 µL digestion buffer, supplemented with 2 µL *Sph*I (10U/µL) and 14 µL dH$_2$O (*see* **Note 23**).
2. Incubate 1 h at 37°C (do not overdigest).
3. 10 Min 65°C.
4. Raise volume with LoTE to 200 µL.
5. Extract with equal volume PC8.
6. Ethanol precipitate the 200 µL sample as described in **Subheading 3.16.3.**
7. Spin directly at full speed for 15 min.
8. Wash twice with 70% ethanol.
9. Resuspend in 80 µL of dH$_2$O (25 ng/µL).

3.16.5.2. ISOLATION OF TEST INSERT

1. Digest an insert-containing plasmid with *Nla*III.
2. Run on agarose gel.
3. Slice out *Nla*III fragment(s).
4. Isolate the fragment from gel using the Qiaquick gel isolation kit.
5. Run 2 µL digested pZero next to 2 µL of the isolated test insert.

6. Dilute the test insert until the concentration is more or less the same as the 2 μL of pZero x *Sph*I.

3.16.5.3. TEST LIGATION

1. Prepare the following ligation reactions: +insert: 1 μL pZero *Sph*I 25 ng/μL, 2 μL Test insert, 1 μL 10X ligbuffer, 1 μL ligase (1 U/μL), and 5 μL dH₂O. (insert: 1 μL pZero *Sph*I 25 ng/μL, 0 μL Test insert, 1 μL 10X ligbuffer, 1 μL ligase (1 U/μL), and 7 μL dH₂O.
2. Incubate 2 h at 16°C.
3. Raise volume to 200 μL with LoTE.
4. Extract with equal volume of PC8.
5. Ethanol precipitate 200 μL sample as described in **Subheading 3.16.4.**
6. Put on ice for 30 min.
7. Spin at full speed for 15 min.
8. Wash four times with 70% ethanol.
9. Air-dry the pellet and dissolve in 6 μL of LoTE.
10. Transform 2–25 μL of DH10B bacteria, by electroporation.
11. Plate 200 μL of bacteria on one big plate (low salt LB/100 μg/mL Zeozin).
12. Incubate overnight 37°C; The +insert ligation should contain at least 10 times more colonies as compared to the –insert ligation.

3.16.5.4. TRANSFORMATION OF DH10B CELLS BY ELECTROPORATION

1. Put white cuvet holder on ice.
2. Mix 25 μL (when using 1-cm cuvet) or 40 μL (when using 2-cm cuvets) of DH10B cells (with 2 μL of ligation product, mix and transfer to cuvet.
3. Be sure the cells are spread on the bottom, without air bubbles.
4. Let stand on ice for 2 min.
5. Dry the cuvet and the cuvet holder with tissues.
6. Place cuvet in holder and holder in machine.
7. Charge the electroporator by pushing the "raise and lower" buttons at the same time, when using 1 cm cuvet: push one time: 1.8 kV, when using 2-cm cuvet: push two times: 2.5 kV.
8. Pulse by pushing the pulse buttons at the same time; keep pushing until the beep.
9. As quickly as possible add 1 mL SOC to the cells and transfer them back to the 1.5-mL tube.
10. Check time constant by pushing the "set volts and actual volts" at the same time (should be around 5 ms).
11. Recover cells 1 h at 37°C shaker (exact).
12. Plate cells (200 μL per big plate; low salt LB/ 100 μL g/mL Zeozin).

3.16.6. Ethidium Bromide Dot Quantitation

For the dot quantification you need a DNA standard which is double-stranded and of about the same length as the DNA you want to quantify. For the dot quantification of the ditags, double strand DNA of about 30-bp is needed. Two complementary primers can be annealed and serve as the standard. Mix 10 μg of each primer in 1X H-buffer (restriction buffer of Boehringer), at a final volume of 20 μL. Heating up to 95°C and then place at 37°C for 30 min. Store at –20°C.

1. Use this DNA to prepare the following standards: 0, 1, 2.5, 5, 7.5, 10, 20 ng/μL, respectively.
2. Use 1 μL of sample DNA to make 1/5, 1/25, and 1/125 dilutions in LoTE.
3. Add 4 μL of each standard or 4 μL of each diluted sample to 4 μL of 1 μg/mL ethidium bromide.
4. Mix well.
5. Place a sheet of plastic wrap on a UV transilluminator.
6. Spot each 8 μL mix on plastic wrap and photograph under UV light.
7. Estimate DNA concentration by comparing the intensity of the sample to the standards.

4. Notes

1. Read the protocol carefully before starting and be aware that digesting and testing of the pZero vector is time-consuming. The SAGE protocol contains many steps and in every step something can go wrong: For first time users, just follow the protocol exact, do not make changes!
2. For each SAGE library aliquot all of the solutions and enzymes; only then can you be sure that you will not contaminate your stock solutions and then contaminate your next SAGE library. It is not possible to stop at every stage of the protocol; but can stop after every precipitation step, after resuspending your product and were the protocol indicates you can stop.
3. Because the SAGE protocol contains many PCR amplification steps, contamination of the pipets or solutions will definitively kill the procedure. So aliquots all solutions, physically separate the steps before and after a PCR reaction to prevent contamination.
4. Use siliconized microcentrifuge tubes until the 130-bp PCR; Aerosol resistant pipet tips are recommended for all procedures; use only sterile, new pipet tips and microcentrifuge tubes; Wear latex gloves while handling reagents and RNA samples to prevent RNase contamination.
5. Be sure to check integrity of RNA by gel electrophoresis, when possible use about 25 μg of total RNA as starting material.
6. If there is a precipitate in the lysis/binding buffer, warm the buffer to 37°C to dissolve the precipitate.
7. During incubation: prepare 400 μL 1X 1st strand buffer; prepare first strand reaction mix.
8. If DTT is opaque/cloudy incubate 10 min at 37°C.
9. During incubation: prepare for second strand synthesis and preheat wash buffer C to 75°C.
10. Add *E. coli* DNA ligase, *E. coli* DNA polymerase, and *E. coli* RNase H at the end.
11. Clumping of the beads will interfere with inactivation and removal of *E. coli* DNA Polymerase. The exonuclease activity of this enzyme may interfere with subsequent

reactions by digesting single stranded overhangs used for subsequent cohesive ligations. It is advisable to inactivate the enzyme immediately after the cDNA synthesis and *Nla*III digestion, by adding wash buffer C and heating.

12. Increase the number of washes, if clumping of the beads occurs.

13. Do not have enough time to complete the next step, store the beads in 1X buffer 4 at 4°C overnight. The next day, proceed to digesting cDNA with *Nla*III (**Subheading 3.3.**).

14. *Nla*III is extremely sensitive to high temperatures. Do not keep the enzyme at room temperature or 4°C for long periods. Use immediately upon removal from –80°C and return the enzyme to –80°C as soon as possible.

15. Always set up the negative control reactions (no template and no ligase) first, and then overlay with 30–40 µL of mineral oil before adding the template to prevent any accidental contamination from the template.

16. To avoid cross-contamination, perform these experiments at different locations. Transfer the negative control to location 1 and the template to the PCR or laminar flow hood in location 2.

17. Digesting the 130-bp ditag with *Nla*III produces a 34-bp ditag. It is important to achieve >80% digestion efficiency with *Nla*III to obtain a good yield of 34-bp ditags.

18. The ditags are small double-stranded DNA fragments. That means that they easily melt and when the DNA is single-stranded you will lose it. So from now on: It is essential to keep the ditags cold!! Perform everything on ice, spin at 4°C, and do not dry in a Speedvac.

19. At least 100-ng of ditags for a concatemer ligation; more optimal is several hundreds of nanograms; with a maximum of 400-ng per ligation reaction. If you are not sure about the amount of ditags you may perform a dot quantification (*see* **Subheading 3.16.6.**).

20. If the efficiency is no longer good, we prepare a fresh digestion. Always check the vector first. Making a new batch of vector is much less work then making a new SAGE library.

21. Be sure the bacteria are resuspended! When plates are stored at 4°C for more than one day, the bacteria should be resuspended either by hand or by robot!

22. Before you start sequencing, the optimal dilutions of terminator mix must be determined; different dilutions of the big dye terminator mix and different amounts of input PCR product should be tested. We routinely dilute the Big Dye Terminators 1:6 and the input 1:6.

23. The pZero vector is digested and tested before it can be used in the concatemer ligation. The digested pZero is unstable at –20°C, so before every concatemer ligation the vector should be tested.

References

1. Velculescu, V. E., Zhang, L., Vogelstein, B., and Kinzler, K. W. (1995) Serial analysis of gene expression. *Science* **270**, 484–487.
2. Lash, A. E., Tolstoshev, C. M., Wagner, L., et al. (2000) SAGEmap: a public gene expression resource. *Genome Res.* **7**, 1051–1060.

3. van Kampen, A. H., van Schaik, B. D., and Pauws, E. (2000) USAGE: a web-based approach towards the analysis of SAGE data. Serial analysis of gene expression. *Bioinformatics* **10,** 899–905.
4. Ruijter, J. M., Van Kampen, A. H., and Baas, F. (2002) Statistical evaluation of SAGE libraries: consequences for experimental design. *Physiolog. Genomics* **11,** 37–44.
5. Baggerly, K. A., Deng, L., Morris, J. S., and Aldaz, C. M. (2004) Overdispersed logistic regression for SAGE: modelling multiple groups and covariates. *BMC Bioinforma.* **5,** 144.
6. Man, M. Z., Wang, X., and Wang, Y. (2000) POWER_SAGE: comparing statistical tests for SAGE experiments. *Bioinformatics* **16,** 953–939.
7. Margulies, E. H. and Innis, J. W. (2000) eSAGE: managing and analysing data generated with serial analysis of gene expression (SAGE). *Bioinformatics* **7,** 650–651.
8. Morris, J. S., Baggerly, K. A., and Coombes, K. R. (2003) Bayesian shrinkage estimation of the relative abundance of mRNA transcripts using SAGE. *Biometrics* **3,** 476–486.
9. Caron, H., van Schaik, B., van der Mee, M., et al. (2001) The human transcriptome map: clustering of highly expressed genes in chromosomal domains. *Science* **291,** 1289–1292.
10. Chen, J. J., Rowley, J. D., and Wang, S. M. (2000) Generation of longer cDNA fragments from serial analysis of gene expression tags for gene identification. *Proc. Natl. Acad. Sci. USA* **97,** 349–353.
11. Polyak, K., Xia, Y., Zweier, J. L., Kinzler, K. W., and Vogelstein, B. (1997) A model for p53-induced apoptosis. *Nature* **389,** 300–305.
12. Saha, S., Sparks, A. B., Rago, C., et al. (2002) Using the transcriptome to annotate the genome. *Nat. Biotechnol.* **5,** 508–512.

5

Gene Expression Profile Analysis of Tumors

Katia Basso and Riccardo Dalla-Favera

Summary

Gene expression profiling is a powerful tool to analyze the complexity of cancer biology. Recent methods allow the generation of gene expression profiles for all known genes in the human genome. The genome-wide analysis of the gene expression patterns of neoplastic and normal cells provides insights into: (1) the identification of previously unknown tumor subtypes; (2) the normal cellular counterparts of tumor cells; (3) the identification of cellular pathways that may be affected by malignant transformation; (4) the identification of new diagnostic markers and potential therapeutic targets. This chapter summarizes experimental approaches addressing these goals using examples from studies on B-cell malignancies.

Key Words: Oligonucleotides arrays; B-cell malignancies; gene expression profiling; gene expression data analysis; molecular classification of tumors; cancer.

1. Introduction
1.1. General Concepts

Gene expression profile analysis has been successfully used in the last few years to investigate various biological and clinical aspects of cancer *(1–7)*. Parallel expression measurements of virtually all genes in the genome produce a set of data that can help to investigate the complexity of the tumor phenotypes.

cDNA or oligonucleotide microarrays are commonly used tools to measure large-scale gene expression. In general, microarrays contain a grid of DNA molecules (cDNA or oligonucleotides) that are complementary to their respective RNA target. The cDNA microarrays require the hybridization of two differentially labeled RNAs (control and test) and thus provide a relative expression measure for each target compared to a reference sample.

From: *Methods in Molecular Biology, vol. 383: Cancer Genomics and Proteomics: Methods and Protocols*
Edited by: P. B. Fisher © Humana Press Inc., Totowa, NJ

Oligonucleotide microarrays contain multiple oligonucleotides for each target and provide an absolute measurement of the abundance of the RNA species represented on the arrays. cDNA microarrays can be produced in house via relatively easy and inexpensive procedures that require the spotting of cDNA on slides. Moreover, it is possible to generate custom microarrays containing only genes of interest (e.g., tissue specific cDNA). Oligonucleotide microarrays are usually commercially available and are produced by either one of two techniques: *in situ* synthesis of oligonucleotides using a photolithographic approach *(8)* or deposition of oligonucleotides on special matrices. These techniques allow the production of high density microarrays containing multiple oligonucleotides for each gene and recently have been made representative of virtually all known genes. Thus, such high density microarrays are powerful tools to generate whole genome gene expression profiles. The variability among arrays generated by *in situ* synthesis of oligonucleotides is extremely low offering a good starting point for highly reproducible experiments. However, gene expression profile analyses are limited by the fact that data generated using cDNA and oligonucleotide microarrays as well as data generated with different oligonucleotide array types or protocols cannot be readily used in the same analysis *(9)*.

1.2. Data Analysis

The approaches currently used to analyze gene expression data can be conceptually divided into two main groups: unsupervised and supervised approaches (**Fig. 1**). Unsupervised analyses allow the identification of components of a dataset that share common expression patterns without using any *a priori* knowledge about the biological identity of each sample. In practice, unsupervised learning is employed to a dataset to detect internal structures or previously unknown relationships among samples, for example, it may identify subgroups within a single tumor entity on the basis of expression profile differences. Supervised approaches are suited to identify genes that are differentially expressed between predefined groups of samples. The supervised learning requires an *a priori* knowledge to identify groups of samples that will be subsequently compared in order to identify genes whose expression levels differ significantly between the groups. Several methods based on unsupervised and supervised analyses have been developed (**Table 1**), and each user should identify the approach that best fits the biological questions that need to be addressed. The results of these analyses are generally lists of genes associated with a certain phenotype. The most challenging part of the data analysis is the assignment of the genes to functional categories and biological pathways in order to recognize the biological processes underlying the single genes. This problem is still largely unresolved mainly

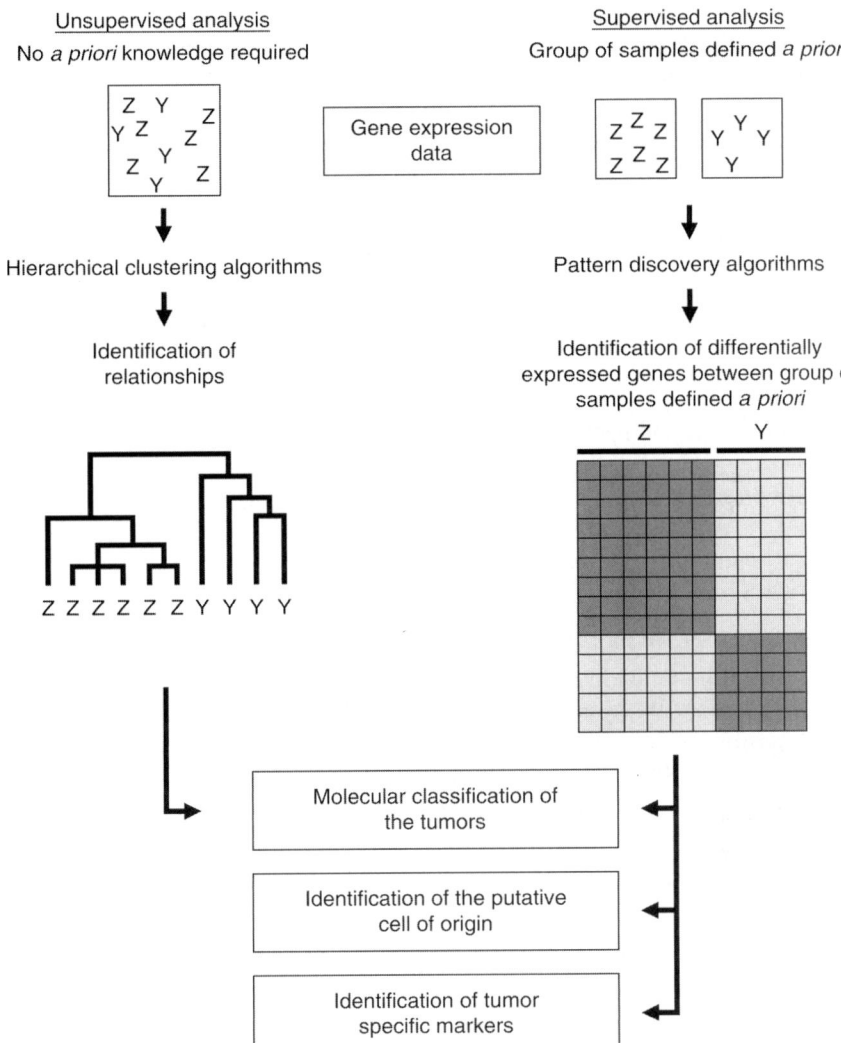

Fig. 1. Schematic representation of the data analysis approaches used to investigate cancer by gene expression profiling.

because of the limited amount of information regarding the function of a sizeable fraction of genes.

1.3. Applications to Cancer

Gene expression profiling of cancer is a first step in the dissection of the tumor phenotype and allows the identification of new diagnostic markers and potential therapeutic targets.

Table 1
List of Freely Available Software Designed to Perform Unsupervised and Supervised Analyses

Software	URL
Genes@Work	http://www.research.ibm.com/FunGen/FGDownloads.htm
dChip	http://www.biostat.harvard.edu/complab/dchip/
Cluster and TreeView	http://rana.lbl.gov/EisenSoftware.htm
GeneCluster 2.0	http://www.broad.mit.edu/cancer/software/genecluster2gc2.html
SAM: Significance analysis for microarrys	http://www-stat.stanford.edu/%7Etibs/SAM/index.html
PAM: Prediction analysis for microarrays	http://www-stat stanford.edu/~tibs/PAM/

1.3.1. Molecular Classification of Tumors

A first question to address when approaching gene expression analysis of a certain tumor type is the definition of its biological identity. The molecular classification of cancer by gene expression profiling was first attempted in a study on acute lymphoblastic leukemia (ALL) and acute myeloid leukemia (AML) that showed the ability to perform class discovery and class prediction using gene expression data *(1)*. ALL and AML could be successfully distinguished and a class predictor able to assign a new case to the correct class was generated. Class discovery was later applied to larger number of tumor subtypes and gene expression analysis was able to confirm the molecular homogeneity of several tumor types (e.g., B-cell chronic lymphocytic leukemia, hairy cell leukemia, and mantle cell lymphoma) as defined by morphologic and immunophenotypic classification *(3,7,10,11)*. However, class discovery by gene expression profiling also showed that not always what is defined as a specific tumor type by the pathologists is a uniform biological entity. For instance, distinct types of diffuse large B-cell lymphoma, the most common non-Hodgkin lymphoma of adults, were recognized by gene expression profiling *(2)*.

The identification of a subset of genes specifically associated with a phenotype (signature genes) allows performing class prediction. A class predictor is a subset of genes whose expression analysis allows assigning unknown cases to a defined tumor phenotype. The accuracy of a predictor generally depends on the number of samples that have been used to identify the signature genes. Predictive signatures can be used in addressing several issues with the common goal of assigning single samples to previously defined subgroups. For instance, class predictors were generated in order to predict survival in diffuse large B-cell lymphoma *(12)* and response to treatment in ALL *(6)*. A main application will be the use of specific gene expression signatures associated with prognosis in order to predict the outcome of a patient already at the time of diagnosis and accordingly develop appropriate therapeutic strategies.

1.3.2. Identification of the Putative Cell of Origin

The identification of the normal counterpart of a tumor cell is essential to reconstruct the aberrant behavior induced by malignant transformation. The relationship of a tumor with normal cells is often addressed by the identification of a few specific markers associated with cellular differentiation. However, it is important to consider the possibility of the aberrant expression of single molecules in the transformed cells. Gene expression profiling of normal cell populations enlarges the number of genes associated with a stage of normal development that can be used to track the normal counterpart of a tumor phenotype. Using gene expression profiles from purified mature B cells representative of preGC, GC and post-GC B cells, a relationship of B-CLL and HCL with post-GC memory B cells was uncovered *(3,7)*. Following the identification of memory B cells as the most related normal cell population to HCL cells it was possible to analyze cellular programs that have been more dramatically affected by the malignant transformation process *(7)*. Overall, the identification of the putative cell of origin represents the starting point for analysis of aberrant phenotypes in the respective tumor cells.

1.3.3. Identification of Tumor-Specific Markers

Gene expression analysis is a powerful tool for the identification of tumor-associated molecules that could represent new diagnostic or prognostic markers as well as therapeutic targets. Ideally, a diagnostic marker should be specifically expressed in the investigated tumor but not in normal cells nor in related tumors. A therapeutic target should not be expressed in the normal cells and it should be susceptible to targeting by pharmacological manipulation of the corresponding cellular pathway or by one of various possible approaches including inhibition by small peptides or Ab-mediated immunotherapy. The comparison between gene expression profiles of the cancer under investigation and other related tumor types and normal cell populations leads to the detection of patterns of genes whose expression is strictly associated to that specific transformed phenotype. Thereby, the larger the representation of normal and transformed cell types related to the tumor under investigation, the more accurate will be the specificity of the selected tumor-associated genes. The identification of prognostic markers requires the availability of gene expression profiles of large panels of cases with similar clinical characteristics, but different prognosis. Prognostic markers are not required to be exclusively expressed in the tumors, but should be differentially expressed in subgroups of the tumor samples that show differences in their outcome. Before considering their potential as diagnostic or prognostic markers or therapeutic targets, the identified genes need to be further validated at the protein level and in a large panel of cases.

2. Materials

The following reagents and materials are recommended by Affymetrix (Santa Clara, CA).

2.1. Sample Preparation

2.1.1. RNA Purification

1. Homogenizer.
2. Trizol Reagent (Invitrogen, Carlsbad, CA).
3. RNeasy Mini Kit (Qiagen, Valencia, CA).
4. UV spectrophotometer.
5. Gel electrophoresis unit with appropriate buffer.
6. Agilent 2100 Bioanalyzer (Agilent Technologies) (Optional, Palo Alto, CA).
7. RNA 6000 Nano LabChip kit (Agilent Technologies) (Optional).

2.1.2. cDNA and Labeled cRNA Synthesis

1. To perform one cycle target labeling:
   a. One-cycle Target Labeling and Control Reagents (Affymetrix).
2. To perform two-cycle target labeling:
   a. Two-cycle Target Labeling and Control Reagents (Affymetrix).
3. MEGAscript High Yield Transcription Kit (Ambion, Austin, TX).

2.1.3. Hybridization, Staining, Washing, and Scanning

1. GeneChip® instrument system (Affymetrix).
2. GeneChip arrays (Affymetrix).
3. Bovine serum albumin solution (50 mg/mL) (Invitrogen).
4. Herring sperm DNA (Promega, Madison, WI).
5. GeneChip Eukaryotic hybridization control kit (Affymetrix).
6. 5 M NaCl RNase-free (Ambion).
7. MES hydrate (Sigma, St. Louis, MO).
8. MES sodium salt (Sigma).
9. EDTA disodium salt, 0.5 M solution (Sigma).
10. DMSO (Sigma).
11. Surfact-Amps 20 (Tween-20), 10% (Pierce Chemical, Rockford, IL).
12. R-Phycoerythrin streptavidin (Molecular Probes).
13. Phosphate-buffered saline, pH 7.2 (Invitrogen).
14. 20X Sodium chloride, sodium phosphate, EDTA (SSPE): 3 M NaCl, 0.2 M NaH$_2$PO$_4$, 0.02 M EDTA (Cambrex Bio Science, Rockland Inc., Rockland, ME).

15. Goat IgG, reagent grade (Sigma).
16. Antistreptavidin antibody (goat), biotinylated (Vector Laboratories, Burlingame, CA).

3. Methods

3.1. Sample Selection

1. The first step before starting a gene expression profiling project is to carefully select the samples that will be used in the study. The samples must be homogeneous regarding biological and manipulation variables. The most commonly available tumor samples are diagnostic specimens, where tumor cells represent a variable part of the tissue. The malignant cell fraction must be evaluated and quantified by standard diagnostic procedures and only samples containing a similar percentage (preferentially >90%) of tumor cells should be analyzed. Moreover, the tumors may reside in different locations and therefore the normal infiltrating cells would be completely different among specimens. The infiltrating cells could represent an extremely small percentage nevertheless in a comparative analysis they will perform as a major player leading to distinguish the different specimens on the base of their nontumor components.
2. The manipulation of samples is a second important variable that needs consideration. The specimens should have been manipulated and stored in the same way. In order to obtain good quality RNA, the samples must be fresh or properly frozen (*see* **Note 1**).
3. Ideally, upon removal the specimens should be immediately placed in physiological solution, frozen in liquid nitrogen as soon as possible and stored in liquid nitrogen or at −80°C. Frequently specimens are included in optimal cutting temperature compound (OCT) before freezing: although it is possible to use these samples, to remove most of the OCT (working on dry ice to avoid thawing of the tissue) before proceeding to the RNA isolation was suggested. If the samples are derived at different institutions and/or have been manipulated differently the gene expression results should be monitored for patterns of expression associated with these variables.

3.2. Sample Preparation

The protocols reported here are for the generation of gene expression profiles using the Affymetrix oligonucleotide GeneChip arrays and platform.

3.2.1. RNA Isolation and Quality Control

1. RNA is isolated from fresh or frozen tissues using Trizol reagent (Invitrogen) and cleaned up using RNeasy columns (Qiagen). If starting from frozen tissue, do not allow the tissue to thaw but immediately transfer the frozen sample from the dry ice to the Trizol reagent and proceed with the homogenization (*see* **Note 2**).
2. RNA must be quantified using a spectrophotometer and its quality checked by agarose gel electrophoresis. Alternatively, it is possible to perform both quantification and quality check using the Agilent 2100 Bioanalyzer and the RNA 6000 Nano LabChip kit (Agilent technologies) following the manufacturer's instructions. If

working with limiting material, the Agilent 2100 Bioanalyzer system that requires as little as 25 ng of RNA to perform accurate quantification and quality check, may be useful.

3. Regardless of the approach used, it is mandatory to quantify and perform a quality check of the RNA. Low quality RNA will lead to biased gene expression data because of the small size RNA fragments used as template in the cDNA synthesis.

3.2.2. cDNA and Biotinylated cRNA Synthesis

Two different approaches are suggested by Affymetrix, the choice of which is depending on the available starting amount of RNA.

1. The first method (one-cycle protocol) requires 1–15 μg of total RNA that are subject to a retrotranscription process using oligo-dT primers containing a T7 sequence in order to obtain a double stranded cDNA that is subsequently used as template for a cRNA synthesis incorporating biotinylated nucleotides. The labeled cRNA is quantified (usually are obtained 40–100 μg) and its quality should be tested by gel electrophoresis or using the Agilent 2100 Bioanalyzer system. Good quality cRNA appears as a smear ranging between 600 and 2000 nt. Fifteen micrograms of biotinylated cRNA are fragmented. The fragmentation protocol is designed in order to break down full-length cRNA to fragments ranging from 35 to 200 bases. The efficiency of fragmentation should be confirmed by gel electrophoresis or using the Agilent 2100 Bioanalyzer system. The fragmented cRNA is dissolved in the hybridization buffer and applied to the microarray.
2. The two-cycle protocol is used for RNA samples whose starting amount is ranging from 10 to 100 ng. It requires an intermediate step of amplification before proceeding with the synthesis of T7 sequence-containing double strand cDNA and labeled cRNA. The cRNA fragments obtained using the two-cycle protocols are shorter (average size 800 nt) than those obtained with the one-cycle protocol. All samples of a gene expression profile study must be prepared starting from the same amount of RNA and using the same protocol (*see* **Note 3**).
3. The reagents are listed in **Subheading 2.** and the detailed protocols are reported in the Affymetrix Expression Analysis Technical Manual (Chapters 1 and 2) downloadable at: http://www.affymetrix.com/support/technical/manual/expression_manual.affx.

3.2.3. Hybridization, Staining, Washing, and Scanning

1. After overnight hybridization (approx 16 h) the microarrays are washed and stained with streptavidin–phycoerythrin solution followed by a signal amplification protocol using antistreptavidin biotinylated antibodies and streptavidin–phycoerythrin solution.
2. The washing and staining procedures are performed automatically using the Fluidics Station and take approx 90 min. Following these procedures, the microarrays can be immediately scanned or stored for several hours at 4°C in the dark.

3. The reagents are listed in **Subheading 2.** and the hybridization, staining, and washing protocols, as well as the procedure for scanning are detailed in the Affymetrix Expression Analysis Technical Manual (Chapter 3) downloadable at: http://www.affymetrix.com/support/technical/manual/expression_manual.affx.

3.3. Gene Expression Data Mining

1. In most cases a DNA microarray facility is in charge of the hybridization and scanning and the user receives the primary gene expression data. The expression data are stored in several file types: two files contain the imaging data (.dat and .cel); the .chp file contains the gene expression values generated by the Affymetrix software (MAS4.0, MAS5.0, GCOS, or the most updated one GCOS1.2); the .rpt file includes the information concerning the quality control of the sample and the hybridization.

2. The signal value generated by the Affymetrix software is calculated using an algorithm that integrates information from approx 11 probe pairs (in the most recent GeneChip arrays) each of them represented by an oligonucleotide perfectly matching the target sequence and a second one carrying a single mismatch in its middle part. The mismatched oligonucleotides allow controlling any nonspecific annealing that may occur. The software generates a "signal value," providing a quantitative measurement of the transcripts, a "detection," that defines each probe set as present, absent, or marginal (neither present nor absent), and a "detection *p*-value" giving an estimation of the presence of each probe set.

3. Before proceeding with the data analysis each sample must be controlled for quality. In particular the following information should be evaluated: 3′/5′ ratio for actin and GAPDH genes; percentage of present, absent and marginal genes; background and scaling factor. The optimal values and thresholds for each one of these parameters are reported in the Affymetrix GeneChip Expression Analysis Data Analysis Fundamentals (Chapter 1) downloadable at: http://www.affymetrix.com/support/technical/manual/ expression_manual.affx

3.4. Data Analysis

The analysis of gene expression data cannot be summarized in a single protocol (*see* **Note 4**). The tools available are numerous and increasing rapidly in number and power. Here, we will provide a general approach for gene expression analysis in tumors referring more to the concepts than to specific methods of analysis. A list of several freely available software is reported in **Table 1**.

3.4.1. Molecular Classification of Tumors

1. In order to investigate the homogeneity or heterogeneity of a tumor type is necessary to apply an unsupervised method that recognizes differences in the gene expression profiles without the bias of any *a priori* class definition.

2. The identification of subgroups within a tumor type should be strongly tested to exclude nonbiological influences (*see* **Subheading 3.1.**) that may be responsible for any clustering. The tumor samples may cluster according to manipulation, quality of RNA, date of hybridization, operator, and so on. These variables must be carefully checked before proceeding in the data analysis.

3. The subclusters of tumor samples identified on exclusion of environmental events are likely to represent biological subtypes of the tumor.

4. The availability of complete information for each sample, including clinical and cytogenetic data, may be essential to associate a specific gene expression pattern to clinical or genetic parameters.

5. Unsupervised analysis is the method of choice for the identification of relationships among different tumor types. The resulting dendrogram gives a visual measure of which tumors share a larger gene expression profile and appear to be more related among each other suggesting a common cellular origin or the acquisition of similar aberrant phenotypes upon malignant transformation. This type of analysis may be strongly affected by the fact that different tumors may reside in different locations and therefore include a different spectrum of infiltrating normal cells. Occasionally, also profiling the normal tissue could help to identify the genes associated with the tumor-infiltrating component. However, the most efficient approach to address this problem is the availability of gene expression profiles for at least a few samples of purified tumor cells.

6. Subsets of genes associated to a tumor type (signature genes) can be used as a "predictor" in order to assign unknown cases to a defined tumor phenotype. A subset of the available tumor samples (training set) is used to identify genes (class predictor genes) that are specifically associated with the tumor type. The remaining samples (testing set) are then tested in order to investigate how efficiently the predictor genes are able to assign the samples to the correct tumor type. The procedure is repeated multiple times using different samples in the training and testing sets. When all samples have been tested the set of selected genes is considered the class predictor. The accuracy of a class predictor increases with the number of samples used to identify the signature genes. Generally, predictive signatures can be used to address several issues with the common of assigning single samples to previously defined subgroups (e.g., clinical, cytogenetic, and molecular groups).

3.4.2. Identification of the Putative Cell of Origin

1. The identification of the normal counterpart of a tumor type also requires the generation of the gene expression profiles of the normal cell populations. This analysis is particularly important in the study of hematological malignancies, where different stages of B- or T-cell development are targeted by malignant transformation, giving rise to aberrant phenotypes that often cannot be easily assigned to a normal counterpart.

2. In order to obtain suitable gene expression profiles, the normal cell populations need to be purified based on the expression of surface markers or isolated from the

tissue by microdissection to assure a uniform population. A supervised approach is applied to identify the gene signatures that distinguish the various normal sub-populations.

3. The tumor samples are then tested for the expression of the signature genes using a classification method in order to define their similarity to each normal population. The classification method allows quantifying this similarity assigning to each sample a score of relatedness to a normal population.

3.4.3. Identification of Tumor-Specific Markers

1. Tumor-specific genes can be identified by both supervised and unsupervised approaches. However, only by a supervised approach it is possible to identify genes that are differentially expressed between the tumor and all other malignant and nonmalignant phenotypes.

2. A gene is defined as specifically associated to a particular tumor type only if it is not expressed in its normal counterpart and in any other related tumor types. The purity of the samples is an important issue in the identification of genes specifically associated to a tumor type. If the normal cells are represented by purified samples and the tumor by whole tissue biopsies, it is difficult to identify tumor-specific candidate genes because it may be impossible to discriminate between the genes expressed in the tumor and those expressed in the contaminating cells. Including in the analysis other tumors showing a similar pattern of infiltrating cells or normal tissues may help to remove genes associated with the infiltrating component.

3. Alternatively, the problem can be addressed by the inclusion of a few purified samples in the tumor panel as this would avoid the identification of genes that are associated only to a subgroup (purified or unpurified) focusing on the common element (tumor component).

4. Candidate tumor specific-genes identified by gene expression analysis need to be validated for protein expression, preferably by immunohistochemistry to further confirm the specific expression of the respective gene products in the tumor cells. Moreover, a large panel of biopsies should be tested before defining a candidate gene as specific and restricted.

4. Notes

1. Formalin-fixed paraffin-embedded tissues: good quality RNA cannot be obtained from fixed tissues (i.e., using formalin). Recently, Affymetrix released the Human GeneChip X3P array that in conjunction with the Paradise™ reagent system (Arcturus, Sunnyvale, CA) has been designed to perform gene expression analysis on formalin-fixed paraffin-embedded tissues. Further information is available at: http://www.affymetrix.com/products/arrays/specific/x3p.affx.

2. RNA quality: the quality of RNA is a major issue in gene expression profiling of tumor samples. The main reason for obtaining low quality RNA resides in the storage procedure. It is absolutely necessary that the tissues are rapidly frozen and never thaw up.

3. cDNA synthesis: one-cycle or two-cycle protocol? The one-cycle protocol requires at least 1μg of total RNA. Alternatively, the two-cycle protocol allows starting from as little as 10 ng of total RNA. The two protocols are equally efficient in generating a sufficient amount of cRNA for the hybridization on the GeneChip array. However, the resulting gene expression data are not directly comparable because of the shorter size of the cRNA obtained using the two-cycle protocol. According to the sample with the lowest amount of RNA available, one of the two protocols has to be chosen and the same amount of RNA has to be used for all samples.

4. Data analysis: the data analysis may be the most critical and time-consuming step in a gene expression profiling project. Although, many programs are available applying all of them may just confuse. Initially, it is suggested to choose one approach and try to become familiar with the data using that particular software, and then eventually interrogate other programs using the same gene expression data set. The data analysis may take longer than the generation of the profiles and may never finish because new tools and knowledge become constantly available to interpret gene expression data.

References

1. Golub, T. R., Slonim, D. K., Tamayo, P., et al. (1999) Molecular classification of cancer: class discovery and class prediction by gene expression monitoring. *Science* **286,** 531–537.
2. Alizadeh, A. A., Eisen, M. B., Davis, R. E., et al. (2000) Distinct types of diffuse large B-cell lymphoma identified by gene expression profiling. *Nature* **403,** 503–511.
3. Klein, U., Tu, Y., Stolovitzky, G. A., et al. (2001) Gene expression profiling of B cell chronic lymphocytic leukemia reveals a homogeneous phenotype related to memory B cells. *J. Exp. Med.* **194,** 1625–1638.
4. Ferrando, A. A., Neuberg, D. S., Staunton, J., et al. (2002) Gene expression signatures define novel oncogenic pathways in T cell acute lymphoblastic leukemia. *Cancer Cell* **1,** 75–87.
5. Shipp, M. A., Ross, K. N., Tamayo, P., et al. (2002) Diffuse large B-cell lymphoma outcome prediction by gene-expression profiling and supervised machine learning. *Nat. Med.* **8,** 68–74.
6. Holleman, A., Cheok, M. H., den Boer, M. L., et al. (2004) Gene-expression patterns in drug-resistant acute lymphoblastic leukemia cells and response to treatment. *N. Engl. J. Med.* **351,** 533–542.
7. Basso, K., Liso, A., Tiacci, E., et al. (2004) Gene expression profiling of hairy cell leukemia reveals a phenotype related to memory B cells with altered expression of chemokine and adhesion receptors. *J. Exp. Med.* **199,** 59–68.
8. Lipshutz, R. J., Fodor, S. P., Gingeras, T. R., and Lockhart, D. J. (1999) High density synthetic oligonucleotide arrays. *Nat. Genet.* **21,** 20–24.

9. Kuo, W. P., Jenssen, T. K., Butte, A. J., Ohno-Machado, L., and Kohane, I. S. (2002) Analysis of matched mRNA measurements from two different microarray technologies. *Bioinformatics* **18,** 405–412.

10. Rosenwald, A., Alizadeh, A. A., Widhopf, G., et al. (2001) Relation of gene expression phenotype to immunoglobulin mutation genotype in B cell chronic lymphocytic leukemia. *J. Exp. Med.* **194,** 1639–1647.

11. Martinez, N., Camacho, F. I., Algara, P., et al. (2003) The molecular signature of mantle cell lymphoma reveals multiple signals favoring cell survival. *Cancer Res.* **63,** 8226–8232.

12. Rosenwald, A., Wright, G., Chan, W. C., et al. (2002) The use of molecular profiling to predict survival after chemotherapy for diffuse large-B-cell lymphoma. *N. Engl. J. Med.* **346,** 1937–1947.

6

cDNA Microarray and Bioinformatic Analysis of Nuclear Factor-κB Related Genes in Squamous Cell Carcinoma

Zhong Chen, Tin-Lap Lee, Xin-Ping Yang, Gang Dong, Amy Loercher, and Carter Van Waes

Summary

Squamous cell carcinomas and several other cancers have been found to exhibit microarray expression profiles that include genes related to nuclear factor (NF)-κB, a signal activated transcription factor that is evolutionarily important in regulating early response gene programs to injury and infection. Inhibition of NF-κB by expression of a dominant negative signal phosphorylation site mutant of inhibitor-κB, IκBαM, under a tetracycline inducible promoter, established the role of NF-κB as an essential molecular switch modulating multiple genes important in the malignant phenotype. Bioinfomatic analysis of the promoter and coding region of IκBαM-modulated genes has enabled identification of new candidates with and without known NF-κB related motifs for validation and functional studies of their relationship to NF-κB. These studies illustrate how microarray data can be used to generate a hypothesis regarding regulation of genes by a specific signal transcription factor, and how genetic mutants and bioinformatic analysis can be used to analyze the relative importance of the regulatory molecule to expression of genes involved in the malignant phenotype.

Key Words: Bioinformatics; cDNA microarray; gene expression profiling; NF-κB; squamous cell carcinoma; transcription.

1. Introduction

Microarray technology and analysis tools developed as a result of the human and murine genome projects have enabled global expression profile comparison among multiple cancers, their intermediate precursors, and normal cells of origin. Annotated bioinformatic databases can now facilitate identification of potential relationships and mechanisms underlying altered expression of genes that cluster together within a pathological subset or throughout certain malignancies.

From: *Methods in Molecular Biology, vol. 383: Cancer Genomics and Proteomics: Methods and Protocols*
Edited by: P. B. Fisher © Humana Press Inc., Totowa, NJ

Sophisticated bioinformatic analysis tools can link together multiple genes that are regulated by common signal and transcription factors, and can identify regulatory motifs for these pathways within sequence of those candidates previously known and unknown. Genetic mutants can be used to analyze and confirm the relative importance of the regulatory molecule hypothesized to be involved in expression of gene programs involved in the malignant phenotype.

Squamous cell carcinomas (SCC) and lymphomas have been noted to exhibit altered microarray expression profiles that include multiple genes related to nuclear factor (NF)-κB, a signal and transcription factor that is evolutionarily important in regulating early response gene programs to injury and infection *(1–3)*. In comparing the differences in gene expression among normal, transformed and metastatic squamous cells in a step-wise murine model, it was observed that coexpression of multiple genes involved in growth, apoptosis, and angiogenesis related to the NF-κB pathway *(1)*. To test the hypothesis that NF-κB activation contributed to expression of these gene clusters, NF-κB was inhibited by expression of a dominant negative signal phosphorylation site mutant of inhibitor-κB, IκBαM, under a tetracycline inducible promoter *(4)*. Microarray comparison of SCC before and after inhibition of NF-κB established the role of NF-κB as an essential molecular switch modulating multiple genes important in the malignant phenotype *(4)*. Bioinfomatic analysis of the promoter and coding region of IκBαM-modulated genes has enabled identification of new candidates with and without known NF-κB related motifs for validation and functional studies of their relationship to NF-κB. These studies illustrate how microarray data can be used to generate a hypothesis regarding regulation of genes by a specific signal transcription factor, and how genetic mutants and bioinformatic analysis can be used to analyze the relative importance of the regulatory molecule to expression of genes involved in the malignant phenotype.

1.1. Multiple Genes Upregulated With Murine SCC Tumor Progression are Related to the NF-κB Signal Pathway

To investigate the molecular mechanisms of SCC tumor progression, a syngeneic murine model was previously developed that includes the spontaneously transformed BALB/c keratinocyte line Pam 212, and rare lymph node and lung metastases (LY and LU) of Pam 212 (*see* **ref. 5** and **Fig. 1**). The metastatic Pam LY and Pam LU cell lines were found to form tumors and metastases at a higher rate than the parental Pam 212 tumor line in vivo *(5)*, consistently related to an increased expression of a repertoire of proinflammatory and proangiogenic factors that are regulated by the transcription factor NF-κB *(6)*. Utilizing a first generation 4000 element murine cDNA microarray to compare the stepwise changes illustrated in **Fig. 1**, a group of genes was

1. Murine squamous cell carcinoma model

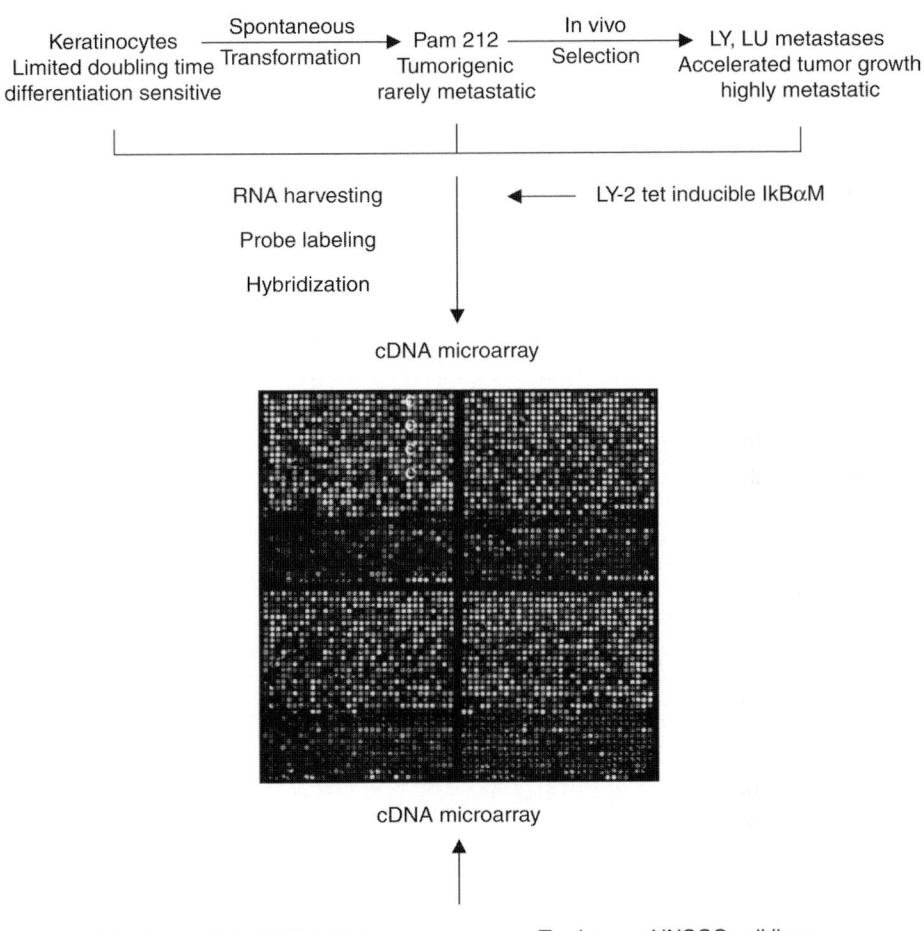

Fig. 1. Diagram of multistage Pam 212 tumor model and identification of differentially expressed genes using microarray analysis. RNA was harvested from primary BALB/c keratinocytes, Pam 212 cells, and metastatic Pam LY, and LU cells, and used for microarray. Illustrative array results were obtained comparing arrays hybridized with cDNA from keratinocytes, Pam 212, and Pam metastatic cell lines. Modified from **ref. *1***, in the public domain as a US Government work.

identified to be upregulated during tumor progression. **Figure 2** shows a cluster analysis of genes differentially upregulated in metastatic LY cells. The functions of the genes involved included immunological, inflammatory, and angiogenesis responses, such as Gro-1 (KC), complement component 3, IL-12B, CSF-1, and

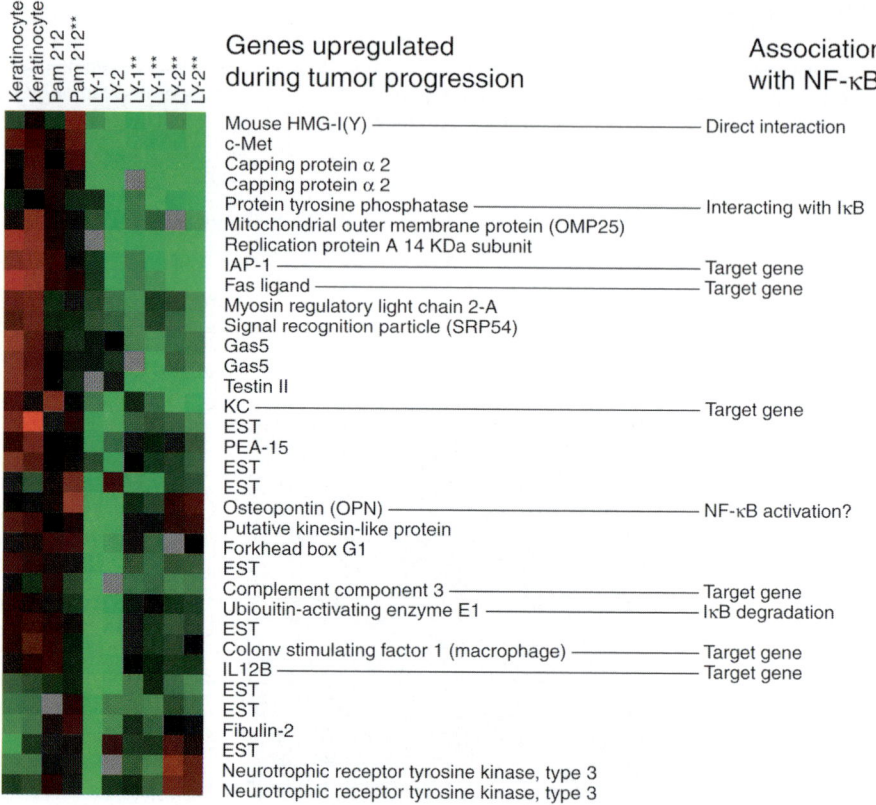

Keratinocyte/Pam212/LY-1, LY-2

Fig. 2. Hierarchical clustering of gene expression data from mouse arrays. Shown is the single dimension clustered gene expression measured from 10 mouse 4K array analyses obtained from 10 samples of normal and malignant keratinocytes. Each row represents a separate cDNA on the microarray and each column the cell source, in which expression is compared. Expression among cDNAs was compared from Pam 212 and LY-1 and LY-2 cells grown in vitro and from whole tumors established from the cell lines in vivo (Pam 212* and LY-1* and LY-2*). Log ratios of hybridized fluorescent cDNA probes from each experiment are depicted according to the color scale. Color schemes for probe labeling are reflected by colored sample names on bottom left. Reproduced from **ref. 1**, in the public domain as a US Government work.

Osteopontin; signal transduction and regulation of gene expression and DNA replication, such as *c*-Met, neurotrophic receptor tyrosine kinase, HMG-1(Y), and replication protein A (14-kd subunit); and modulation of cell cycle and apoptosis, such as cIAP-1, Fas ligand, phosphoprotein enriched in astrocytes, 15-KD (PEA-15), and ubiquitin-activating enzyme E1. PubMed search revealed that many of the genes are targets or involved in regulation of the NF-κB signal transduction pathway *(1)*.

1.2. Tetracycline Inducible IκBαM Suppresses NF-κB-Regulated Genes in Murine SCC

The diversity of genes differentially expressed among the various stages of tumor progression was further examined and determined the effect of conditionally inhibiting NF-κB in Pam LY-2 cells in a subsequent study using a second generation 15,000 element cDNA microarray *(4)*. Because NF-κB nuclear translocation and activation involves signal phosphorylation and degradation of Inhibitor-κBα, the effect of expressing an IκBα phosphorylation mutant (IκBαM) under control of a tetracycline (doxycycline; Dox) inducible promoter on the gene clusters and expression profile was examined. **Figure 3** shows the cluster analysis for 308 genes (horizontal rows) that exhibited >2-fold difference in expression between Keratinocytes, Transformed Pam 212, or Metastatic Pam LY-2 cells, and Pam LY-2 cells expressing IκBαM following Dox. Three dominant patterns were detected that clustered according to the step-wise differences in phenotype between keratinocytes, Pam 212, Pam LY-2 cells. Cluster 1 genes showed similar expression in normal and transformed Pam 212 cells, but showed reduced expression in metastatic LY-2 cells. Cluster 2 genes showed lower expression in normal cells, but increased expression in transformed and metastatic cells. Cluster 3 showed increased expression primarily in LY-2 metastatic cells. Overall, 141 genes in the upper cluster showed a decrease, and 167 genes in the Clusters 2 and 3 were increased in association with transformation, tumor progression, and activation of NF-κB. Remarkably, **Fig. 3** reveals that expression of IκBαM in Pam LY-2 results in restoration of the pattern of expression observed for keratinocytes for many of the genes up- and downregulated between keratinocytes and Pam LY2 metastatic cells. The list of genes and the fold-difference in expression between LY-2 and keratinocytes, and LY-2 and LY-2-cells expressing IκBαM might be found in **ref. 4**, and at http://www.nidcd.nih.gov/research/scientists/vanwaesc.asp. Of the genes upregulated in LY-2 cells, hybridization of 115/167 (69%) decreased by less than twofold toward normal levels after expression of IκBαM. Fully 88/141 (62%) genes downregulated in LY-2 cells were restored by more than twofold back toward the level observed in keratinocytes following expression of IκBαM. The genes were also classified according to putative function and published associations with NF-κB as determined by search of NCBI PubMed and LocusLink. Analysis of the sequence of the promoter region was also performed using TFSearch, and genes containing NF-κB concensus promoter sites were identified *(4)*. This bioinformatic analysis confirmed that many genes detected as upregulated in LY2 cells had previously been associated with NF-κB; that these and additional previously unidentified genes contained NF-κB promoter sites.

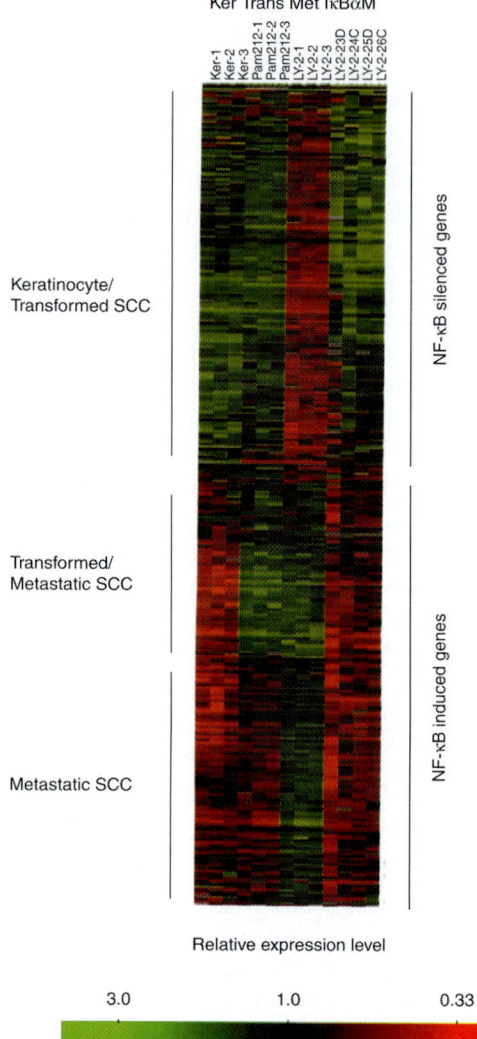

Fig. 3. Cluster analysis of genes related to tumor progression and NF-κB in squamous cell carcinoma. cDNA was prepared from mRNA from three independent preparations of Keratinocytes, Pam 212, and Pam LY-2 cells, and 4 Pam LY-2 clones expressing IκBαM following culture with dox for 72 h. Three hundred and eight of 15,000 cDNA microarray elements exhibited mean signal differences of ≥2-fold (99% confidence interval) for triplicate comparisons between cDNAs from normal keratinocytes (Ker), transformed Pam 212 (Trans) and/or metastatic Pam LY-2 (Met) cells. Three dominant patterns were detected that clustered according to the step-wise differences in phenotype between KER, Pam 212, Pam LY-2 cells. Cluster 1 genes showed similar profiles in keratinocytes and transformed Pam 212 cells (green), but showed reduced expression in metastatic LY-2

1.3. Differentially Expressed Genes Associated With Human Head and Neck Squamous Cell Carcinomas

Concurrent with the study in the murine SCC tumor model, we compared gene profiles of 10 human head and neck squamous cell carcinomas (HNSCC) cell lines with normal human keratinocytes utilizing a 23,000 element human cDNA microarray, developed by the National Human Genome Research Institute. A total of 969 differentially expressed genes were identified in the tumor group, and differed significantly from keratinocyte group by the *F*-test ($p < 0.01$). Genes were classified category according to gene ontology and function (DAVID, NIAID, NIH, **ref. 7**). The upregulated genes in the data set included many of the gene families identified previously in the murine tumor progression model, such as growth cytokines (i.e., human IL-8, a homolog of murine Gro 1), signal transduction (i.e., YAP1 and PI3KR3), cell cycle and apoptosis (i.e., cyclin D1 and CIAP-1), DNA replication (PCNA), and adhesion and migration (TIMP2 and BMP7). Within human HNSCC, the cluster and PCA analysis also identified two subgroups, as shown in **Fig. 4**. One group showed upregulation of the cluster of genes involving in early injury responses and cell death, such as interleukin (IL)-6, IL-8, Yes-associated protein (YAP)1, allograft inflammatory factor 1, baculoviral IAP-containing 2 (C-IAP1) or genes related to differentiation, like keratin 8. As in the murine model, many genes in this cluster have previously been reported to be regulated by NF-κB, or contain putative NF-κB binding sites by the computation prediction performed by Genomatix software suite. Inhibition of NF-κB p65 by RNAi confirmed that several of these genes are upregulated by NF-κB (Lee et al., manuscript in preparation). These results are consistent with evidence that alterations in NF-κB are prevalent in HNSCC, and suggest that NF-κB activation and expression of related genes might be important in a subset of HNSCC.

In summary, the microarray studies in the murine model and a broader panel of human SCC lines reveal clusters of genes that are related to specific pathways by promoter analysis, and which can be validated by genetic inhibition with dominant negative mutants or RNAi targeted at the candidate regulatory molecule.

2. Materials

2.1. Murine SCC Cell Lines, Transfection, and In Vivo Tumor Growth

1. 37°C, 5% CO_2 cell culture incubator.
2. Murine SCC cell lines: murine SCC line Pam 212 is a spontaneously transformed cell

---

Fig. 3. *(Continued)* cells (red). Cluster 2 genes showed lower expression in normal cells (red), but increased expression in transformed and metastatic cells (green). Cluster 3 showed low intensity in keratinocytes and transformed cells (red) and increased signal intensity primarily in LY-2 metastatic cells (green). Reproduced from **ref. 4**, in the public domain as a US Government work.

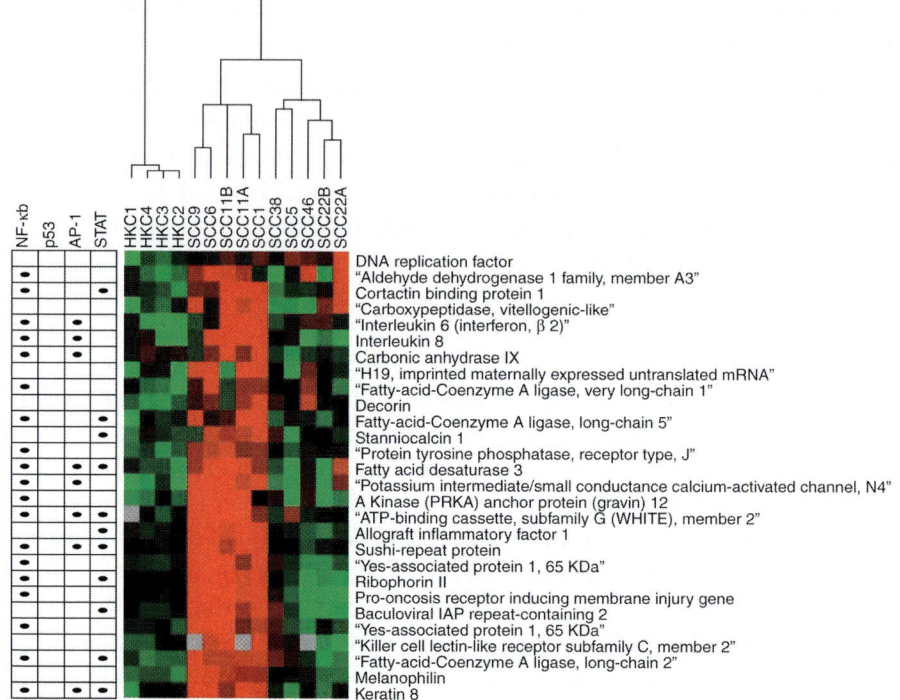

Fig. 4. Differentially expressed genes and their promoter analysis. Unsupervised cluster analysis was performed and two clusters (A and B) were identified from a hierarchical clustering dendrogram of 969 genes that exhibited a >2-fold difference between 10 UM-SCC cell lines and normal kerintinocytes. Differentially expressed genes from cluster B with analysis of transcription factor binding sites in their promoters are shown. The putative transcription factor binding sites of NF-κB, p53, AP-1, and STAT were listed on the left of the hierarchical clustering dendrogram. Red indicates upregulated genes; green indicates low gene expression.

line derived from neonatal BALB/c skin keratinocytes in vitro, which forms squamous cell carcinomas but rarely metastasizes in vivo when subcutaneously inoculated in normal BALB/c mice (*5*). Metastatic LY and LU lines were isolated from lymph node and lung metastases of Pam 212 tumor implants in BALB/c mice (*5*).

3. Growth medium for murine SCC cell lines: Eagle's minimal essential media (EMEM) plus 10% heat inactivated fetal bovine serum (FBS), penicillin, streptomycin, and glutamine (Invitrogen).

4. Murine primary keratinocytes were isolated from the skin of male BALB/c neonates as described previously (*1*) and grown either in EMEM plus 10% FBS (thus designated as high Ca^{++}) or same media with calcium concentration reduced to 0.02 m*M* (low Ca^{++}).

5. BALB/c male mice, 4–6 wk old.
6. Tet-inducible IκBαM expression system (Clontech Laboratories, Inc., Palo Alto, CA).
7. Lipofectamine plus transfection reagents and Opti-MEM medium (Invitrogen).
8. G418 (Gentamicin, Invitrogen).
9. Doxycycline (Sigma).

2.2. Murine cDNA Microarray

1. Murine cDNA microarray. The first generation of murine 4K element cDNA microarray was developed by the Division of Clinical Sciences and Microarray Facility, Advanced Technology Center, NCI, NIH. The 15K murine cDNA microarray was developed by the Radiation Oncology Sciences Program, NCI, using a cDNA library from the National Institute of Aging (Mm-ROSP-NIA15K) (http://nciarray.nci.nih.gov). Each of the 15K murine array slides contains 15,398 spots with about 8490 known genes.
2. Trizol (Invitrogen).
3. Cy3 and Cy5-dUTP (Amersham Pharmacia Biotech Inc, Piscataway, NJ).
4. SuperScript II RT (Invitrogen).
5. Microcon column (Millipore, Bedford, MA).
6. Mouse COT-1 DNA (Invitrogen).
7. PolyA (Amersham Biosciences).
8. 20X SSC (sodium chloride, 3 M; sodium citrate, 0.3 M pH 7.0) and 10% sodium dodecyl sulfate (SDS).
9. 65°C Hybridization chamber/water bath.

2.3. HNSCC Cell Lines and Normal Primary Keratinocytes

1. HNSCC cell lines were from the University of Michigan series (UM-SCC, **ref. 8**, Dr. T. E. Carey, University of Michigan, Ann Arbor, MI). The UM-SCC cell lines were derived from eight patients with SCC of the upper aerodigestive tract. The 10 UM-SCC cell lines selected were obtained from patients with tumors distributed among stages I–IV, and oral, pharyngeal, and laryngeal sites. Cell lines established from isolates of two time-points or anatomical sites were designated with an alphabetical suffix (i.e., "A" or "B").
2. Growth medium for UM-SCC cell lines. EMEM supplemented with 10% heat inactivated fetal bovine serum and penicillin/streptomycin/glutamine (Invitrogen).
3. Human normal keratinocytes (HKC, Cascade Biologics Inc., Portland, OR).
4. Growth medium for human normal keratonocytes. Keratinocyte serum-free medium 154CF containing 0.08 mM of calcium chloride and supplemented with human keratinocyte growth supplements (HKGS, Cascade Biologics Inc., Portland, OR). The final concentration of HKGS in the complete medium are: 0.2% (v/v) of bovine pituitary extract (BPE), 5 µg/mL of bovine insulin, 0.18 µg/mL of hydrocortisone and 5 µg/mL of transferring, and 0.2 ng/mL of human recombinant EGF. All normal human keratinocytes used for experiments were used within ≤ 5 passages.

2.4. Human cDNA Microarray

1. The human cDNA microarray spotted with 23K elements were provided by the microarray core facility of National Human Genome Research Institute (NHGRI, NIH, Bethesda, MD). Each of the human array slides contains approx 23,220 cDNA spots, including approx 12,270 human known genes, amplified from human sequence-verified clone sets of named genes and ESTs purchased from Research Genetics (acquired by Invitrogen).
2. All other reagents and instruments for RNA isolation, cDNA labeling and hybridization, and image scanning are the same as described in **Subheading 2.2.**, murine cDNA microarrays *(2–10)*, except human COT-1 DNA (Invitrogen) was used.

2.5. Instruments and Softwares for cDNA Microarray Data Collection

1. GenePix 4000 microarray scanner with GenePix Pro software (Axon Instruments, Union City, CA).
2. mAdb (microarray data base, NCI, NIH).
3. GeneSpring (Silicon Genetics, Redwood City, CA).
4. ArraySuite 2.1 extension (Y. Chen, NHGRI) in the IPLab program (Scanalytics, Inc., Fairfax, VA).

2.6. Softwares for cDNA Microarray Data Analysis

1. BRB Array Tool 2.0 (NCI, NIH, http://linus.nci.nih.gov/BRB-ArrayTools.html).
2. Principle component analysis (PCA) was carried by using Partek Pro 5.1 software (Partek Inc., St. Louis, MO).
3. Cluster and Treeview software (Stanford University).
4. Significance analysis for microarray (SAM; Stanford University, http://www-stat.stanford.edu/~tibs/SAM/index.html).
5. Bioperl (www.bioperl.org).
6. MatchMiner (http://discover.nci.nih.gov/matchminer/html/index.jsp).
7. Database for Annotation, Visualization, and Integrated Discovery (DAVID) from NIAID, NIH (http://apps1.niaid.nih.gov/david/).
8. GoMiner, (http://discover.nci.nih.gov/gominer/).
9. Transcription Factor Search, a software for searching transcription factor binding sites (http://www.cbrc.jp/research/db/TFSEARCH.html, Japan).
10. Genomatix software suite (Munich, Germany).

3. Methods

3.1. In Vitro Growth and Transfection of Murine SCC Cell Lines and In Vivo Growth of Murine SCC Tumors

1. LY-2 cells were cultured in complete EMEM medium and cotransfected with the Tet-On plasmid system containing the tet-responsive transcriptional activator that binds the Tet Responsive Element and the gentamicin resistance gene (neo), together with the tetR plasmid that binds the Tet responsive element in

the absence of dox that reduces background levels of transcription. Fourteen clones with low background and high fold induction of luciferase reporter activity were selected and pooled as controls (LY-2P) and for transfection by the response plasmid containing IκBαM. After G418 selection, colonies were expanded for screening for Tet inducible expression of IκBαM by Western blot. The inhibition of NF-κB in clones was compared with control LY-2P cells by NF-κB luciferase and electromobility shift assay, as described in detail elsewhere *(4)*.

2. To grow tumors from cell lines, 5×10^6 Pam 212 and LY cells in log growth phase were injected subcutaneously into male BALB/c mice. Tumors were allowed to grow larger than 1 cm in diameter, harvested by grossly stripping off the stroma tissues, snap frozen, and stored in −80°C.

3.2. RNA Isolation, Reverse Transcription, and Labeling of cDNA and Hybridization of Murine SCC Cell Lines and Tumors

1. For initial microarray studies in the murine SCC tumor model with the 4K mouse array, total RNA was isolated from primary BALB/c keratinocytes, Pam 212, LY-1, and LY-2 cells or tumors using Trizol reagent (*see* **Note 1**).
2. For subsequent studies of the effects of IκBαM on gene expression, RNA was isolated from LY-2P and the four inducible LY-2 IκBαM clones (23D, 24C, 25D, 26C) with and without dox treatment (**ref. 4**, *see* **Note 1**).
3. To make fluorescence-labeled cDNA by reverse transcription, 50–100 µg total RNA was incubated in a cocktail containing Cy3- or Cy5-dUTP and SuperScript II RT. Labeled cDNAs were purified using a Microcon column. The appropriate Cy3- and Cy5-labeled cDNAs were combined, along with 2 µL (20 µg) mouse COT-1 DNA, 1 µL (8–10 µg) polyA, 2.6 µL 20X SSC, and 0.45 µL 10% SDS in a final volume of 15 µL. After denaturation, labeled cDNAs were added to processed cDNA array chips, which were then placed in hybridization chambers and incubated overnight (10–16 h) at 65°C. The next day, slides were washed for 1 min in 1X SSC, 1 min in 0.2X SSC, 10 s in 0.05X SSC, and then spin dried. Fluorescence images were captured by a GenePix 4000 microarray scanner with GenePix Pro software. The photomultiplier tube voltage for both channels was adjusted between 750 and 890 V range to give the same overall intensity between both channels and the raw image was saved in TIFF format (*see* **Notes 2–4**).

3.3. Growth of Human UM-SCC Cell Lines and Normal Keratinocytes

1. Human UM-SCC cell lines and primary normal keratinocytes were grown in the medium as described in **Subheading 2.3.**

3.4. RNA Isolation, Reverse Transcription, and Labeling of cDNA and Hybridization of Human SCC Cell Lines and Normal Keratinocytes

1. The protocols are the same as described for murine cell lines (*see* **Subheading 3.2.**, **Notes 1–4**).

3.5. Data Collection, Filtering, and Quality Control

1. Raw array images were analyzed using the ArraySuite 2.1 extension (Y. Chen, NHGRI) in the IPLab program (Scanalytics, Inc., Fairfax, VA), to calibrate relative ratios and develop confidence intervals for their significance *(9)*. Fluorescence intensities of local background of each spot were subtracted for both dyes (Cy3 and Cy5). Spots were visually checked and filtered based on RQuality score provided in the software (with 0 as the lowest quality and 1 as the highest quality). Implementation of the quality metric enables unified and universally applicable data filtering before downstream data analysis. Any spots with RQuality score <0.5 were excluded. For each spot, the calibrated ratio (the median of the pixel-by-pixel ratios of pixel intensities that have the median background intensity subtracted) was used in subsequent analysis. Expression outliers were determined using ±3.0 SD and 99% confidence interval as cutoff. For the purpose of analysis for known genes, the resulting gene list was further filtered to exclude EST and hypothetical clones (*see* **Notes 5–8**).

3.6. Cluster Analysis, Finding Differentially Expressed Genes, Gene Ontology, and Promoter Analysis

1. Principle component analysis (PCA). In our investigations, PCA was performed to identify groups of normal and cancer cells which have similar and different expression patterns. PCA is a mathematic technique used to reduce the dimensionality of a data set or data matrix, and to visualize the global segregation of the microarray data. It describes the data set in terms of its variance. Each principal component describes a percentage of the total variance of a data set and takes into consideration weights that each variant contributes to this variance. The first principal component of a data set describes the greatest amount of variance of the data set. The coefficients of the principal components quantify the loading or weight of each variant to that amount of variance. In our laboratory, PCA analysis was performed on normalized data by Partek Pro 5.1 software (Partek Inc., St. Louis, MO). In addition, Genespring and TIGR array suite also include PCA function analysis (*see* **Notes 9** and **10**).

2. Hierarchical clustering. Hierarchical clustering was carried out on a gene list where the average ratio difference for 10 HNSCC tumor cell lines to normal kerintocyte were more than twofold and had a *t*-test score at $p < 0.01$. The result file including data for 969 genes were analyzed by cluster software based on clustering algorithm of Eisen et al. *(11)* and the expression maps of clustered genes was visualized using Java treeview. This software is especially useful for data visualization. In addition, BRB array tool developed by NCI/NIH is also a powerful tool, and it indicates the distance on a hierarchical tree, particularly useful if looking for a refined classification. (http://linus.nci.nih.gov/BRB-ArrayTools.html). Genespring also includes a clustering tool (*see* **Notes 9** and **10**).

3. Finding differentially expressed genes. For the microarray studies in the murine tumor progression model, the array elements exhibiting mean signal differences of

≥2-fold (99% confidence interval) for triplicate comparisons between cDNAs from normal keratinocytes, transformed Pam 212 and/or metastatic Pam LY-2 cells were scored as differentially expressed with tumor progression. Array elements exhibiting reversion of mean differences of ≥2-fold (99% confidence interval) for comparisons between cDNAs from metastatic Pam LY-2P and four LY-2 IκBαM clones +/− doxycycline were scored as NF-κB modulated genes (*see* **Note 10**).

4. For studies of multiple human HNSCC cell lines, analysis of differential gene expression included statistical comparison. Many microarray software packages can identify differentially expressed genes by various statistical methods. These include BRB array tool and SAM (http://www-stat.stanford.edu/~tibs/SAM/index.html). To analyze the differential expression between the tumor and normal cells, as two classes, statistical analysis of the data from the hierarchical clustering was performed using BRB-Array Tools 2.0 (NCI, NIH), where a class comparison tool was used. Univariate *F*-tests were performed on the two classes. This tool also computes a global permutation test that excludes genes that differ significantly owing to chance alone. Genes were considered to be significantly differentially expressed when $p < 0.001$ after test of 2000 permutations. For sequence identification and ID conversion, softwares available include Bioperl (www.bioperl.org), and MatchMiner (http://discover.nci.nih.gov/matchminer/html/index.jsp). These softwares allow the batch-translation among files with many types of gene and protein identifiers (*see* **Note 10**).

5. Gene Ontology and promoter analysis. Gene Ontology analysis was carried out by DAVID from NIAID, NIH (http://apps1.niaid.nih.gov/david/). It is a web-based, client/server application that allows users to access a relational database of functional annotation. Functional annotations are derived primarily from LocusLink at the National Center for Biotechnology Information (NCBI, NIH). DAVID uses LocusLink accession numbers to link gene identifiers to biological annotations including gene names and aliases, functional summaries, gene ontologies, protein domains, and biochemical and signal transduction pathways. Annotation pedigrees are provided through direct links to the primary sources of annotation, which also provide additional gene specific information. Another software is GoMiner, which leverages the gene ontology for the biological interpretation of microarray data (http://discover.nci.nih.gov/gominer/). To detect transcription factor binding sites in the promoter region on selected clusters, promoter analysis was performed by Genomatix software suite (Munich, Germany) using the default matrix index (*see* **Note 10**).

6. To study the importance of NF-κB in modulation of gene expression in the murine tumor progression model, the sequence from 2 kb 5′ to the coding region and coding region for genes found to be differentially expressed by microarray was examined for NF-κB related binding motifs. The sequence data were generated through the use of the Celera Discovery System and public databases. Sequences were analyzed for NF-κB binding site motifs using Transcription Factor Search, a software for searching transcription factor binding sites (http://www.cbrc.jp/research/db/TFSEARCH.html, Japan).

4. Notes

1. Quality of RNA. The most important step in the microarray procedure that a research scientist can control is the quality of the RNA. Only the intact RNA with a 260/280 ratio ≥1.8 is strongly suggested to be used in microarray experiments. If a reference RNA will be used to cohybridize with different biological samples, a pool of the reference RNA which is enough for the entire study should be prepared ahead of time to minimize the variability caused by different preparation of reference RNA.

2. Experimental reagents. In order to minimize the noise introduced by reagents used in experiments, it is strongly recommended to purchase all reagents, especially enzymes (reverse transcriptase, Cy3- and Cy5-labeled nucleotide), in adequate amounts for the entire study. We have experienced in our experiments that Cy3- and Cy5-labeled nucleotide from different sources or different batches from the same source gave different labeling efficiencies. Cy3- and Cy5-labeled nucleotides were purchased from Amersham Pharmacia. In our experiments, the direct labeling method was used. Because of the nature of Cy3 and Cy5 (Cy5 is a bigger molecule than Cy3), the efficiency of Cy3 labeling is higher than Cy5 labeling. In order to have closer amount of RNA labeled by Cy3 and Cy5, double the amount of RNA was used to label with Cy5 than with Cy3, to yield a similar intensity of fluorescent labeling.

3. Quality of microarray chips. Custom cDNA microarrays produced by noncommercial academic facilities such as used in our studies are subject to more variability and variation in quality. In order to minimize the noise introduced by printing of the microarrays, it is better to use the arrays from the same batch for one study. Several array chips from the each batch should be picked and checked for quality before the experiment. Typically, this includes every 10th slide to select batches with sufficient uniformity that can be used together. Batches that show evidence of widespread loss of spots from insufficient probe or pipetor function are not used. The authors have used Panomer™ 9 Random Oligodeoxynucleotides, Alexa Fluor 546 Conjugate (similar to Cy3; Molecular Probes) to assess the overall quality of cDNA microarray spotting using the fluorescence dye staining procedure. Briefly, the oligo was diluted in hybridization buffer normally used for microarray at final concentration of 7.5 µM, heated to 90°C for 2 min, cooled by centrifugation for 5 min, applied in adequate amount of solution onto slid and covered with cover slip. The slides were hybridized at room temperature for 3–5 min, then washed, dried, and scanned using the channel corresponding to the oligo used (same as Cy3).

4. Procedures. It is observed that the labeling efficiency of cDNA probes varies from different methods. It is advisable to stay with same labeling protocol throughout the entire study. The same rule should be applied to hybridization and all other steps of microarray experiments. Technical handling also attributes to variables. Make sure to preform all initial labeling reactions in an RNase-free environment. It is also very important to place the cover slip over the hybridization solution on

the microarray in such a way that no bubble will be formed and no leakage of hybridization solution will occur. Bubble formation during the hybridization will leave the bubble area unhybridized with high background, in the edge of the bubble. The leakage of hybridization solution will lead to drying of slides and cause extremely high background, which cannot be reduced by washing steps. It is also advisable that the same person should perform all experiments through out a study and have all experiments done within a short period of time. This will reduce variables caused by different individual, different experiment, and deterioration of microarrays.

5. Scanning equipment and its settings. Different scanner or same scanner with different PMT setting can lead to considerable variation of the expression levels extracted from the array. It is better to choose a scanner with auto-calibration of PMT for both Cy3 and Cy5 channels so that bias will be eliminated.

6. Image collection and quality control. In addition to the IPLab with array extension suite (GIPO file, NHGRI), softwares available for microarray image collection and quality control include Genpix (GAL file, Axon Instruments), mAdb (microarray database, NCI, NIH) and Genespring (Silicon Genetics). It is important to follow the manufacture's or developer's procedure to flag, filter and normalize the data using the built-in algorithm, for example, RQuality and statistical criteria. RQuality score is important to test the quality of the microarray hybridization to enable unified and universally applicable data filtering before downstream data analysis. It is important to avoid input of result data without filtering and flagging. The integrity of result file has to be preserved even when some columns or rows may be blank. It is important to build and keep data files in the excel format with complete information, which will be the input resource for the subsequent analyses.

7. Processing microarray data. With the advent of enormous data generated by high-throughput technologies, the need for biological data analysis and management has been elevated to be a challenge to biologists. Microarray data handling and analyses have become an important issue, and continue to evolve. The key to getting a reasonable sense of the quality of microarray experiment is done by preprocessing of raw data by normalization and filtering by different algorithms and criteria. The old adage of computing, garbage in is garbage out, is certainly appropriate in dealing with the large amounts of data involved with microarrays. Therefore, one should accurately calibrate the signal and background based on standards, controls, and replicates. Such data-handling work requires computational knowledge to accomplish, and may pose difficulties to biologists. However, thanks to the increasing popularity of microarray experiments, many free and commercial software packages are now available to help analyze microarray data sets to accomplish different bioinformatics tasks. Some are in a form of packaged suites while others are in individual package for specific task, such as gene ontology, signaling pathway analysis, and gene prediction. The data analysis tools for microarray we have utilized are listed in **Table 1**. It is important to clearly define the biological questions before applying the appropriate software for analysis.

8. Formatting and managing microarray data. One important issue when performing microarray analysis is the file format. Information obtained from online databases is usually in the flat file format, which is similar to text file that can be opened by text editors available in almost all computer operating systems. One example is the sequence information retrieved from Genebank. Although such flat files can be read easily without installation of specific applications, it is not a good portable file format for long-term storage of large data sets and might pose problems if the software used requires a different file format. Therefore, before performing analysis, researchers should seek a standard data file format that could be imported into the software they are going to use. This may be accomplished by using Microsoft Excel or other spreadsheet programs to save the files in tab-delimited text format. It offers a very easy way to manage the data, including filtering, sorting and calculation. One limitation of Excel program is that it could not handle any table with more than 65,536 rows or 256 columns. This usually happens when handling very large data sets like SAGE tag data. When the capacities of the spreadsheet program are not sufficiently robust, database programs like Microsoft Access or Filemaker Pro can be used. Programming tools can also be used for huge data analysis. There are actually several programming languages that are especially suited for the data file formatting and other text file manipulations, such as Perl and Awk. These program languages are very powerful and easy to manipulate text information.

9. Cluster analysis. Cluster analysis allows visualization and analysis of the results of complex microarray experiments. Through the analysis, data are significantly reduced to increase the signal-to-noise ratio for more sensitive discovery. This can be done using the criteria of fold-difference likely to be of biological relevance (e.g., twofold), without or with statistical criteria, and unsupervised or supervised comparison. The unsupervised comparison does not have any training criteria. The data will be clustered according to the differences in expression. In supervised comparison, the data classification requires supervised learning, i.e., to use a set of microarray data to train the software to specify what is to be learned (by classes, such as whether certain gene expression patterns are related to tumor classification or outcomes). After the training, the software algorithm can be used to analyze unknown experimental data sets the same way, to classify the unknowns using the criteria. Examples of this have been published in HNSCC by others *(10)*. There are many clustering methods available, including PCA, hierarchical clustering, self-organizing maps (SOMs), and *k*-means clustering.

10. Appropriate usage of softwares. Another issue on microarray analysis is how to find the appropriate softwares for the analysis. There are many software packages that are commercially available, totally free, or free for academic researchers, as listed in **Table 1**. Researchers should define the outline of analysis first and find the software that suits their needs. We have experience with quite a few software packages and observed some compatibility problems. These may be as a result of hardware conflicts or software problems. In some instances this can be overcame by updating the system and obtain the most recent version of the program. Usually the functionality and features of free softwares are usually limited compared with

Table 1
Summary of Commonly Used Software for Microarray Analysis

Suites	Company/institutes	License	URL
Genespring	Silicon Genetics	Commercial	www.sigenetics.com
GeneSight	BioDiscovery	Commercial	www.biodiscovery.com/ genesight.asp
Vector Xpression	Informax	Commercial	http://register.informaxinc.com/
Expressionist	Gene Data	Commercial	http://www.genedata.com/
BRB Arraytools	NCI/NIH	Free	http://linus.nci.nih.gov/BRB-ArrayTools.html
MultiExperiment Viewer	TIGR	Free	http://www.tigr.org/ software/tm4/mev.html
Clustering and dendrogram			
Cluster and TreeView	Stanford University	Free	http://rana.lbl.gov/ EisenSoftware.htm
GeneCluster	Broad Institute	Free	http://www.broad.mit.edu/ cancer/software/ genecluster2/gc2.html
Motif/promoter identification			
Genomatrix	Genomatix Software	Commercial	http://www.genomatrix.de/
Eukaryotic Promoter Database	EMBL	Free	http://www.epd.isb-sib.ch/
Neural Network Promoter Prediction	U.C. Berkeley	Free	http://www.fruitfly.org/ seq_tools/promoter.html
Pathway analysis/construction			
Pathwayassist	Ariadne Genomics	Commercial	http://www.ariadnegenomics. com/products/pathway.html
GenMAPP	U.C. San Francisco	Commercial	http://www.genmapp.org/
Pubgene	Pubgene	Free	http://www.pubgene.org/
KEGG Pathway Database	Kyoto University	Free	http://www.genome.jp/kegg/ pathway.html
Gene onotology			
DAVID	NIAID/NIH	Free	http://apps1.niaid.nih.gov/ david/
SOURCE	Stanford University	Free	http://source.stanford.edu/ cgi-bin/source/sourceSearch
GoMiner	NCI/NIH	Free	http://discover.nci.nih.gov/ gominer/

those commercial ones. Researchers may need to find other alternatives to complement their needs. There are a number of commercial softwares that are useful. GeneSpring is a general analysis tool specifically tailored for DNA microarray data analysis. The intended groups of users are biologists, who actually perform the experiments. The program contains various data preprocessing, clustering, and pathway construction tools. Gene annotation can also be retrieved directly from web-based databases. GeneSpring can also be expanded with user-made scripts or programs (Java APIs). It is good comprehensive microarray analysis tool, but a bit limited if the work is focused on editing biological pathways. Another problem is its performance on Macintosh is slower and less stable owing to the JAVA runtime problem for Mac OS X. Another commercial software, Vector Xpression, provides integration with their Vector NTI Advance (sequence analysis application suite) and Vector PathBlazer software (bioinformatics tool, providing the ability to build, visualize, and analyze biological pathways), which creates a dynamic analysis environment and ease of data integration.

Acknowledgments

Reading and comments on this chapter by Drs. James F. Battey and James B. Mitchell are appreciated.

References

1. Dong, G., Loukinova, E., Chen, Z., et al. (2001) Molecular profiling of transformed and metastatic murine squamous carcinoma cells by differential display and cDNA microarray reveals altered expression of multiple genes related to growth, apoptosis, angiogenesis, and the NF-kappaB signal pathway. *Cancer Res.* **61,** 4797–4808.
2. Davis, R. E., Brown K. D., Siebenlist, U., and Staudt, L. M. (2001) Constitutive nuclear factor kappaB activity is required for survival of activated B cell-like diffuse large B cell lymphoma cells. *J. Exp. Med.* **194,** 1861–1874.
3. Hinz. M., Lemke, P., Anagnostopoulos, I., et al. (2002) Nuclear factor kappaB-dependent gene expression profiling of Hodgkin's disease tumor cells, pathogenetic significance, and link to constitutive signal transducer and activator of transcription 5a activity. *J. Exp. Med.* **196,** 605–617.
4. Loercher, A., Lee, T. L., Ricker, J. L., et al. (2004) Nuclear factor-kappaB is an important modulator of the altered gene expression profile and malignant phenotype in squamous cell carcinoma. *Cancer Res.* **64,** 6511–6523.
5. Chen, Z., Smith, C.W., Kiel, D., and Van Waes, C. (1997) Metastatic variants derived following in vivo tumor progression of an in vitro transformed squamous cell carcinoma line acquire a differential growth advantage requiring tumor-host interaction. *Clin. Exp. Metastasis* **15,** 527–537.
6. Dong, G., Chen, Z., Kato, T., and Van Waes, C. (1999) The host environment promotes the constitutive activation of nuclear factor-kappaB and proinflammatory cytokine expression during metastatic tumor progression of murine squamous cell carcinoma. *Cancer Res.* **59,** 3495–3504.

7. Dennis, G., Jr., Sherman, B. T., Hosack, D. A., et al. (2003) DAVID: database for Annotation, Visualization, and Integrated Discovery. *Genome Biol.* **4,** 9R60. 1-11 (http://genomebiology.com/2003/4/9/R60).
8. Krause, C. J., Carey, T. E., Ott, R. W., Hurbis, C., McClatchey, K. D., and Regezi J. A. (1981) Human squamous cell carcinoma. Establishment and characterization of new permanent cell lines. *Arch. Otolaryngol.* **107,** 703–710.
9. Chen, Y., Dougherty, E. R., and Bittner, M. L. (1997) Ratio-based decisions and the quantitative analysis of cDNA microarray images. *J. Biomed. Optics* **2,** 364–374.
10. Chung, C. H., Parker, J. S., Karaca, G., et al. (2004) Molecular classification of head and neck squamous cell carcinomas using patterns of gene expression. *Cancer Cell* **5,** 489–500.
11. Eisen, M. B., Spellman, P. T., Brown, P. O., and Botstein, D. (1998) Cluster analysis and display of genome-wide expression patterns. *Proc. Natl. Acad. Sci. USA* **95,** 14,863–14,868.

7

Gene Profiling Uncovers Retinoid Target Genes

Yan Ma, Qing Feng, Ian Pitha-Rowe, Sutisak Kitareewan, and Ethan Dmitrovsky

Summary

Decades of hypothesis-driven research have identified candidate targets for cancer therapy and chemoprevention. Recently, genomic, proteomic, and tissue-based microarray approaches have made possible another scientific approach. This is one that interrogates comprehensively the complex profile of mRNA or protein expression present in normal, preneoplastic, or malignant cells and tissues. This in turn can uncover critical targets for cancer pharmacology and also lead to a better understanding of the known or novel networks of gene expression that play a rate-limiting role in carcinogenesis. This chapter addresses the use of mRNA expression profiling to uncover candidate target genes active in cancer pharmacology by citing as an example how this has already proven useful to reveal that retinoids (natural and synthetic derivatives of vitamin A) signal through pathways, which promote tumor cell differentiation, induce growth suppression, trigger apoptosis or affect other growth regulatory pathways. Pathways involved in the regulation of protein stability will be highlighted as these play a critical role in mediating pharmacological effects of the retinoids in cancer therapy or chemoprevention.

Key Words: Cancer therapy; cancer chemoprevention; microarray; gene profiling; retinoid target genes; retinoids.

1. Introduction

The transformation process that causes normal cells to become malignant is of a chronic and multistep nature. These include the initiation, promotion, and progression steps. The initiation phase is typically viewed as the earliest and one associated with cellular damage and changes in gene expression. The promotion phase subsequently leads to expansion of altered cells or tissues that form preneoplastic lesions that precede invasive malignant disease. Additional genetic and epigenetic changes result in critical cellular and tissue changes that cause invasive or metastatic malignant disease to become clinically evident.

From: *Methods in Molecular Biology, vol. 383: Cancer Genomics and Proteomics: Methods and Protocols*
Edited by: P. B. Fisher © Humana Press Inc., Totowa, NJ

Cancer chemoprevention is a strategy that uses natural or synthetic agents to reverse, suppress, or prevent carcinogenesis *(1–4)*. Multiple mechanisms are engaged by chemopreventive agents that can target the initiation step. Examples include the scavenging of reactive oxygen species and inhibiting depletion of antioxidant defenses that would prevent DNA oxidative damage. Regulation of phase I enzymes responsible for formation of reactive carcinogenic metabolites can reduce formation of these species *(2,4)*. Modulation of oxidative enzymes often produces products of lower carcinogenic potential than the parent compound. Induction of detoxification (phase II) enzymes and pathways that conjugate carcinogens also can affect carcinogenesis *(3,4)*. The scavenging of reactive, carcinogenic intermediates occurs through direct biochemical interactions that affect DNA repair mechanisms *(4)*. Agents that target promotion or progression steps include those that enhance detoxification of a tumor promoter *(4)* and those that inhibit inflammation, proliferation, angiogenesis, and those that induce cellular differentiation or promote apoptosis *(5–8)*.

Diverse natural and synthetic compounds exhibit chemopreventive properties. For example, antioxidant actions are exerted by certain vitamins. Other examples of candidate chemopreventive agents include polyphenols, certain isoflavonoids, other antioxidant agents, and agents that inhibit cyclooxygenase or lipoxygenase dependent pathways to reduce inflammation or proliferation *(9–14)*. Isothiocyanates and indole-3-carbinol modulate phase II enzymes and promote apoptosis *(15,16)*. Certain unsaturated fatty acids also affect the carcinogenesis process *(17)*.

Clinical work has supported the validity of cancer chemoprevention using selective estrogen receptor modulation with tamoxifen to reduce risk of breast cancer *(18)*. Selective cyclooxygenase-2 inhibition reduces polyp formation in familial adenomatous polyposis and nonsteroidal anti-inflammatory drugs (NSAIDs) reduce incidence of colon cancer, as reviewed in **ref. *19***. Retinoids, natural and synthetic derivatives of vitamin A, are active in cancer therapy and chemoprevention *(7,8,20)*. The precise mechanisms engaged in these clinical responses need to be understood so as to broaden clinical use of these agents. Retinoids suppress several steps in carcinogenesis. This chapter addresses gene-profiling studies of retinoid target genes that have clinical relevancy.

1.1. Retinoid Nuclear Receptors

The retinoids have diverse and important roles in development, proliferation, differentiation, vision, immune response, and other pathways involved in cellular and tissue homeostasis. Wolbach and Howe *(21)* first found in 1925 that vitamin A-deficient rodents developed squamous metaplasia. Subsequent epidemiological studies and in vitro as well as animal model experiments provided a strong rationale for clinical use of retinoids. Retinoids are active in the treatment of certain nonneoplastic, preneoplastic, and malignant diseases as well as in chemoprevention, as reviewed in **ref. 20**. These are quite active in treatment of acute promyelocytic leukemia (APL), juvenile chronic myelogenous leukemia, mycosis fungoides, Kaposi's sarcoma, and high-risk neuroblastoma *(7,8,20)*. Antiproliferative, differentiation-, or apoptosis-inducing mechanisms appear to signal retinoid therapeutic or chemopreventive effects *(20,22,23)*. Retinoids are useful tools to uncover other pharmacological targets. For instance, the epidermal growth factor receptor (EGFR) is a retinoid target gene *(23)*. The EGFR pathway can be targeted by other pharmacological agents *(7)*.

Retinoids exert biological effects through nuclear retinoid receptors. Two classes of nuclear retinoid receptors exist. These are the retinoic acid receptors (RARs) (classical retinoid receptors) and the retinoid X receptors (RXRs), which are examples of nonclassical retinoid receptors *(20)*. There are three RARs: RARα, RARβ, and RARγ, and three RXRs: RXRα, RXRβ, and RXRγ. Multiple isoforms can exist for these receptors. RARs can heterodimerize with RXRs, whereas RXRs heterodimerize with other nuclear receptors including the thyroid hormone receptor (TR), vitamin D receptor (VDR), peroxisomal proliferator activated receptor (PPARγ), among other receptors, as reviewed in **ref. 20**.

Retinoid receptors have several functional domains with varying degrees of homology between family members. These include the DNA binding domain, ligand binding domain, and amino-terminal region, which is the least conserved domain *(20)*. Unliganded nuclear retinoid receptors can bind to DNA response elements present in the promoter region of target genes as well as associate with inhibitory corepressors. Corepressors mediate inhibitory transcriptional effects by recruiting histone deacetylase complexes (HDACs) that alter chromatin structure. Stimulatory coactivators are recruited after ligand treatment. Coactivators can enhance transcriptional activation through histone acetyl transferase (HAT) activity that affects chromatin structure and transcription of target genes *(20)*.

Ligand-binding domains of RARs and RXRs can be independently pharmacologically targeted *(20)*. For example, 9-*cis*-retinoic acid is a bifunctional retinoid that activates both RAR and RXR pathways, while all *trans*-retinoic acid (RA) activates only the classical RAR pathway. These agents have been extensively studied during induced tumor cell differentiation, and growth suppression. Some nonclassical retinoids appear to act through retinoid receptor independent pathways. An example is fenretinide (4HPR) that preferentially signals apoptosis *(20,22)*. This chapter emphasizes analyses of retinoid target genes uncovered through gene expression profiling experiments designed to identify therapeutic or chemopreventive effects of the classical retinoid, RA.

1.2. Retinoid Target Genes

Preclinical models to assess retinoid response with microarray-based technologies are useful to identify the network of retinoid target genes that signal retinoid biological effects. Functional assessment of candidate retinoid target genes has typically involved in depth laboratory studies to assess each target gene. Considerable time is needed to explore the function of individual retinoid target genes. Here, we will describe some approaches taken for retinoid target gene validation. Examples of retinoid target genes that were identified through gene profiling will be discussed along with how each was validated functionally.

1.3. RIP140

Receptor interacting protein 140 (RIP140) was initially found as a candidate retinoid target gene in human embryonal carcinoma cells through microarray expression profiling *(24,25)*. RIP140 mRNA levels were increased within 3 h of RA-treatment and protein levels were increased within 6 h of RA-treatment *(24)*. *De novo* protein synthesis was not required for induction of RIP140 and an upstream promoter domain was found to contain a near consensus direct repeat 5 retinoic acid response element (RARE) that is activated by RA-treatment, suggesting that RIP140 was activated directly by this retinoid treatment *(24)*.

The precise role of RIP140 in mediating retinoid effects is under study. However, prior evidence revealed that RIP140 suppressed transcriptional activity of ligand-bound nuclear receptors *(26–31)*. Several different approaches extended this prior work and revealed that RIP140 acted as a negative regulator of retinoid transcriptional signaling and retinoid-mediated differentiation in embryonal carcinoma cells *(24,32)*. Expression of RIP140 inhibited RA-dependent activation of an RARE containing reporter plasmid *(24)*. Subsequent studies revealed small interfering RNA (siRNA) mediated repression of RIP140 enhanced retinoid-induced RARE reporter activity and retinoid-mediated induction of endogenous RA target gene expression *(32)*. This also accelerated retinoid-induced embryonal carcinoma cell differentiation *(32)*. Taken together, these investigations provided strong evidence that retinoid induction of RIP140 inhibited the degree of retinoid signaling and thereby delayed RA-dependent biologic effects. These findings also highlighted a potential negative-feedback regulatory mechanism for this corepressor that could play an important role not only in retinoid signaling, but also in cross-talk with estrogenic signaling in breast carcinogenesis *(24,32)*.

1.4. Tumor Necrosis Factor-Related Apoptosis-Inducing Ligand

Altucci et al. *(33)* uncovered tumor necrosis factor-related apoptosis-inducing ligand (TRAIL) as a retinoid target gene active in APL through studies of retinoid-induced apoptosis in the NB4 APL cell line. Prolonged treatment of this APL line with an RARα agonist induced apoptosis, along with increased expression of TRAIL *(33)*. TRAIL can cause apoptosis of APL and other cancer cell lines *(33)*. Retinoid treatment of an RA-resistant NB4 APL cell line did not trigger apoptosis or induce TRAIL expression in these cells *(33)*. However, both retinoid sensitive and resistant APL cell lines underwent apoptosis on TRAIL treatment, even in the absence of retinoid treatment *(33)*. Inhibiting TRAIL receptor activity reduced retinoid-mediated apoptosis *(33)*. Blast cells isolated from APL cases induced TRAIL expression after prolonged RA treatment and these cells were also sensitive to retinoid as well as TRAIL-mediated apoptosis *(33)*. Interestingly, there is evidence that retinoid treatment enhanced expression of the death receptors 4 and 5 (DR4 and DR5, respectively), thereby sensitizing tumor cell lines to TRAIL-dependent apoptosis *(34)*.

1.5. Approaches to Study Retinoid Target Genes

Retinoid target genes share several features. Expression of these target genes are regulated by retinoid treatments. The induced target gene expression in turn plays a key functional role in retinoid-response. Diverse models, techniques, and approaches have been used to examine specific target gene function. Findings reveal an impressive network of retinoid-mediated regulation of target gene expression as well as complex functional roles for the target gene products in mediating retinoid effects. Retinoid-mediated changes in gene regulation have been initially highlighted through analyses of mRNA- or protein-based expression arrays, as reviewed in the **ref. 20**. Additional studies are needed to define further the function of candidate retinoid target genes. Available approaches include molecular and cellular biological strategies applied after retinoid treatments such as kinetic studies of gene expression, immunoblot analyses, transient transfection reporter assays, stable transfection experiments, examination of retinoid sensitive as well as resistant cell lines, and analyses of gene expression profiles in clinical specimens along with linkage of the results to clinical databases.

Kinetic studies of gene expression can provide insight into the retinoid regulation of a target gene. Although retinoid treatment might affect mRNA as well as protein expression, the relationship between the time-course of this regulation and the specific biological or biochemical effect can provide a clue regarding the potential function of an individual target gene. For example, earlier work revealed that augmented expression of RIP140 by RA-treatment directly affects aspects of retinoid response *(24)*, but induction of TRAIL by RA-treatment was shown to require expression of interferon regulatory factor 1 *(35)*. The time to peak induction of expression might provide insights into target gene function. By observing expression changes that occurred before retinoid-induced apoptosis of NB4 APL cells, Altucci et al. *(33)* were able to highlight specific changes in target gene expression involved in triggering this apoptosis.

Validated preclinical models, including analyses of retinoid responsive and resistant cell lines, are useful to confirm a relationship between changes in target gene expression and biological response. Microarray analyses have revealed differences in key target genes involved in mediating retinoid biological effects in different cellular contexts *(25,36,37)*. For this reason, it is useful to assess the functional impact of target gene expression in multiple models, including well-characterized cell lines, experimental animal models as well as in clinical specimens.

Functional validation of retinoid target genes would determine the precise role played by each of these species in a particular cell context. In the examples already discussed, several different experimental approaches were undertaken to assess target gene function. Evidence for an inhibitory role of RIP140 in retinoid signaling was found by reducing RIP140 levels after RA-treatment and by exogenously expressing RIP140 in a relevant cell context. When RIP140 levels were repressed, retinoid-mediated differentiation and gene induction were each enhanced *(32)*. Conversely, exogenous expression of RIP140 inhibited retinoid-mediated reporter activity *(24)*. Different experiments were used to establish a role for TRAIL in retinoid-mediated apoptosis *(33)*. These and other studies have proven useful to establish that retinoid target genes themselves are able to signal key aspects of retinoid response. Knock-down of these species can affect retinoid response *(20,32)*. Retinoid resistance owing to deregulation of a target gene should be overcome by exogenous expression of the specific target gene *(20)*.

Retinoid target gene identification and functional validation will be influenced by evolving techniques. For example, comprehensive proteomic screens might identify candidate retinoid target genes that were not highlighted by prior mRNA microarray analyses. Although the techniques and tools described in this chapter would continue to play a role in target gene identification, the ability to follow retinoid-mediated changes in posttranslational modification and protein processing would become important in uncovering the complex network of changes in gene expression that confer retinoid biological or clinical effects.

Functional validation of candidate retinoid target genes is currently a time-intensive process. As such, it is desirable to limit the number of candidate target genes that require this investigation. As screening approaches involving gain and loss of function studies mature, it would become possible to determine more rapidly than is currently possible the roles played by each retinoid target gene in conferring retinoid responses. The role of a specific retinoid target gene, UBE1L, which appears active not only in differentiation therapy, but also in chemoprevention will be discussed in the **Subheading 1.6.**

1.6. Dual Roles for Retinoid Target Genes

Microarray expression analysis has become an indispensable tool in biomedical research to assess mRNA expression profiles. It is a methodology routinely used to discern gene expression patterns under various experimental and clinical conditions *(24,36,38–41)*. The complex network of relationships present between expressed gene products might not be readily discerned, if only individually species were examined. The expression profiles of species induced during differentiation therapy of APL following RA-treatment illustrates this point *(37,42)*.

Identifying those species that regulate RA-mediated differentiation in APL should provide insight into the mechanisms responsible for this leukemic cell differentiation program. This approach should provide a basis for identifying molecular pharmacological targets or pathways involved in retinoid response. Using DNA microarray-based techniques, several candidate RA-target genes were highlighted as prominently induced after RA-treatment of the NB4 APL cell line *(37)*. Several of these species were deregulated in their expression in RA-resistant APL cells *(37,42,43)*.

Among these RA-induced genes, UBE1L (ubiquitin-activating enzyme-E1 like protein) was selected for further study. Ealier work revealed that RA-treatment activated the proteasome-dependent degradation of PML/RARα, the oncogenic translocation product found in APL *(44)*. RA-treatment also caused proteasomal degradation of cyclin D1 during induced chemoprevention and tumor cell differentiation *(45,46)*. UBE1L mRNA and protein were each prominently increased within 24 h of RA-treatment in the RA-sensitive NB4-S1 APL cell line *(42)*. In contrast, this was not observed in RA-resistant NB4-R1 APL cells after RA-treatment *(42)*. Augmentation of UBE1L protein expression in NB4-S1 cells after RA-treatment was inversely associated with PML/RARα protein expression (*see* **Fig. 1**) in these APL cells *(42)*. Reduction of PML/RARα protein expression by UBE1L was confirmed in cotransfection experiments establishing that UBE1L directly triggered repression of PML/RARα *(42)*.

RA is a pharmacological agent that is approved by the Federal Drug Administration (FDA) for differentiation-based treatment of APL, as reviewed in **ref. 20**. Retinoids are also clinically active in other therapeutic and chemopreventive settings, as previously reviewed in **refs. 20** and **47**. Earlier work revealed that RA-treatment of BEAS-2B immortalized human bronchial epithelial (HBE) cells prevented carcinogenic transformation of these cells *(48)*. In a follow-up study employing affymetrix-based microarray DNA analysis, UBE1L was found to be augmented in RA-chemoprevented HBE cells, as compared with carcinogen-transformed HBE cells *(49)*. This finding indicated that induced UBE1L expression could play an important role in RA-response beyond APL. Microarray analyses *(49)* were performed using cDNA isolated independently from immortalized (BEAS-2B) HBE cells and from HBE cells (BEAS-2B$_{NNK}$) that were transformed with the carcinogen N-nitrosamine-4-(methylnitrosamino)-1-(3-pyridyl)-1-butanone (NNK). Findings were compared

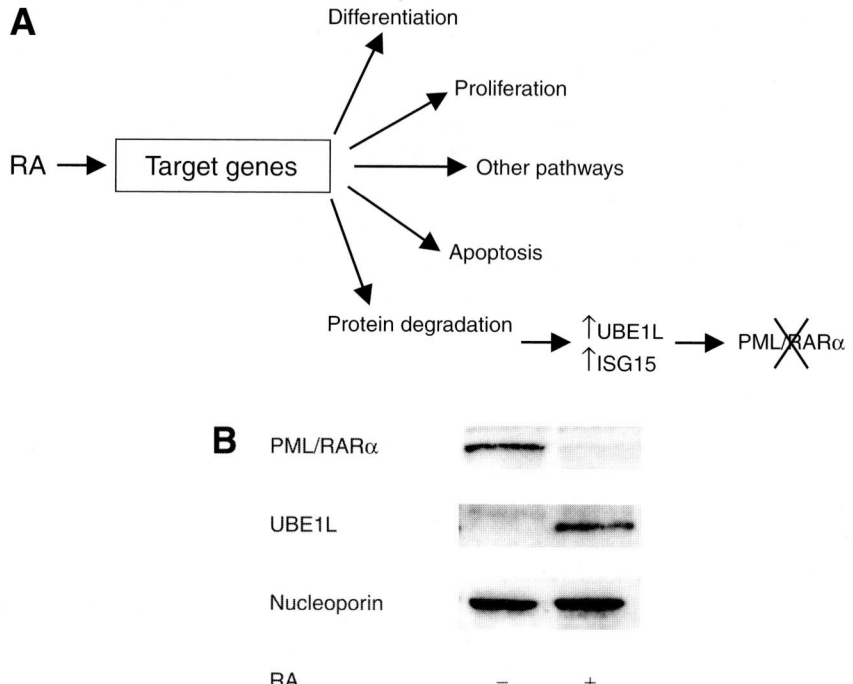

Fig. 1. Effects of all-*trans* retinoic acid (RA) on the oncogenic protein found in acute promyelocytic leukemia (APL), PML/RARα. (A) RA has been shown to affect the stability of oncogenic proteins, such as PML/RARα and cyclin D1. An inverse relationship exists between UBE1L and PML/RARα expression. A model exists *(59)* whereby RA-mediated induction of UBE1L and ISG15 (displayed in this figure as ↑UBE1L and ↑ISG15, respectively) triggered degradation of PML/RARα (depicted by the "X") and thereby can overcome the cellular differentiation block found in APL. Whether this is mediated through an ISG15-dependent conjugation of PML/RARα is the subject of current work. (B) RA-mediated induction of UBE1L was directly related to repression of PML/RARα protein. Immunoblot analyses of PML/RARα and UBE1L before and after RA-treatment. NB4-S1 APL cells were treated with RA (1 μ*M*) (+) or with dimethyl sulfoxide (DMSO) vehicle control (−) for 48 h before immunoblot analyses were performed. Nucleoporin expression served as a loading control for total protein.

with the gene profile evident in chemoprevented (BEAS-2B$_{NNK\ RA}$) HBE cells that were treated with RA immediately before exposure to NNK *(49)*.

Analyses of these microarrays indicated that the expression profile of immortalized HBE cells resembled that of RA chemoprevented HBE cells (as shown in **Fig. 2**), but were distinct from the profile of the transformed HBE cells. Notably, only a small cluster of species was prominently induced during

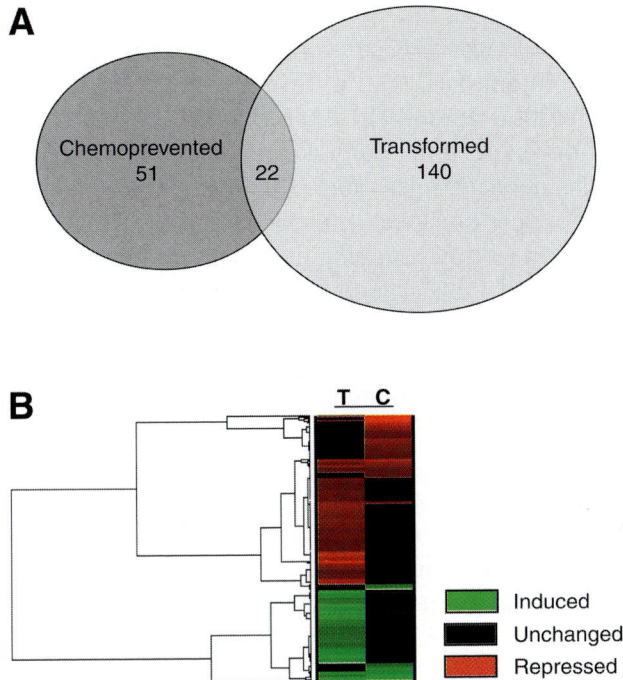

Fig. 2. Global gene expression analysis of regulated species in immortalized BEAS-2B human bronchial epithelial (HBE) cells and derived cell lines. (**A**) Those species induced 2.5-fold in chemoprevented as compared to carcinogen-transformed HBE cell lines relative to the parental BEAS-2B line were depicted as the numbers displayed in this figure. (**B**) The cluster analysis of the 2.5-fold augmented species in the indicated HBE cell lines. The symbol "T" is meant to designate those species increased in the transformed as compared to the parental BEAS-2B HBE cell line and "C" represents those species augmented in the chemoprevented as compared with the immortalized BEAS-2B HBE cell line.

this RA-mediated chemoprevention *(49)*. These species were categorized into two groups. The first group included those that were novel or previously recognized retinoid target genes and the second group was unexpectedly interferon-stimulated genes (ISGs) *(49)*. Some species were members of both groups. It is interesting to note that unlike those species that were identified after prolonged RA-treatment, acute RA-treatment uncovered a distinct casette of retinoid target genes *(36)*. Thus, prolonged RA-treatment induced a different gene expression profile as compared to acute RA-treatment of HBE cells *(36,49)*. These chronically induced species included the ISGs: 2′-5′-oligoadenylate synthetase 1 (OAS1), myxovirus resistance 1 (MX1), retinoic acid-induced gene, G protein (RIG-G), UBE1L, midkine (MK), and interferon-stimulated protein, 15 kDa

(ISG15), among others *(49)*. Examples of previously recognized retinoid-target genes augmented in HBE cells after prolonged RA-treatment included the inhibitor of DNA binding 1 (ID1), insulin-like growth factor-binding protein 6 (IGFBP6), and E74-like factor 3 (ELF3) *(36,49)*.

UBE1L was of interest to study since its expression was reported as repressed in lung cancer cell lines *(50,51)*. This implicated a role for UBE1L as a potential tumor suppressor gene in these cells. These in vitro findings were confirmed and extended by showing that UBE1L immunoblot expression was lower in subsets of lung cancers than adjacent normal lung tissues *(49)*. Immunohistochemical studies of the normal human bronchial epithelium demonstrated abundant UBE1L expression that was inversely related to the low levels of cyclin D1 immunohistochemical expression detected in these tissues *(49)*. Cyclin D1 is a recognized chemoprevention marker in bronchial epithelial cells *(45,49,52)*. Enhanced cyclin D1 expression has been associated with carcinogenesis in diverse cellular contexts *(53–55)*. It was hypothesized that UBE1L might directly affect the stability of cyclin D1 protein in normal as compared to preneoplastic or transformed pulmonary epithelial cells or tissues. Transient transfection assays performed in HBE cells confirmed that cyclin D1 expression was repressed by UBE1L in a manner that depended on the transfected UBE1L dosage *(49)*.

The precise mechanisms responsible for UBE1L-dependent repression of PML/RARα and cyclin D1 need to be determined. UBE1L is an activating enzyme for the ubiquitin-like protein, ISG15 *(43,56)*. It is postulated that target proteins can be degraded through proteasome-dependent or related mechanisms that employ UBE1L as an activating enzyme. Notably, UBE1L and the ubiquitin-related ISG15 species were found to be coordinately regulated by RA- and interferon-treatments and were also coexpressed in many adult and fetal cells or tissues *(43)*. Yet, some cancer cell lines and *de novo* cancers exhibited UBE1L repression *(43)*. UBE1L and ISG15 can associate in vivo *(43)*. Taken together, these finding indicate that UBE1L-ISG15 represent a pharmacologically induced pathway that appears to have distinct targets *(20,42,47)*. This pathway is a potential molecular pharmacological target. The protease UBP43 serves as a deconjugase for UBE1L and ISG15 and this would in turn regulate the biological consequences of their conjugation *(57,58)*. This possibility was confirmed by the development of UBP43 knock-out mice that resulted in accumulation of ISG15 target protein conjugates, following interferon treatment *(58)*.

DNA microarray analyses have uncovered UBE1L as a retinoid target gene in several cellular contexts, as shown in **Figs. 2** and **3** and as reviewed elsewhere in **ref. 59**. The fact that UBE1L associates with ISG15 and is coordinately expressed with ISG15 in diverse cell contexts to affect stability of proteins that regulate carcinogenesis (such as PML/RARα and cyclin D1) indicates that a

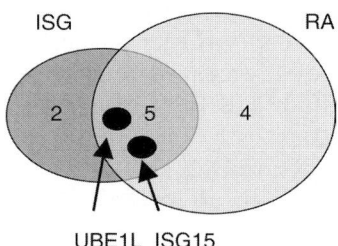

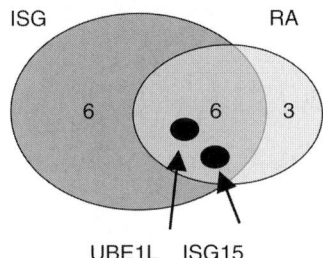

Fig. 3. Summary of species prominently induced (fourfold) by microarray analyses during all-*trans*-retinoic acid (RA)-induced chemoprevention of BEAS-2B human bronchial epithelial (HBE) cells. This included previously recognized retinoid target genes as well as interferon-stimulated genes (ISGs). Two species prominently induced during this RA-mediated chemoprevention included UBE1L and ISG15, as displayed in this figure. The numbers in this figure refer to the number of species augmented fourfold in each subgroup.

novel molecular therapeutic pathway might exist. Whether UBE1L, or its physical partner, ISG15, or the deconjugase UBP43 are pharmacological targets in cancer therapy or chemoprevention needs to be determined.

2. Materials

1. Serum-free LHC-9 medium (used to culture immortalized BEAS-2B or RA-refractory BEAS-2B-R1 HBE cells) (Biofluids, Rockville, MD).
2. RPMI 1640 medium (Invitrogen, Carlsbad, CA) (used to culture independently the RA-sensitive NB4-S1 and the RA-resistant NB4-R1 APL cell lines was also purchased).
3. Fetal bovine serum (Gemini BioProducts, Calabrasas, CA).
4. Penicillin (Gemini BioProducts).
5. Streptomycin (Gemini BioProducts).
6. Glutamine (Cellgro by Mediatech Inc, Herndon, VA).

7. RA (Sigma, St. Louis, MO).
8. Dimethyl sulfoxide (DMSO) (control for RA treatments) (Sigma).
9. Nitrocellulose membranes (Schleicher and Schuell Bioscience, Inc., Keene, NH).
10. Polyclonal rabbit antibody (used as independent primary antibodies) that recognized the amino terminus of UBE1L *(42)* or the carboxyl terminus *(42)*, respectively (*see* **Notes 2** and **3**).
11. Rabbit polyclonal anti-RARα antibody (Santa Cruz Biotechnology, Santa Cruz, CA).
12. Goat anti-actin antibody (used as a control for loaded protein) (Santa Cruz Biotechnology).
13. Anti-nucleoporin antibody (used as a loading control) (Santa Cruz Biotechnology).
14. Anti-rabbit antisera (Amersham Biosciences, Piscataway, NJ).
15. Anti-goat antisera (Santa Cruz Biotechnology).
16. Chemiluminescence system for immunoblot experiments (Amersham Biosciences).
17. RNeasy® Protect Mini Kit (used to isolate total cellular RNA) (Qiagen, Valcencia, CA).
18. DNA-free kit (used to remove contaminating DNA from the total RNA) (Ambion, Austin, TX).
19. Superscript II reverse transcriptase (used to synthesize cDNA from RNA) (Invitrogen, Carlsbad, CA).
20. Taq polymerase (used to perform polymerase chain reactions [PCR]) (Invitrogen).
21. Gelstar nucleic acid stain (used to stain PCR products that were resolved on 1% agarose gels) (Biowhittaker Molecular Applications, Rockland, ME).
22. Storm™ (Molecular Dynamics, Sunnyvale, CA).
23. ImageQuant Software (Amersham Biosciences Corp.).
24. Trizol reagent (used to isolate total RNA for gene profiling experiments) (Life Technologies, Gaithersburg, MD).
25. Affymetrix HU6800 microarray (derived from unigene 18, GenBank, and TIGR).
26. Customized 8600 microarray (Affymetrix, Santa Clara, CA).
27. Affymetrix Microarray Suite 4.0 software (used to process the primary microarray data) (Affymetrix).
28. Effectene transfection reagent (used for transfection experiments) (Qiagen) was used.
29. RIPA cell lysis buffer:

    a. 0.01 M Tris-HCl (pH 7.4).
    b. 5 mM EDTA.
    c. 0.15 M NaCl.
    d. 1% Triton X-100.
    e. 0.01% Sodium deoxycholate.
    f. 0.1% SDS.
    g. 0.4% Aprotinin.
    h. 0.1 mM Sodium vandate, with 1 mM dichlorodiphenyltrichloroethane (DTT).
    i. Protease inhibitors.

30. Formaldehyde (Ambion, Austin, TX) (used for Northern analysis).
31. ^{32}PdCTP (Amersham Biosciences) (used for Northern analysis).

32. Hybridization solution (used for Northern analysis):
    a. 0.5 M Sodium phosphate (pH 7.0).
    b. 7% SDS.
    c. 1% Bovine serum albumin.
    d. 1 mM EDTA.
33. 0.1X SSC (used for Northern analysis):
    a. 15 mM NaCl.
    b. 1.5 mM sodium acetate.
34. Nonfat milk (Nestle, Glendale, CA) (used for Western analysis).
35. Coomassie (Bio-Rad, Hercules, CA).

3. Methods

3.1. Cell Lines, Culture Conditions, and RA-Treatments (see Notes 1, 2, 4, and 5)

1. Immortalized (BEAS-2B) HBE cells, NNK-carcinogen transformed (BEAS-2B$_{NNK}$) HBE cells, and RA-refractory (BEAS-2B-R1) HBE cells were each cultured in serum-free LHC-9 medium supplemented with 100 U/mL penicillin, 100 µg/mL streptomycin, and 2 mM L-glutamine, as previously described in **refs. *45*** and ***49***.
2. The chemoprevented (BEAS-2B$_{NNK\ RA}$) HBE cells were treated with RA (1 µM) and were cultured in LHC-9 media, as previously reported *(48)*.
3. The RA-sensitive NB4-S1 and RA-resistant NB4-R1 APL cell lines were each cultured in RPMI medium supplemented with 10% fetal bovine serum along with 100 U/mL penicillin, 100 µg/mL streptomycin, and 2 mM L-glutamine.
4. All cell lines were cultured at 37°C in 5% CO_2 using a humidified incubator.

3.2. Semiquantitative RT-PCR Assays (see Note 2)

1. Logarithmically growing NB4-S1 and NB4-R1 cells were each seeded in the indicated medium at a density of 3×10^5 cells/mL.
2. These cell lines were either supplemented with RA (1 µM) or with the vehicle, DMSO.
3. Total cellular RNA was extracted using either TRI reagent or the RNeasy Protect Kit with contaminating DNA removed as per the manufacturer's procedures.
4. The cDNA was isolated by synthesis from 5 µg of total RNA that was primed with random hexamers and with Supsript II reverse transcriptase (RT).
5. PCR amplifications were accomplished with Taq polymerase and expression patterns of the unregulated species actin or glyceraldehyde 3-phosphate dehydrogenase (GAPDH) were compared with that of the RA-regulated species, UBE1L and ISG15, as described *(42,43)*.
6. These RT-PCR semiquantitative expression studies were performed with oligonucleotides that have been described previously in **refs. *36,42*,** and ***49***.

3.3. Northern and Western Analyses

1. Northern analyses were performed using 10 µg of total cellular RNA that were size-fractionated on a 1% agarose-formaldehyde gel before transfer to a nitrocellulose

membrane that was incubated overnight at 65°C in hybridization solution with the desired ^{32}PdCTP-radiolabeled probe before stringent washings at between 56 and 60°C in a 0.1X SSC, 0.1% sodium dodecyl sulfate (SDS) solution before autoradiography by exposure to a phosphorimager screen with the image analyzed by Storm^{TM} and Image Quant Software (*see* **Notes 1**, **2**, and **5**).

2. For these expression studies, a ^{32}PdCTP-radiolabeled probe (by random-priming) for actin was hybridized to confirm that similar amounts of total RNA were added to each lane.

3. For immunoblot analyses, size-fractionation of total cellular protein was accomplished using SDS-polyacrylamide (PAGE) gels before transfer to nitrocellulose membranes. Parental BEAS-2B, RA-refractory BEAS-2B-R1, transformed BEAS-2B$_{NNK}$, and chemoprevented BEAS-2B$_{NNKRA}$ HBE cells as well as NB4-S1 and RA-resistant NB4-R1 APL cells were individually harvested and washed twice with ice-cold phosphate-buffered saline and then lysed with ice-cold RIPA cell lysis buffer.

4. The protein concentrations were measured in duplicate using Bradford assays. Following transfer to nitrocellulose filters, blocking was accomplished with 5% nonfat milk before incubation overnight with the primary antibody.

5. Filters were then washed and incubated with the desired secondary antibody and detection was with the described chemiluminescence system.

6. As loading controls to assure comparable amounts of total protein per lane, expression of actin or of nucleoporin was examined.

7. Gene expression was assessed using the Personal Densitometer SI^{TM} (Molecular Dynamics, Sunnyvale, CA) and ImageQuant Software.

3.4. Gene Expression Profiles

1. Total RNA was isolated individually from BEAS-2B or its derived chemoprevented and carcinogen-transformed HBE cell lines that were either treated with or without DMSO as a vehicle or with RA (1 μM) using Trizol reagent to generate cDNA (*see* **Notes 4** and **5**).

2. The Affymetrix Hu6800 microarray and the customized 8600 microarray contained 6095 and 7976 unique species, respectively, with a total of 11,069 distinct species displayed on these two microarrays. These arrays were hybridized, washed, and scanned as per the recommendation of the manufacturer.

3. The obtained data were assessed using Affymetrix Microarray Suite 4.0 software to measure average difference values and to assess signal intensity for each probe set.

4. By use of these microarrays that were normalized using the global scaling method recommended by Affymetrix, it was possible to score fold changes in these different cellular contexts.

5. Those species induced by RA at 2.5- or 4-fold or more were examined further, as shown in **Figs. 2** and **3**.

6. Highlighted species were validated independently by use of RT-PCR, Northern, or Western analyses as well as by examination of species displayed on both arrays.

7. For selected novel species (such as UBE1L) for which antisera did not previously exist, polyclonal antisera were generated for use in immunoblot as well as immunohistochemical assays *(42,43)*.

3.5. Functional Validation of Retinoid Target Genes

1. Once a candidate RA regulated target gene has been identified, then its functional role in retinoid response needs to be determined. Two experiments were conducted to address this. The first examined the relationship between UBE1L and PML/RARα expression in exponentially growing BEAS-2B cells that do not express PML/RARα (*see* **Note 1**). The second one examines the effect of UBE1L on cyclin D1 expression in BEAS-2B cells (*see* **Note 5**). These cells were independently cotransfected using the Effectene transfection reagent along with expression plasmids containing full length UBE1L (pSG5-UBE1L), an inactive UBE1L (pSG5-UBE1L-T), the hemagglutinin (HA)-tagged PML/RARα expression vector (pCMX-PML/RARα), or a HA-tagged cyclin D1 expression vector (pCMV-Cyclin D1). Insertless vectors served as controls as did cotransfection of the pCMV-β-galactosidase reporter plasmid that permitted β-galactosidase activity measurements along with total protein measurements. Coomassie staining of gels served as an additional loading control.

4. Notes

1. The data presented in this chapter indicate that microarray analyses have proven useful to identify the E1-like activating enzyme, UBE1L, as a retinoid target gene in APL *(42)*. Following this observation, as summarized in **Fig. 1A**, molecular genetic and biochemical assays that included use of siRNA strategies, cotransfection assays, and immunoblot analyses were used to reveal that the in vivo partner of UBE1L, ISG15, was also prominently induced by RA-treatment of APL cells *(43)*. Notably, an inverse relationship was shown to exist between UBE1L and PML/RARα protein expression, as indicated in **Fig. 1B**. This inverse relationship indicated that UBE1L might directly repress PML/RARα expression. This possibility was confirmed by cotransfection experiments demonstrating that UBE1L overexpression would, in a dose-dependant manner, repress expression of PML/RARα, while the classical E1 was unable to trigger this effect *(42)*.

2. Taken together, these findings underscore how useful it is to "mine" data from microarray experiments by independently confirming gene profiling findings at the RNA level of expression by use of real-time or semiquantitative RT-PCR assays and Northern analyses as well as through Western analyses to detect protein expression of the highlighted retinoid target genes. In the case of UBE1L, to validate the array data, two independent polyclonal antibodies that recognized UBE1L were derived including one generated using a carboxyl-terminal peptide and the other using an amino-terminal peptide *(42)*. In this way, blocking peptides could be used to confirm the specificity of the obtained immunoreactive species on immunoblot

analyses. Because two independent polyclonal antibodies were generated, the immunoblot results obtained from one antibody could be confirmed by the other.

3. The results obtained by immunoblot assays can be extended to the cellular or tissue levels of expression by adapting the antibodies for use in immunohistochemical assays suitable for analyses of clinical specimens including those that were paraffin-embedded and formalin-fixed. For UBE1L, each of the derived anti-UBE1L antisera were adapted for this purpose and high levels of UBE1L immunohistochemical expression were found in those cells and tissues that would be expected to exhibit abundant UBE1L expression such as in RA-treated as compared with RA-untreated NB4-S1 APL cells (data not shown) as well as in pulmonary macrophages and the histologically normal human bronchial epithelium *(49)*. The controls needed to verify these immunohistochemical results include use of blocking peptides to inhibit immunohistochemical staining, as well as use of preimmune antisera to confirm that nonspecific straining of tissues was not present *(49)*.

4. Another way to confirm and extend the results obtained by microarray studies is to explore gene profiling patterns of the desired cluster or subset of species in other cell contexts. **Figures 2** and **3** address this by assessing the expression profiles of UBE1L and its partner, ISG15, in NNK-carcinogen transformed HBE cells as well as in RA-chemoprevented HBE cells. When these cells were compared by assessing prominently induced species in chemoprevented against immortalized and in chemoprevented vs transformed HBE cells, UBE1L, and ISG15 were found to be prominently induced as were several RA target genes and ISGs, as summarized in **Fig. 3**.

5. Notably, this UBE1L-dependent pathway was shown to be active in a previously unrecognized chemoprevention mechanism in that UBE1L and cyclin D1 cotransfection led to repression of cyclin D1 expression in a manner that depended on the expression of transfected UBE1L. Another consequence of this analysis in lung cancer chemprevention was the finding of the relatively similar gene profiles of the chemoprevented as compared with the immortalized HBE cells, as displayed in **Fig. 2**.

6. In summary, this chapter addressed use of mRNA expression profiling to uncover candidate target genes active in cancer pharmacology by citing as an example how these technologies have already revealed key pathways that signal effects of retinoids in cancer therapy and chemoprevention. Highlighted target genes and associated pathways include those that promote tumor cell differentiation, induce growth suppression, trigger apoptosis, regulate protein stability, or affect other key regulatory pathways. Notably, gene profiling experiments identified UBE1L and its physical partner, ISG15, as playing mechanistic roles in retinoid differentiation and chemopreventive response. Future work should determine whether these or other species are themselves pharmacological targets for cancer therapy or chemoprevention.

Acknowledgments

This work was supported by the National Institutes of Health grants RO1-CA087546 (E.D.), RO1-CA111422 (E.D.), and RO1-CA62275 (E.D.), a research grant from Pfizer (E.D.), a Samuel Waxman Foundation Cancer Research Award (E.D.), and the Oracle Giving Fund (E.D.). IPR was supported by the National Institutes of Health Grant T32-CA09658. We thank Ms. Ann Frost for expert assistance in the preparation of this manuscript. We thank Drs. Jean Beebe and Tom Turi as well as their colleagues at Pfizer Global Research and Development, Groton, CT for helpful consultation.

References

1. Sporn, M. B., Dunlop, N. M., Newton, D. L., and Smith, J. M. (1976) Prevention of chemical carcinogenesis by vitamin A and its synthetic analogs (retinoids). *Fed. Proc.* **35,** 1332–1338.

2. Windmill, K. F., McKinnon, R. A., Zhu, X., Gaedigk, A., Grant, D. M., and McManus, M. E. (1997) The role of xenobiotic metabolizing enzymes in arylamine toxicity and carcinogenesis: functional and localization studies. *Mutat. Res.* **376,** 153–160.

3. Talalay, P. (1989) Mechanisms of induction of enzymes that protect against chemical carcinogenesis. *Adv. Enzyme Regul.* **28,** 237–250.

4. Stoner, G. D., Morse, M. A., and Kelloff, G. J. (1997) Perspectives in cancer chemoprevention. *Environ. Health Perspect.* **105 Suppl. 4,** 945–954.

5. Andela, V. B. (2004) Functional antagonism between NF-κB and nuclear receptors: implications in carcinogenesis and strategies for optimal cancer chemopreventive interventions. *Curr. Cancer Drug Targets* **4,** 337–344.

6. Sun, S. Y., Hail, N., Jr., and Lotan, R. (2004) Apoptosis as a novel target for cancer chemoprevention. *J. Natl. Cancer Inst.* **96,** 662–672.

7. Dragnev, K. H., Stover, D., and Dmitrovsky, E. (2003) Lung cancer prevention: the guidelines. *Chest* **123,** 60S–71S.

8. Dragnev, K. H., Rigas, J. R., and Dmitrovsky, E. (2000) The retinoids and cancer prevention mechanisms. *Oncologist* **5,** 361–368.

9. Moyers, S. B. and Kumar, N. B. (2004) Green tea polyphenols and cancer chemoprevention: multiple mechanisms and endpoints for phase II trials. *Nutr. Rev.* **62,** 204–211.

10. Conney, A. H. (2003) Enzyme induction and dietary chemicals as approaches to cancer chemoprevention: the Seventh DeWitt S. Goodman Lecture. *Cancer Res.* **63,** 7005–7031.

11. Jang, M., Cai, L., Udeani, G. O., et al. (1997) Cancer chemopreventive activity of resveratrol, a natural product derived from grapes. *Science* **275,** 218–220.

12. Dong, Z. (2003) Molecular mechanism of the chemopreventive effect of resveratrol. *Mutat. Res.* **523–524,** 145–150.

13. Barnes, S. (2004) Soy isoflavones—phytoestrogens and what else? *J. Nutr.* **134,** 1225S–1228S.

14. Birt, D. F., Hendrich, S., and Wang, W. (2001) Dietary agents in cancer prevention: flavonoids and isoflavonoids. *Pharmacol. Ther.* **90,** 157–177.

15. Conaway, C. C., Yang, Y. M., and Chung, F. L. (2002) Isothiocyanates as cancer chemopreventive agents: their biological activities and metabolism in rodents and humans. *Curr. Drug Metab.* **3,** 233–255.
16. Murillo, G. and Mehta, R. G. (2001) Cruciferous vegetables and cancer prevention. *Nutr. Cancer* **41,** 17–28.
17. Rose, D. P. and Connolly, J. M. (1999) Omega-3 fatty acids as cancer chemopreventive agents. *Pharmacol. Ther.* **83,** 217–244.
18. Fisher, B., Costantino, J. P., Wickerham, D. L., et al. (1998) Tamoxifen for prevention of breast cancer: report of the National Surgical Adjuvant Breast and Bowel Project P-1 Study. *J. Natl. Cancer Inst.* **90,** 1371–1388.
19. Rao, C. V. and Reddy, B. S. (2004) NSAIDs and chemoprevention. *Curr. Cancer Drug Targets* **4,** 29–42.
20. Freemantle, S. J., Spinella, M. J., and Dmitrovsky, E. (2003) Retinoids in cancer therapy and chemoprevention: promise meets resistance. *Oncogene* **22,** 7305–7315.
21. Wolbach, S. B. and Howe, P. R. (1925) Tissue changes following deprivation of fat-soluble A vitamin. *J. Exp. Med.* **42,** 753–777.
22. Dmitrovsky, E. (2004) Fenretinide activates a distinct apoptotic pathway. *J. Natl. Cancer Inst.* **96,** 1264–1265.
23. Lonardo, F., Dragnev, K. H., Freemantle, S. J., et al. (2002) Evidence for the epidermal growth factor receptor as a target for lung cancer prevention. *Clin. Cancer Res.* **8,** 54–60.
24. Kerley, J. S., Olsen, S. L., Freemantle, S. J., and Spinella, M. J. (2001) Transcriptional activation of the nuclear receptor corepressor RIP140 by retinoic acid: a potential negative-feedback regulatory mechanism. *Biochem. Biophys. Res. Commun.* **285,** 969–975.
25. Freemantle, S. J., Kerley, J. S., Olsen, S. L., Gross, R. H., and Spinella, M. J. (2002) Developmentally-related candidate retinoic acid target genes regulated early during neuronal differentiation of human embryonal carcinoma. *Oncogene* **21,** 2880–2889.
26. Lee, C. H. and Wei, L. N. (1999) Characterization of receptor-interacting protein 140 in retinoid receptor activities. *J. Biol. Chem.* **274,** 31,320–31,326.
27. Treuter, E., Albrektsen, T., Johansson, L., Leers, J., and Gustafsson, J. A. (1998) A regulatory role for RIP140 in nuclear receptor activation. *Mol. Endocrinol.* **12,** 864–881.
28. Chuang, F. M., West, B. L., Baxter, J. D., and Schuafele, F. (1997) Activities in Pit-1 determine whether receptor interacting protein 140 activates or inhibits Pit-1/nuclear receptor transcriptional synergy. *Mol. Endocrinol.* **11,** 1332–1341.
29. Subramaniam, N., Treuter, E., and Okret, S. (1999) Receptor interacting protein RIP140 inhibits both positive and negative gene regulation by glucocorticoids. *J. Biol. Chem.* **274,** 18,121–18,127.
30. Miyata, K. S., McCaw, S. E., Meertens, L. M., Patel, H. V., Rachubinski, R. A., and Capone, J. P. (1998) Receptor-interacting protein 140 interacts with and

inhibits transactivation by peroxisome proliferator-activated receptor alpha and liver-X-receptor alpha. *Mol. Cell Endocrinol.* **146**, 69–76.

31. Eng, F., Barsalou, A., Akutsu, N., et al. (1998) Different classes of coactivators recognize distinct but overlapping binding sites on the estrogen receptor ligand binding domain. *J. Biol. Chem.* **273**, 28,371–28,377.
32. White, K. A., Yore, M. M., Warburton, S. L., et al. (2003) Negative feedback at the level of nuclear receptor coregulation. Self-limitation of retinoid signaling by RIP140. *J. Biol. Chem.* **278**, 43,889–43,892.
33. Altucci, L., Rossin, A., Raffelsberger, W., Reitmair, A., Chomienne, C., and Gronemeyer, H. (2001) Retinoic acid-induced apoptosis in leukemia cells is mediated by paracrine action of tumor-selective death ligand TRAIL. *Nat. Med.* **7**, 680–686.
34. Sun, S. Y., Yue, P., Hong, W. K., and Lotan, R. (2000) Augmentation of tumor necrosis factor-related apoptosis-inducing ligand (TRAIL)-induced apoptosis by the synthetic retinoid 6-[3-(1-adamantyl)-4-hydroxyphenyl]-2-naphthalene carboxylic acid (CD437) through up-regulation of TRAIL receptors in human lung cancer cells. *Cancer Res.* **60**, 7149–7155.
35. Clarke, N., Jimenez-Lara, A. M., Voltz, E., and Gronemeyer, H. (2004) Tumor suppressor IRF1 mediates retinoid and interferon anticancer signaling to death ligand TRAIL. *EMBO J.* **23**, 3051–3060.
36. Ma, Y., Koza-Taylor, P. H., DiMattia, D. A., et al. (2003) Microarray analysis uncovers retinoid targets in human bronchial epithelial cells. *Oncogene* **22**, 4924–4932.
37. Tamayo, P., Slonim, D., Mesirov, J., et al. (1999) Interpreting patterns of gene expression with self-organizing maps: methods and application to hematopoietic differentiation. *Proc. Natl. Acad. Sci. USA* **96**, 2907–2912.
38. Almstrup, K., Hoei-Hansen, C. E., Wirkner, U., et al. (2004) Embryonic stem cell-like features of testicular carcinoma in situ revealed by genome-wide expression profiling. *Cancer Res.* **64**, 4736–4743.
39. McElwaine, S., Mulligan, C., Groet, J., et al. (2004) Microarray transcript profiling distinguishes the transient from the acute type of megakaryoblastic leukemia (M7) in Down's syndrome, revealing PRAME as a specific discriminating marker. *Br. J. Haematol.* **125**, 729–742.
40. Oh, T. J., Kim, C. J., Woo, S. K., et al. (2004) Development and clinical evaluation of a highly sensitive DNA microarray for detection and genotyping of human papillomaviruses. *J. Clin. Microbiol.* **42**, 3272–3280.
41. Weisz, A., Basile, W., Scafoglio, C., et al. (2004) Molecular identification of ERalpha-positive breast cancer cells by the expression profile of an intrinsic set of estrogen regulated genes. *J. Cell Physiol.* **200**, 440–450.
42. Kitareewan, S., Pitha-Rowe, I., Sekula, D., et al. (2002) UBE1L is a retinoid target that triggers PML/RARα degradation and apoptosis in acute promyelocytic leukemia. *Proc. Natl. Acad. Sci. USA* **99**, 3806–3811.
43. Pitha-Rowe, I., Hassel, B. A., and Dmitrovsky, E. (2004) Involvement of UBE1L in ISG15 conjugation during retinoid-induced differentiation of acute promyelocytic leukemia. *J. Biol. Chem.* **279**, 18,178–18,187.

44. Zhu, J., Gianni, M., Kopf, E., et al. (1999) Retinoic acid induces proteasome-dependent degradation of retinoic acid receptor alpha (RARalpha) and oncogenic RARalpha fusion proteins. *Proc. Natl. Acad. Sci. USA* **96,** 14,807–14,812.

45. Langenfeld, J., Kiyokawa, H., Sekula, D., Boyle, J., and Dmitrovsky, E. (1997) Posttranslational regulation of cyclin D1 by retinoic acid: a chemoprevention mechanism. *Proc. Natl. Acad. Sci. USA* **94,** 12,070–12,074.

46. Spinella, M. J., Freemantle, S. J., Sekula, D., Chang, J. H., Christie, A. J., and Dmitrovsky, E. (1999) Retinoic acid promotes ubiquitination and proteolysis of cyclin D1 during induced tumor cell differentiation. *J. Biol. Chem.* **274,** 22,013–22,018.

47. Kitareewan, S., Pitha-Rowe, I., Ma, Y., Freemantle, S. J., and Dmitrovsky, E. (2004) The retinoids and cancer prevention mechanisms, in *Cancer Chemoprevention Volume 1: Promising Cancer Chemopreventive Agents,* (Kelloff, G. J., Hawk, E. T., and Sigman, C. C., eds.), Humana Press, Totowa, NJ, pp. 277–288.

48. Langenfeld, J., Lonardo, F., Kiyokawa, H., et al. (1996) Inhibited transformation of immortalized human bronchial epithelial cells by retinoic acid is linked to cyclin E down-regulation. *Oncogene* **13,** 1983–1990.

49. Pitha-Rowe, I., Petty, W. J., Feng, Q., et al. (2004) Microarray analyses uncover UBE1L as a candidate target gene for lung cancer chemoprevention. *Cancer Res.* **64,** 8109–8115.

50. Kok, K., Hofstra, R., Pilz, A., et al. (1993) A gene in the chromosomal region 3p21 with greatly reduced expression in lung cancer is similar to the gene for ubiquitin-activating enzyme. *Proc. Natl. Acad. Sci. USA* **90,** 6071–6075.

51. Carritt, B., Kok, K., van den Berg, A., et al. (1992) A gene from human chromosome region 3p21 with reduced expression in small cell lung cancer. *Cancer Res.* **52,** 1536–1541.

52. Boyle, J. O., Langenfeld, J., Lonardo, F., et al. (1999) Cyclin D1 proteolysis: a retinoid chemoprevention signal in normal, immortalized, and transformed human bronchial epithelial cells. *J. Natl. Cancer. Inst.* **91,** 373–379.

53. Koziczak, M., Holbro, T., and Hynes, N. E. (2004) Blocking of FGFR signaling inhibits breast cancer cell proliferation through downregulating of D-type cyclins. *Oncogene* **23,** 3501–3508.

54. Chung, D. C. (2004) Cyclin D1 in human neuroendocrine: turmorigenesis. *Ann. NY. Acad. Sci.* **1014,** 209–217.

55. Barbieri, F., Lorenzi, P., Ragni, N., et al. (2004) Overexpression of cyclin D1 is associated with poor survival in epithelial ovarian cancer. *Oncology* **66,** 310–315.

56. Yuan, W. and Krug, R. M. (2001) Influenza B virus NS1 protein inhibits conjugation of the interferon (IFN)-induced ubiquitin-like ISG15 protein. *EMBO J.* **20,** 362–371.

57. Malakhov, M. P., Malakhova, O. A., Kim, K. I., Ritchie, K. J., and Zhang, D. E. (2002) UBP43 (USP18) specifically removes ISG15 from conjugated proteins. *J. Biol. Chem.* **277,** 9976–9981.

58. Malakhova, O. A., Yan, M., Malakhov, M. P., et al. (2003) Protein ISGylation modulates the JAK-STAT signaling pathway. *Genes Dev.* **17,** 455–460.
59. Pitha-Rowe, I., Petty, W. J., Kitareewan, S., and Dmitrovsky, E. (2003) Retinoid target genes in acute promyelocytic leukemia. *Leukemia* **17,** 1723–1730.

8

Complete Open Reading Frame (C-ORF) Technique

Rapid and Efficient Method for Obtaining Complete Protein Coding Sequences

Dong-chul Kang and Paul B. Fisher

Summary

Several approaches, generally referred to as rapid amplification of cDNA ends, are currently used as a means of obtaining full-length cDNA clones by PCR. However, these protocols are not infallible and in specific instances they have proven unsuccessful, emphasizing a need for further refinement. A novel method, the complete open reading frame (C-ORF) technique, is presently described, which has proven successful in cases, where standard rapid amplification of cDNA ends (RACE) has not worked. In C-ORF, the 5′ PCR primer site is provided by a degenerative stem–loop annealing primer, which consists of a stem-loop structure and a 3′ random 12-mer. degenerative stem–loop annealing primer is designed to anneal at random sites of the first strand cDNA, while promoting second strand synthesis from the end of given cDNA. Although this technique manifests weak sequence preference for GC-rich regions, in practice it has been successfully applied to clone both known and unknown genes with varying regions of GC-rich content. C-ORF does not use additional enzymes other than reverse transcriptase and *Taq* polymerase making it a cost-effective and relatively simple method that should be of general utility for gene cloning in multiple laboratories.

Key Words: cDNA analysis; expressed sequence tags; open reading frame cloning; protein sequence determination; PCR; C-ORF; *PCTA-1*; *mda-5*; *mda-9*; *PSGen-13*.

1. Introduction

Obtaining full-length cDNA is a prerequisite for functional gene analysis. With the completion of the human genome project, focus is now shifting toward establishing databases of experimentally verified full-length sequences of all expressed genes as well as producing a unified annotation of their gene products. A recent report indicates the cloning of approx 21,000 full-length cDNA sequences *(1)*. However, the UniGene database for all reported cDNAs currently

From: *Methods in Molecular Biology, vol. 383: Cancer Genomics and Proteomics: Methods and Protocols*
Edited by: P. B. Fisher © Humana Press Inc., Totowa, NJ

lists approx 50,000 unique human gene clusters, which exceeds the number of identified full-length clones by twofold, although the number of unique transcripts may be overestimated by occasionally recognizing a 5′ sequence and a 3′ sequence of a single clone as incorrectly representing two distinct transcripts. In these contexts, full-length cDNA cloning represents the ongoing effort.

Traditionally, full-length cDNAs could be obtained by repeated screening of a cDNA library. Recent advances in constructing full-length-enriched cDNA libraries significantly decrease the burden of repeated screening *(2–4)*. However, the screening of cDNA libraries is expensive, tedious, and time-consuming. This makes this approach less attractive for cloning small numbers of cDNA species. This limitation has resulted in the extensive use of PCR-based cloning methods, called rapid amplification of cDNA ends (RACE) *(3,5–7)* (*see* **Note 1**). In addition, PCR-based RACE requires a considerably smaller quantity of RNA samples than construction of a cDNA library. Currently, full-length-enriched cDNA libraries are primarily used in mass analysis of random cDNAs for database construction, while RACE is applied to cloning one or two cDNAs of precise interest, thereby satisfying the requirement of most research laboratories.

RACE, first developed by Frohman et al. *(6)* and its variants (suffixed also with RACE) *(3,5,7)*, eliminates the tedious nature of library screening and significantly reduces the amount of starting material and the cost of cDNA cloning. A major huddle of cDNA cloning by PCR resides in obtaining the 5′-end of a cDNA. Unlike the 3′-end of an mRNA that contains a defined sequence, such as the poly A tail, the 5′-ends of mRNAs are heterogeneous and do not have defined sequence motifs. Thus, various protocols have been developed to provide definitive sequence information for the 5′-ends of cDNAs. Among these approaches are homopolymeric tailing by terminal deoxynucleotidyl transferase *(6)*, ligation of oligoribonucleotides with RNA ligase *(3,7)*, and oligocapping by chemical modification and template switching in reverse transcription reactions *(4,5)*. In RACE, cDNAs tagged with a specific 5′ sequence are amplified by PCR with a gene specific primer and a tagged sequence primer. Successful applications of these methods have been documented in numerous reports. In addition, methods such as oligonucleotide ligation *(3)*, oligocapping *(4)*, and template switching *(2)* have been implemented to enrich the proportion of full-length cDNAs in a cDNA library.

In practice; however, RACE is not as simple and straightforward as the theory behind the technique would suggest *(8)*. TdT-mediated addition in original RACE often ends short and requires repeated application to obtain full-length cDNAs, most likely because of the existence of intragenic primer sites similar to the homopolymer tail. Moroever, the RNA ligation-mediated protocol requires complicated sequential enzymatic reactions with relatively low efficiencies, but which are critical in the success of this approach *(3)*. The SMART protocol is by far the simplest among the RACE techniques, but it is primarily dependent on sequence context *(5)*. Thus, although currently available RACE methods significantly ameliorate the labor and cost of cDNA library screening, improvements in this protocol that make it simpler and more reliable will greatly increase the applications of this approach for most research laboratories.

In this chapter, a simplified RACE method, called the complete open reading frame (C-ORF) approach is described, which will have wide applications for full-length cDNA cloning (*see* **Note 2**). Instead of generating a universal primer site with TdT-tailing or anchor ligation, a degenerate stem and loop annealing primer (dSLAP) that is designed to anneal the 3′-end of the first strand cDNA, which is used to start second strand cDNA synthesis from the annealed site (**Fig. 1A**). dSLAP also provides a universal adaptor primer site for subsequent PCR reactions. The protocol includes reverse transcription, second strand cDNA synthesis with dSLAP and PCR amplification with a gene specific primer and an adaptor primer (**Fig. 1B**). The C-ORF protocol requires RT, Taq polymerase, and dSLAP only. The procedure was tested with known genes including *ISG*-56 (1.5 kb) *(9)* and melanoma differentiation associated gene-9 (*mda*-9)/syntenin (2 kb) *(10–12)* and also successfully applied to clone full coding region of a novel gene melanoma differentiation associated gene-5 (*mda*-5) (3 kb) *(13,14)*, human polynucleotide phosphorylase (*hPNPase^{old-35}*) *(15–19)*, progression elevated gene-28 (*PEGen*-28) (1.2 kb) *(20)*, progression elevated gene-42 (*PEGen*-42) (1.5 kb) *(20)*, prostate carcinoma tumor antigen (*PCTA*)-1 (3.5 and 6 kb) *(21,22)* and astrocyte elevated gene-1 (*AEG*-1) *(23,24)* (*see* **Notes 3** and **4**). With its labor-saving, cost-efficiency, and improved reliability, the C-ORF method is applicable for performing both single and multiple cDNA cloning projects simultaneously. Modified dSLAPs were also tested for cDNA cloning and this primer was found to be more effective for longer products (~5–8 kb) in comparison with the original dSLAP primer *(25)*.

2. Materials

1. Total RNA—2 μg/each sample, prepared either by Qiagen RNeasy column or phenol extraction followed by ethanol precipitation.
2. Reverse transcriptase—Superscript® RT II (Invitrogen).
3. dNTP mix—10 m*M* each.
4. RnaseOUT® (Invitrogen)—optional.
5. Gene specific primer 1.
6. RNase A and RNase H—can be replaced by alkaline hydrolysis of RNA.
7. GlassMax® (Invitrogen)—can be replaced by Qiagen PCR purification kit.
8. KlenTaq™ (BD Science).
9. dSLAP-5′ GTC TCG AGT TTA AAC ACT TTC TGG TCG ACT AGT GTT TAA ACT CGA GAC N$_{12}$ 3′.

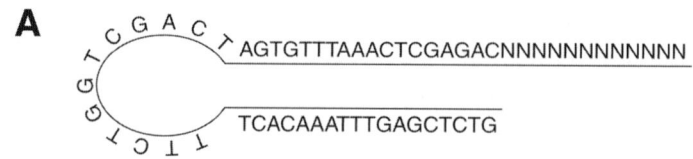

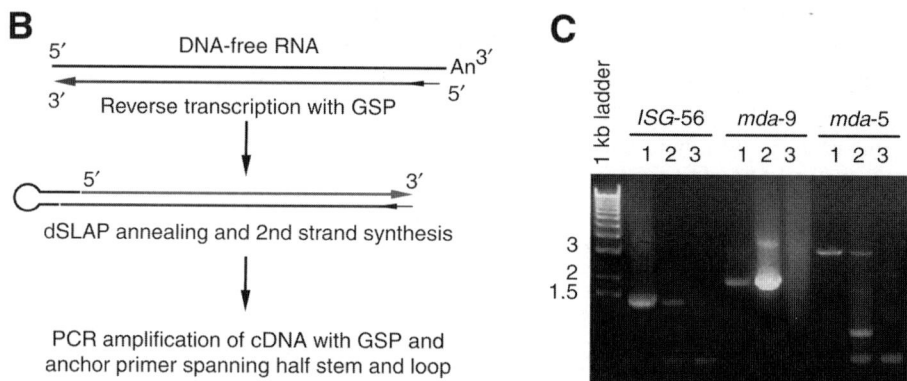

Fig. 1. Complete open reading frame technique. **(A)** Degenerate stem-loop annealing primer (dSLAP) used in C-ORF. The stem-loop structure is designed to prevent degenerate region from annealing to internal site and efficient extension thereby by steric hindrance. Adaptor primer sequence is in bold face. **(B)** Schematic diagram of C-ORF (details in text). **(C)** Representative C-ORF application. C-ORF products of ISG-56, *mds*-9/syntenin and *mda*-5 are resolved in 1% agarose gel (lane 2). The size of C-ORF products if compared with RT-PCR product of each gene with common 3′ nested primer and 5′ primer from reported sequence (lane 1). Nested PCR of C-ORF with anchor primer only (lane 3). (From Kang and Fisher *[25]*.)

10. dNTP mix—10 m*M* each.
11. 2.3 PCR amplification.
12. KlenTaq (BD Science).
13. dNTP mix—10 m*M* each.
14. Gene specific primer 1 and 2 for nested PCR.
15. Adaptor primer-5′ TTC TGG TCG ACT AGT GTT TAA ACT CGA GAC 3′.
16. Mixture 1:
    a. 4 µL 5X Superscript RT II buffer.
    b. 1 µL 100 m*M* DTT.
    c. 1 µL 10 m*M* dNTP mix.
    d. 5 U RNaseOUT® (optional).
    e. Up to 9 µL DW.
17. Mixture 2:
    a. 10–16 µL Cleaned first strand cDNA.

b. 2 pmol dSLAP.

c. 2 µL 10X KlenTaq buffer.

d. up to 20 µL DW.

18. Mixture 3:

a. 0.25 µL Advantage cDNA polymerase mix.

b. 0.5 µL 10X KlenTaq buffer.

c. 0.5 µL 10 mM dNTPs.

d. 3.75 µL DW.

19. Mixture 4:

a. 5 µL Second strand synthesis reaction.

b. 2 µL 10X KlenTaq buffer.

c. 0.5 µL 10 mM dNTPs.

d. 5 pmol gene specific primer 1.

e. 10 pmol anchor primer.

f. 0.25 µL Advantage cDNA polymerase mix.

g. Up to 25 µL DW.

20. Mixture 5:

a. 0.5 µL Primary PCR product.

b. 5 µL 10X KlenTaq buffer.

c. 1 µL 10 mM dNTPs.

d. 10 pmol Nested gene specific primer (GSP2).

e. 10 pmol Anchor primer.

f. 0.5 µL Advantage cDNA polymerase mix.

g. Up to 50 µL DW.

3. Methods

3.1. Preparation of the First Strand cDNA

3.1.1. Reverse Transcription

1. Take 2 µg total RNA, 2 pmol of Gene specific primer, and up to 10 µL of distilled water (DW). Incubate at 70°C for 5 min. Quench on ice (*see* **Notes 5–11**).
2. Add mixture 1. Incubate at 45°C for 5 min.
3. Add 1 mL of Superscript RT II (200 U) and incubate 45°C for additional 60 min.

3.1.2. Purification of First Strand cDNA With GlassMax

1. Treat with RNase H (2.2 U) and RNase A (0.5 µg) for 30 min at 37°C.
2. Purify cDNA with GlassMax as in the manufacturer's protocol.

3.2. Second Strand Synthesis With dSLAP

Assemble 20 µL annealing reaction mixture 2.

1. Incubated at 95°C for 1 min.
2. Gradually cooled at 5°C/min to 42°C and keep for additional 5 min.

3. Add the 5 µL polymerase mixture 3 to the annealing reaction while the 42°C incubation.
4. Incubate for 30 min at 68°C.

3.3. PCR Amplification

3.3.1. Primary PCR 25 µL Reaction

Perform PCR as follows (*see* mixture 4).

1. One cycle of 95°C for 1 min.
2. Twenty seven cycles of 95°C for 30 s, 59°C for 1 min (depending on primer) and 68°C for specified period (1 min for 1 kb target).
3. One cycle of 5 min at 68°C.

3.3.2. Nested PCR 50 µL Reaction

Perform PCR as follows (*see* mixture 5)

1. One cycle of 95°C for 1 min.
2. Twenty-seven cycles of 95°C for 30 s, 59°C for 1 min (depending on primer) and 68°C for specified period (1 min for 1 kb target).
3. One cycle of 5 min at 68°C.

Single primer reaction with GSP2 only and anchor primer only was performed with primary PCR reactions to distinguish potential C-ORF artifacts. Resolve in 1% agarose gels and purify bands with gel purification kit (Qiagen). Direct sequencing with either anchor primer or GSP2 or TA cloning.

4. Notes

1. RACE techniques have significantly simplified full-length cDNA cloning procedures. This methodology has permitted many research laboratories to obtain full-length cDNAs without the enormous investment of time and energy required to produce cDNA libraries. There are three widely used RACE methods that differ in 5′-end PCR primer site-homopolymeric tailing, ligation of oligoribonucleotide and template switching *(3,5–7)*. However, none of these methods is without problems and success is often contingent upon the sequence of the cDNA being analyzed.
2. C-ORF is a novel RACE method that has proven effective in obtaining a full open reading frame for many cDNAs and in most cases with a single application using common molecular biology reagents *(25)*. Unlike other RACE methods, C-ORF employs a dSLAP primer that is designed to anneal to cDNAs with random 12 nucleotides, but to prevent the second strand synthesis from starting from inside annealing sites because of its stem-loop structural motif **(Fig. 1A)**. Annealing and second strand extension in C-ORF does not require additional enzymatic reactions such as TdT and RNA ligase that are frequently employed in the RACE approach.

3. Initial tests of the C-ORF approach resulted in the full open reading frames of *ISG*-56 (1.5 kb) *(9)* and *mda*-9/syntenin (2 kb) *(10–12)* in a single step **(Fig. 1C)**. Subsequently, the utility of C-ORF and proof-of-principle was further demonstrated by cloning full open reading frames of previously unknown genes, including *mda*-5 *(13,14)* **(Fig. 1C)**, *hPNPase*$^{old-35}$ *(15–19)* and 5′-extension and identification of splice variants of PCTA-1 *(21,22)*. dSLAP, which can anneal to random sites of a cDNA through its random sequence, does not promote arbitrary second strand synthesis. Rather, the C-ORF products in these applications form discrete bands of full-length size, which strongly suggests the preferential second-strand synthesis from dSLAP anneals near the end of the cDNA. Nonspecific bands, as observed in other PCR-based cloning methods, were also apparent in the C-ORF reactions in comparison with RT-PCR. Simultaneously running a single primer PCR reaction with either an anchor primer only or a gene specific primer only readily distinguished these PCR artifacts using C-ORF **(Subheading 3.3.2., Fig. 1B)**.

4. Although cDNAs cloned by a single round of C-ORF contained the full open reading frame, they ended short by 34 nucleotides (nts) for *ISG*-56, 27 nts for *mda*-9/syntenin and 26 nts for *mda*-5, respectively. Premature ending of the cDNA clone could be resulting from either incomplete reverse transcription reaction caused by RNA secondary structure or annealing preference of degenerate sequence in dSLAP to specific cDNA sequences, such as GC-rich regions. GC content of dSLAP-annealed sequences in 10 C-ORF products was averaged as 70.8%, which implied mild sequence-preference of dSLAP annealing sites **(Table 1)**. However, the GC content does not appear to be the primary determinant of dSLAP annealing and second strand extension. For example, in the case of ISG-56, the GC content of dSLAP annealing site of the C-ORF product is 58.3%, whereas ISG-56 has two internal 12-bp GC-rich stretches of 75% and 83% GC content, respectively. The dSLAP can potentially anneal to all of the GC-rich sites, but the C-ORF product of ISG-56 derived from the lowest GC content site (58.3%). Thus, the stem-loop structure of dSLAP clearly promotes preferential extension of second strand from the dSLAP annealed to the end of first-strand cDNA. In general, the optimal fragment selected for further analysis is a specific band of longest size based on the target cDNA size that can be readily predicted from Northern blot analysis **(Subheading 3.3.2.)**.

5. Efficiency in reverse transcription impacts greatly on the outcome of any RACE application **(Subheading 3.1.1.)**. Premature termination of RT often results from extensive secondary structure at the 5′-UTR of mRNA. Increasing RT reaction temperature to dissolve secondary structure of mRNA should prove helpful to facilitate first-strand cDNA synthesis to the end of the transcript. Superscript RT II (a RNase H-minus MMLV-RT) was found effective up to 48°C. A thermostable RT (~55–65°C) can be another option when even higher temperature is required.

6. Removal of cDNA-associated RNA is definitely helpful in dSLAP annealing and second strand extension **(Subheading 3.1.2.)**. To simplify the protocol and to reduce the number of enzyme reactions, alkaline RNA hydrolysis followed by neutralization can be employed. RNA can be hydrolyzed by addition of 2.2 μL of

Table 1
Sequences of 5′ End of Various C-ORF Products

Name	5′ sequence	G/C score	G/C (%)
mda-5	GCG CGC CGGC CT	11/12	91.7
ISG-56	TGC AGA ACGG CT	7/12	58.3
mda-9	GGC GGC GGCG GC	12/12	100
PCTA-1A	TGG AGG CCTG GA	8/12	66.7
PCTA-1B	GCC AGT GCCT CA	8/12	66.7
PCTA-1C	CGATGT TGG CC TT	7/12	58.3
hPNPase^{old-35}	CGG AGG ACCA AT	7/12	58.3
PSGen 12	GCG GTG GTGA CG	9/12	75
PEGen 28	GTG TGG TGTG TC	7/12	58.3
PEGen 42	GGC GTT GCGA CG	9/12	75
G/C score	899 486 7986 74	85/120	70

5 *M* NaOH and incubation for 30 min at 55°C. The alkaline hydrolysis reaction is neutralized with 72 µL of 1% HAC. The remaining cDNA can be purified with the Qiagen PCR purification kit.

7. The C-ORF strategy is based on annealing the dSLAP to reverse transcribed cDNA during second strand cDNA synthesis. The hairpin structure of dSLAP that could form bulky loop structures is hypothesized to prevent the random sequences from extending into the middle of reverse transcribed cDNA because of stearic hindrance. To ensure formation of the dSLAP structure prior to association with target cDNA, the length of the stem is designed to be longer than the degenerate sequences (18 vs 12 nt) and denatured annealing mixture cools gradually to the annealing temperature (5°C/min) during second strand cDNA synthesis (**Subheading 3.2.**). Standard PCR units (MJ Minicyler was used) control temperature parameters in this process. The effect of annealing temperature on C-ORF performance was further investigated. Despite dependence on gene species, the annealing temperature up to 50°C was as effective as 42°C, and increasing annealing temperature in some cases resulted in reduction of band complexity. The addition of preannealed dSLAP reagent did not significantly enhance PCR yield as compared with the standard C-ORF protocol.

8. Second strand cDNA synthesis appears to depend on the dSLAP concentration (**Subheading 3.2.**). In the evaluation of dSLAP concentration effect, it was found that the concentration of the D-SLAP reagent should be higher than 20 n*M*. The concentration ratio of anchor primer to gene specific primer in the primary PCR has also been tested (**Subheading 3.3.1.**). Although both 0.2 µ*M* anchor-0.2 µ*M* GSP1 and 0.4 µ*M* anchor-0.2 µ*M* GSP1 produced an appropriate sized PCR product, the C-ORF product yield was generally higher with 0.4 µ*M* anchor-0.2 µ*M* GSP1. An additional anchor primer may be required for efficient PCR amplification because significant amount of dSLAP (one-fifth of the second strand cDNA synthesis reaction) is carried over to the primary PCR reaction.

9. Since the stem-loop structure of dSLAP facilitates second strand cDNA synthesis from the end or near the end of the cDNA, we thought that a bulkier structure like a cloverleaf degenerate structure could be more effective than dSLAP in inhibiting second strand cDNA synthesis from internal priming sites, thereby alleviating the GC preferences. Thus, cloverleaf annealing primers (dCLAP) containing 13 random oligonucleotides (3′-NNNNNNNNNNNNNAGAGCTCACAGCTGAAGCA GCTGACTAGCACCTAGTGTAGAATACATCTTGAGCTAT-5′), were designed and tested for cDNA cloning **(Subheading 3.2.)**. C-ORF with dCLAP generated bands of the expected size, not only for shorter transcripts (ISG-56 and *mda*-9) but also for larger transcripts (PCTA-1/pA and fibronectin), although this approach did increase band complexity with increasing target size. Comparison of dCLAP with dSLAP demonstrated that dCLAP was more effective for large size targets such as fibronectin (>6 kb).

10. C-ORF can also be applied to cloning the 3′-end of a cDNA. An oligo dT stem-loop annealing primer (tSLAP: 3′-TTTTTTTTTTTTCAGAGCTCAAATTGTGA TCAGCTGGTCTTFCACAAATTTGAGCTCTG-5′) was designed by replacing oligo dT for the random sequence of the dSLAP. RT reaction with tSLAP followed by PCR amplification (primary and nested) with adaptor primer resulted in cognate cDNA fragment both in a test case application and in practical applications **(Subheading 3.1.)**. These results indicate that the primers based on a stem-loop structure (dSLAP, dCLAP, and tSLAP) can be effectively used for both 5′- and 3′-end cloning of cDNAs.

11. A significant number of expressed sequence tags (ESTs) identified in various experimental systems remain to be cloned as full-length cDNAs. C-ORF can be used with other RACE protocols to extend partial sequence information and to develop complete cDNAs that will be useful for defining functional relevance of potentially physiologically important genes. The C-ORF approach described in this chapter, as evidenced by numerous proof-of-practice examples, provides a very useful approach for full-length gene cloning. Although C-ORF in its current form does not appear applicable for identifying the transcription start site of a cDNA, C-ORF is quite effective in obtaining biologically meaningful sequence data, in most cases within a single application. The elegance of C-ORF resides in the simplicity of this protocol and its efficiency and cost-effectiveness. Employing the described procedure, primers including dSLAP, adaptor, and gene-specific primers are the only additional materials needed in addition to standard molecular biological reagents for full-length cDNA cloning. In principle, C-ORF can be applied to any genetic system, mammal, plant, yeast, bacteria, and so on, requiring complete gene identification. Thus, C-ORF offers a viable and very proficient option for those interested in rapidly and expeditiously obtaining full-length cDNA sequences to begin the ultimate and important task of defining gene function.

Acknowledgments

The present study was supported in part by National Institutes of Health Grants CA035675, CA097318, CA098712, P01 CA104177, GM068448, and

P01 NS31492; the Samuel Waxman Cancer Research Foundation; and the Chernow Endowment. P. B. F. is the Michael and Stella Chernow Urological Cancer Research Scientist in the Departments of Pathology and Urology, College of Physicians and Surgeons of Columbia University and a SWCRF Investigator.

References

1. Ota, T., Suzuki, Y., Nishikawa, T., et al. (2004) Complete sequencing and characterization of 21,243 full-length human cDNAs. *Nat. Genet.* **36,** 40–45.
2. Zhu, Y. Y., Machleder, E. M., Chenchik, A., Li, R., and Siebert, P. D. (2001) Reverse transcriptase template switching: a SMART approach for full-length cDNA library construction. *Biotechniques* **30,** 892–897.
3. Maruyama, K. and Sugano, S. (1994) Oligo-capping: a simple method to replace the cap structure of eukaryotic mRNAs with oligoribonucleotides. *Gene* **138,** 171–174.
4. Carninci, P., Kvam, C., Kitamura, A., et al. (1996) High-efficiency full-length cDNA cloning by biotinylated CAP trapper. *Genomics* **37,** 327–336.
5. Matz, M., Shagin, D., Bogdanova, E., et al. (1999) Amplification of cDNA ends based on template-switching effect and step-out PCR. *Nucleic Acids Res.* **27,** 1558–1560.
6. Frohman, M. A., Dush, M. K., and Martin, G. R. (1988) Rapid production of full-length cDNAs from rare transcripts: amplification using a single gene-specific oligonucleotide primer. *Proc. Natl. Acad. Sci. USA* **85,** 8998–9002.
7. Liu, X. and Gorovsky, M. A. (1993) Mapping the 5′ and 3′ ends of Tetrahymena thermophila mRNAs using RNA ligase mediated amplification of cDNA ends (RLM-RACE). *Nucleic Acids Res.* **21,** 4954–4960.
8. Hooft van Huijsduijnen, R. (1998) PCR-assisted cDNA cloning: a guided tour of the minefield. *Biotechniques* **24,** 390–392.
9. Wathelet, M., Moutschen, S., Defilippi, P., et al. (1986) Molecular cloning, full-length sequence and preliminary characterization of a 56-kDa protein induced by human interferons. *Eur. J. Biochem.* **155,** 11–17.
10. Lin, J. J., Jiang, H., and Fisher, P. B. (1996). Characterization of a novel melanoma differentiation associated gene, *mda-9*, that is down-regulated during terminal cell differentiation. *Mol. Cell. Different.* **4,** 317–333.
11. Lin, J. J., Jiang, H., and Fisher, P. B. (1998). Melanoma differentiation associated gene-9, *mda-9*, is a human gamma interferon responsive gene. *Gene* **207,** 105–110.
12. Sarkar, D., Boukerche, H., Su, Z. Z., and Fisher, P. B. (2004) *mda-9*/Syntenin: Recent insights into a novel cell signaling and metastasis associated gene. Pharm. Ther. **104,** 101–115.
13. Kang, D. -C., Gopalkrishnan, R. V., Wu, Q., Jankowsky, E., Pyle, A. M., and Fisher, P. B. (2002) mda-5: an interferon-inducible putative RNA helicase with double-stranded RNA-dependent ATPase activity and melanoma growth-suppressive properties. *Proc. Natl. Acad. Sci. USA* **99,** 637–642.
14. Kang, D. -C., Gopalkrishnan, R. V., Lin, L., et al. (2004) Expression analysis and genomic characterization of human melanoma differentiation associated gene-5, *mda-5*: a novel type I interferon apoptosis-inducing gene. *Oncogene* **23,** 1789–1800.

15. Leszczyniecka, M., Kang, D. -C., Sarkar, D., et al. (2002) Identification and cloning of human polynucleotide phosphorylase, *hPNPase^{old-35}*, in the context of terminal differentiation and cellular senescence. *Proc. Natl. Acad. Sci. USA* **99,** 16,636–16,641.

16. Leszczyniecka, M., Kang, D. -C., Su, Z. -Z., Sarkar D., and Fisher, P. B. (2003) Expression regulation and genomic organization of human polynucleotide phosphorylase, *hPNPase^{old-35}*, a type I interferon inducible early response gene. *Gene* **316,** 143–156.

17. Sarkar, D., Leszczyniecka, M., Kang, D. -C., et al. (2003) Downregulation of *Myc* as a potential target for growth arrest induced by human polynucleotide phosphorylase (*hPNPase^{old-35}*) in human melanoma cells. *J. Biol. Chem.* **278,** 24,542–24,551.

18. Leszczyniecka, M., DeSalle, R., Kang, D. -C., and Fisher, P. B. (2004) The origin of polynucleotide phosphorylase domains. *Mol. Phylogenet. Evol.* **31,** 123–130.

19. Sarkar, D., Lebedeva, I. V., Emdad, L., Kang, D. -C., Baldwin, A. S. Jr., and Fisher, P. B. (2004) Human polynucleotide phosphorylase (*hPNPase^{old-35}*): a potential link between aging and inflammation. *Cancer Res.* **64,** 7473–7478.

20. Kang, D. -C., LaFrance, R., Su, Z. -Z., and Fisher, P. B. (1998) Reciprocal subtraction differential RNA display: an efficient and rapid procedure for isolating differentially expressed gene sequences. *Proc. Natl. Acad. Sci. USA* **95,** 13,788–13,793.

21. Su, Z. Z., Lin, J., Shen, R., Fisher, P. E., Goldstein, N. I., and Fisher, P. B. (1996) Surface-epitope masking and expression cloning identifies the human prostate carcinoma tumor antigen gene *PCTA-1* a member of the galectin gene family. *Proc. Natl. Acad. Sci. USA* **93,** 7252–7257.

22. Gopalkrishnan, R. V., Roberts, T., Tuli, S., Kang, D. -C., Christiansen, K. A., and Fisher, P. B. (2000) Molecular characterization of prostate carcinoma tumor antigen-1, *PCTA-1*, a human galectin-8 related gene. *Oncogene* **19,** 4405–4416.

23. Su, Z. -Z., Kang, D. -C., Chen, Y., et al. (2002) Identification and cloning of human astrocyte genes displaying elevated expression after infection with HIV-1 or exposure to HIV-1 envelope glycoprotein by rapid subtraction hybridization, RaSH. *Oncogene* **21,** 3592–3602.

24. Su, Z. -Z., Chen, Y., Kang, D. -C., et al. (2003) Customized rapid subtraction hybridization (RaSH) gene microarrays identify overlapping expression changes in human fetal astrocytes resulting from human immunodeficiency virus-1 infection or tumor necrosis factor-alpha treatment. *Gene* **306,** 67–78.

25. Kang, D. -C. and Fisher, P. B. (2005) Complete open reading frame (C-ORF) technology: simple and efficient technique for cloning full-length protein-coding sequences. *Gene* **353,** 1–7.

9

Chromatin Immunoprecipitation Assays

Molecular Analysis of Chromatin Modification and Gene Regulation

Piyali Dasgupta and Srikumar P. Chellappan

Summary

Gene expression pattern in cancer cells differ significantly from their normal counter parts, owing to mutations in oncogenes and tumor suppressor genes, their downstream targets, or owing to increased proliferation, and altered apoptotic potential. Various microarray based techniques have been widely utilized to study the differential expression of genes in cancer in recent years. Along with this, attempts have been made to study the transcriptional regulatory mechanisms and chromatin modifications facilitating such differential gene expression. One of the widely used assays for this purpose is the chromatin immunoprecipitation (ChIP) assay, which enables the analysis of the association of regulatory molecules with specific promoters or changes in histone modifications in vivo, without overexpressing any component. This has been of immense value, because ChIP assays can provide a snapshot of the regulatory mechanisms involved in the expression of a single gene, or a variety of genes at the same time. This review article outlines the general strategies and protocols used to carry out ChIP assays to study the differential recruitment of transcription factors, based on the experience in studying E2F1 and histone modifications as well as other published protocols. In addition, the use of ChIP assays to carry out global analysis of transcription factor recruitment is also addressed.

Key Words: Acetylation; ChIP; E2F; histone modification; immunoprecipitation; microarrays.

1. Introduction

Chromatin immunoprecipitation assays provide an unbiased glimpse into the changes occurring on chromatin in vivo, in response to extracellular signals *(1–4)*. It is clear that differential gene expression correlates with the recruitment of specific and general transcription factors on relevant promoters; in addition, covalent modification of histones in the promoter region also contributes significantly to the process *(5–12)*. Indeed, the pattern of acetylation and methylation of histones

From: *Methods in Molecular Biology, vol. 383: Cancer Genomics and Proteomics: Methods and Protocols*
Edited by: P. B. Fisher © Humana Press Inc., Totowa, NJ

is so vital to modulating gene expression, it is thought that these processes code for the gene expression profiles *per se (13–20)*. Although modifications of histones and recruitment of transcription factors are very dynamic events, it is possible to freeze the process chemically and examine the changes *(1,21–23)*. ChIP assays provide an efficient, cost effective means to achieve this. The lab has utilized ChIP assays to study the recruitment of E2F1, prohibitin or transcriptional corepressors to promoters, in response to specific signaling events *(14,24–27)*. Studies from the laboratories of Doug Dean, Peggy Farnham, and others have established many of the protocols currently in use *(2,7,21,23,28–35)*.

Generally, ChIP assays can be done on tissue culture cells or from tissue samples. ChIP assays involve multiple steps, each of which might need to be standardized for optimum results. The basic assay protocol used to study histone modifications is as follows; this has been modified in certain situations. Generally, the first step in the ChIP assay is to crosslink the DNA to the associated proteins by treating the cells with formaldehyde *(22,30)*; in certain situations, investigators have bypassed this step to prevent chemical modifications of the chromatin *(9,34–36)*. The second step is to isolate the DNA–protein complexes; the isolated DNA is then fragmented into 1–2 kb fragments, either by sonication *(22)* or treatment with micrococcal nuclease *(9,19,23)*, in the third step. The DNA fragments are then immunoprecipitated using specific antibodies of interest—antibodies to transcription factors, modified histones, coactivators or repressors, and so on, which is the fifth step in the ChIP assay. The immunoprecipitated DNA–protein complexes are decrosslinked in the sixth step; last, this is followed by PCR using specific primers to the promoters of interest in the seventh step. Alternately, the DNA can be used for probing microarrays of promoter regions to get an idea of global changes in factor recruitment and histone modifications. Protocols for doing this are described elsewhere in detail *(8,37–40)*.

2. Materials

1. 37% Formaldehyde (molecular biology reagent grade) (Sigma-Aldrich, cat. no. F-8775).
2. 1 M Glycine:
   a. 3.75 g Tissue culture grade glycine (Fischer Biotech, cat. no. BP381-1).
   b. 50 mL Water.
3. 20X Phosphate-buffered saline (PBS):
   a. 160 g NaCl.
   b. 4 g KCl.
   c. 2.88 g Na_2HPO_4.
   d. 4.8 g KH_2PO_4.
   e. 1 L Milli-Q water.
   f. Adjust the pH of the buffer to 6.4 to obtain 20X PBS stock solution.
   g. Take 50 mL of this buffer and make it up to 1 L with Milli-Q water to obtain 1X PBS.

4. Bovine serum albumin (BSA) (10 mg/mL):
   a. 100 mg of BSA (Sigma-Aldrich, cat. no. A3059-50G).
   b. 10 mL of Milli-Q water.
   c. Aliquot in 100 µL and store at −20°C.
5. 10 mg/mL Solution salmon sperm DNA (Sigma-Aldrich, cat. no. D7696) (use 10 µL for preclearing steps).
6. 10 mg/mL Solution herring sperm DNA (Sigma-Aldrich, cat. no. D7290) (use 10 µL for preclearing).
7. Lysis buffer:
   a. 44 mM Tris-HCl, pH 8.1.
   b. 1% Sodium dodecyl sulfate (SDS).
   c. 10 mM Ethylenediaminetetra-acetic acid (EDTA).
   d. 10 mM Sodium butyrate (for histone acetylation ChIP assays).
   e. Add the following protease inhibitors fresh to the lysis buffer just before use.
   f. 1 mM Phenyl methyl sulfonyl chloride (PMSF).
   g. 10 µg/mL Aprotinin.
   h. 10 µg/mL Leupeptin.
   i. 10 µg/mL Pepstatin.
8. Lysis buffer 2 *(31)* (http://genomecenter.ucdavies.edu/farnham/farnham/protocols/chips.html):
   a. 5 mM Piperazine-N-N′-bis (2-ethane sulfonic acid) (PIPES), pH 8.0.
   b. 85 mM KCl.
   c. 0.5% NP-40.
   d. Protease inhibitors.
9. Lysis buffer 3 *(41)*:
   a. 50 mM Tris, pH 8.1.
   b. 150 mM NaCl.
   c. 5 mM EDTA.
   d. 0.1% Sodium deoxycholate.
   e. 1% Triton X-100.
   f. Protease inhibitors.
10. Lysis buffer 4 *(42)* (http://www.biochem.northwestern.edu/ibis/morimoto/Protocols/):
    a. 50 mM Tris-HCl, pH 8.0.
    b. 150 mM NaCl.
    c. 1% SDS.
    d. 2% Triton X-100.
    e. 1 mM EDTA.
    f. Protease inhibitors.
11. Lysis buffer 5 *(36)*:
    a. 250 mM sucrose.
    b. 10 mM Tris-HCl, pH 7.4.
    c. 10 mM Sodium butyrate.
    d. 4 mM Magnesium chloride.
    e. 0.1 M PMSF.

    f. 0.1 *M* Benzamidine.

    g. 0.1% w/v Triton X-100.

    h. Protease inhibitors.

12. 100 m*M* PMSF:

    a. 0.174 g PMSF (Sigma-Aldrich).

    b. 10 mL Absolute ethanol.

    c. Aliquot 1 mL into microcentrifuge tubes and store at −20°C. (Remember that the half-life of 1 m*M* PMSF [after it has been added to the lysis buffer] is 30 min, so add it just before use.)

13. 5 mg/mL Aprotinin, leupeptin, or pepstatin (Sigma-Aldrich). (Each of the inhibitors are dissolved in water at a concentration of 5 mg/mL, aliquoted in 100 μL and stored at −20°C. Use at 1:500 dilution.)

14. Dilution buffer:

    a. 16.7 m*M* Tris-HCl, pH 8.1.

    b. 250 m*M* NaCl.

    c. 0.01% SDS.

    d. 1% Triton X-100.

    e. 10 m*M* sodium butyrate (for histone acetylation assays).

15. Wash buffer 1:

    a. 16.7 m*M* Tris-HCl, pH 8.1.

    b. 375 m*M* NaCl.

    c. 0.01% SDS.

    d. 1% Triton X-100.

    e. 10 m*M* sodium butyrate (for histone acetylation assays).

16. Wash buffer 2 *(31)* (http://genomecenter.ucdavies.edu/farnham/farnham/protocols/chips.html):

    a. 100 m*M* Tris-HCl, pH 9.0 (8.0 for monoclonal antibodies).

    b. 500 m*M* LiCl.

    c. 1% NP-40.

    d. 1% Sodium deoxycholate.

17. Wash buffer 3 *(36)*.

    a. 250 m*M* Sucrose.

    b. 10 m*M* Sodium butyrate.

    c. 4 m*M* Magnesium chloride.

    d. 0.1 m*M* PMSF.

18. Low salt wash buffer *(31,43)*:

    a. 20 m*M* Tris-HCl, pH 8.1.

    b. 150 m*M* NaCl.

    c. 0.1% SDS.

    d. 1% Triton X-100.

    e. 2 m*M* EDTA.

19. High salt wash buffer:

    a. 20 m*M* Tris-HCl, pH 8.1.

    b. 500 m*M* NaCl.

    c. 0.1% SDS.

    d. 1% Triton X-100.

    e. 2 mM EDTA.

20. LiCl wash buffer:

    a. 10 mM Tris-HCl, pH 8.0.

    b. 0.25 mM LiCl.

    c. 1% NP-40.

    d. 1% Deoxycholate.

    e. 1 mM EDTA.

21. Elution buffer:

    a. 0.1 M NaHCO$_3$.

    b. 1% SDS.

    c. 5 mM NaCl.

    d. 10 mM Sodium butyrate (for histone acetylation assays).

22. Resuspension buffer:

    a. 50 mM Tris-HCl, pH 8.0.

    b. 2 mM EDTA.

    c. 0.2% Sarkosyl.

23. IP Elution buffer:

    a. 50 mM NaHCO$_3$.

    b. 1% SDS.

24. Proteinase K (25 mg/mL):

    a. 50 mg proteinase K (Fisher Biotech, cat. no. BP-1700-100).

    b. 2 mL Milli-Q water.

    c. Aliquot into 100 µL and store at −20°C.

25. 5X PK buffer:

    a. 50 mM Tris-HCl, pH 7.5.

    b. 25 mM EDTA.

    c. 1.25% SDS.

    d. 3 M Sodium acetate, pH 5.2.

    e. 20.4 g Sodium acetate.

    f. 50 mL Water.

    g. Adjust the pH to 5.2.

26. Phenol–chloform–isoamyl alcohol (Fisher Biotech, cat. no. BP1752-100).

27. 1X TE:

    a. 10 mM Tris-HCl, pH 8.1.

    b. 1 mM EDTA.

28. 20 mg/mL glycogen (Roche, cat. no. 901393).

29. RNase A (10 mg/mL):

    a. 0.1 g RNase A (Fisher Biotech, cat. no. BP2539-250).

    b. 10 mL Water.

    c. Store at 4°C.

30. 5 M NaCl:

   a. 58.4 g NaCl.
   b. 200 mL of Milli-Q water.
   c. Micrococal nuclease (MNase) Enzyme.
   d. MNase (Amersham Biosciences, cat. no. E70196Y).
   e. Milli-Q water containing 20–50% glycerol at a concentration of 1 mg/mL.
   f. Aliquots (10–20 µL) are stored at −20°C and used only once after thawing on ice.
31. MNase digestion buffer:
   a. 0.32 *M* Sucrose.
   b. 50 m*M* Tris-HCl, pH 7.5.
   c. 1 m*M* CaCl$_2$.
   d. 4 m*M* MgCl$_2$.
   e. 0.1 m*M* PMSF.

3. Methods

3.1. Chromatin Immunoprecipitation (ChIP) Assay Protocol

3.1.1. Preparation of ChIP Cell Lysate

1. ChIP assays to analyze multiple factors or modifications will require a substantial amount of cells as the starting material. It is recommended that at least 2×10^7 adherent or suspension cells be used per sample (*see* **Subheading 4.**, **Note 1**) *(26,31,32,41)*. Add formaldehyde directly added to the tissue culture media at a final concentration of 1%. This corresponds to 280 µL of the commercially available 37% formaldehyde per 10 mL of tissue culture media. Incubate the cells with the formaldehyde on an orbital shaker for 20 min at room temperature for efficient crosslinking of DNA to the protein.

2. Terminate the crosslinking reaction by adding glycine to a final concentration of 0.125 *M*, using the 1 *M* stock solution. Incubate the cells on the orbital shaker for additional 5 min.

3. If the assay is performed on adherent cells, scrape the cells in 5 mL of 1X ice-cold PBS and transfer a fresh 15-mL centrifuge tube. In case of nonadherent cells, collect the cells along with the media and transfer to a 15-mL tube. Centrifuge the cells at 800*g* (1000 rpm) for 5 min in a table-top centrifuge with a swing-bucket rotor (ThermoForma, General purpose centrifuge) to collect the cells. Wash the cell pellet twice with 10 mL of ice-cold PBS. Resuspend the pellet carefully in 1 mL 1X cold PBS and transfer it to an Eppendorf tube. Use a sheared 1-mL pipet tip (a tip with the end cut off) for the resuspending the cells. Spin at 800*g* (3000 rpm) for 5 min in a refrigerated table-top microcentrifuge at 4°C (Beckman-Coulter, Microfuge R or an equivalent model cat. no. 368830).

4. Aspirate the supernatant carefully and resuspend the cell pellet in ice-cold lysis buffer containing protease inhibitors (for 75 µL cell pellet, add 200 µL lysis buffer). Mix well using a pipetman to obtain a uniform homogenate; avoid clumping of the cells (*see* **Subheading 4.**, **Notes 2** and **3**). Do not vortex the lysate. The efficiency of lysis can be checked by Trypan blue staining; if cells are not lysed they can be

homogenized on ice using a Dounce homogenizer (Fisher Scientific, cat. no. NC 9357430) with B type pestle. Certain studies have used a glass bead disruption method to prepare ChIP lysates *(41)*, as described in **Subheading 3.3.1.**

5. Sonicate the lysate to shear the chromatin below an average length of 1-kb. The time and the number of pulses vary according to cell type, sonicator used and extent of crosslinking. However, we have found for most cells, three cycles of sonication at power 4 for 30 s on a Fisher Sonic Dismembrator (Model 100), followed by incubation on ice for 30 s, is sufficient to shear the chromatin. As an alternative to sonication, digestion with micrococcal nuclease *(9,23,34–36)* can be also done to cleave the chromatin into oligonucleosomes (refer to Alternate Protocols, **Subheading 3.3.2.**). This method is referred to as "native chromatin immunoprecipitation." However, this technique is incompatible with formaldehyde crosslinking because crosslinked chromatin is inefficiently cleaved by nucleases. The protocol for cell lysis and nuclei preparation in native ChIP assays is described in detail elsewhere *(9,23,36)*.

6. Spin at 20,100*g* (15,000 rpm) in a table-top refrigerated microcentrifuge for 15 min. Carefully transfer the supernatant into a new microcentrifuge tube. At this point the chromatin can be snap frozen in liquid nitrogen and stored at −70°C.

3.1.2. Setting Up the Immunoprecipitation (IP) Reaction

1. If previously frozen, thaw out the ChIP lysate on ice. Save 20% of the lysate as the input control. Keep the input control aside and do not use it for the immunoprecipitation reaction. The rest of the lysate is equally divided into three microcentrifuge tubes. Set up the following immunoprecipitation reactions:
   a. Irrelevant antibody.
   b. Positive-control antibody.
   c. The antibody that is being used for the assay.
2. Use 3 µg of antibody per immunoprecipitation reaction. Add 300 µL of dilution buffer and rotate the mixture overnight at 4°C. Preclearing ChIP lysates is a good method to reduce nonspecific backgrounds and eliminate false-positives during the assay (for preclearing protocols *see* **Subheading 3.2.2.**). Although it may not be necessary to preclear in all cases, it should be resorted to in case of nonspecific background bands are observed in PCR.
3. ChIP assays can be performed with Protein A (Amersham Bioscences, cat. no. 17-0872-05) Protein G (Amersham Bioscences, cat. no. 17-0618-01), ProteinA/G beads (Santa Cruz Biotechnology, cat. no. sc-2003) or Staph A (American Type Culture Collection) cells. The choice of beads depends on the isotype of the antibody to be used for the immunoprecipitation. A comparison of the relative affinities of antibodies to Protein A or Protein G is described in **ref. 44**. Add the 120 µL of 1:1 Protein G slurry or 400 µL of Protein A slurry (for preparation of beads or Staph A, refer to support protocols, **Subheading 3.2.1.**). Thereafter, make up the volume to 700 µL with dilution buffer and rotate for 2 h at 4°C. If you are using Staph A beads, incubate each sample with 10 µL of preblocked Staph A beads *(29,31,32,45)* (*see* **Subheading 3.2.2.1.**). Rotate on a nutator at 4°C for 15 min.

3.1.3. Extraction of Immunoprecipitated DNA

1. Spin the tubes from **step 3, Subheading 3.1.2.** at 560*g* (2500 rpm) at room temperature for 30 s, in a table-top microcentrifuge at room temperature. Carefully aspirate the supernatant and discard.
2. Efficient washing is critical to reduce background. Wash the beads three times with dilution buffer and twice with wash buffer. It has been found that transferring the beads to a new tube after the first wash helps reduce nonspecific bands. Remove the supernatant and discard each time; care should be taken not to lose beads during the wash steps (*see* **Subheading 4., Note 8**). Carefully remove all the supernatant using a gel-loading tip in the final wash.
3. Add 300 µL of elution buffer. Rotate for 15 min at room temperature.
4. Spin the tubes at 8000*g* (10,000 rpm) for 2 min at room temperature, in a table-top microcentrifuge.
5. Transfer the supernatant to fresh tubes and incubate 65°C overnight to reverse formaldehyde crosslinking. The tube containing the input DNA (**Subheading 3.1.2., step 1**) should also be incubated at 65°C overnight.
6. Add 0.1 vol of 3 *M* sodium acetate, pH 5.2, followed by two and a half volumes of ice-cold 100% ethanol to the tubes and incubate at −70°C for 10 min to precipitate the DNA.
7. Centrifuge the samples at 16,000*g* (14,000 rpm) for 15 min at 4°C, in a refrigerated microcentrifuge.
8. Wash the pellet with 2.5 vol of ice cold 70% ethanol. Spin the tubes at 16,000*g* (14,000 rpm) for 15 min at 4°C. Air-dry the pellet.
9. Resuspend the DNA in 200 µL of Milli-Q water. Add 50 µg of proteinase K (2 µL from a stock solution of 25 mg/mL proteinase K (Fischer Biotech, cat. no. BP-1700-100) to each sample and incubate at 37°C for 30 min.
10. Extract the DNA by adding equal volumes of phenol–chloform–isoamyl alcohol (Fischer Biotech, cat. no. BP1752-100). Vortex well to mix the layers. Spin the tubes at 14,000 rpm for 15 min at 4°C, in a refrigerated microcentrifuge to separate the layers. Collect the upper layer carefully using a pipetman, transfer to a fresh tube and repeat **steps 14–16** to extract the DNA. Alternately, use a QiaQuick spin column (Qiagen Corporation) to extract DNA and elute the DNA in 50 µL of TE.
11. Resuspend the immunoprecipitated DNA samples in 50 µL water. The input DNA is resuspended in 250 µL water. The DNA of interest is amplified by PCR using specific primers to the sequence of interest (*see* **Subheading 4., Notes 6** and **7**). Use 5 µL of immunoprecipitated DNA for PCR analysis; 1 µL of the input DNA would be sufficient for PCR. PCR conditions vary according to the primer sites used; generally 30 cycles of amplification will result in visible bands. Analyze the DNA fragments by agarose gel electrophoresis.

Note: If the immunoprecipitation reaction has been set up using Staph A beads, DNA extraction protocols from Staph A beads are described in **Subheading 3.2.1.**

3.2. Support Protocols for ChIP Assay

3.2.1. Preparation of Protein A, G, A/G Beads, or Staph A Cells

Appropriate beads are chosen depending on the isotype of the antibodies used and their affinity for protein A, protein G, protein A/G. Protocols from the Farnham lab use Staph A cells for immunoprecipitation reaction. Protein G or protein A is used for most applications in the laboratory.

1. Protein A: weigh out 60 mg of protein A and resuspend it in 2 mL of dilution buffer. Incubate for 30 min on ice. Subsequently, use 400 µL of the above slurry for each ChIP reaction.
2. Protein G: Protein G slurry (Amersham Biosciences) containing 20% ethanol was used. The required volume of beads are taken and washed three times with dilution buffer to remove the ethanol preservative. Subsequently, the beads are resuspended in equal volume of dilution buffer and 120 µL of beads are used per sample. The same procedure can be also used for preparation of protein A/G beads.
3. Staph A cells: resuspend 1 g of Staph A cells in 10 mL of 1X resuspension buffer *(21,29,31)*. Centrifuge at 8000g (10,000 rpm) for 5 min at 4°C. This step is repeated once again. Thereafter resuspend in 3 mL of 1X PBS containing 3% SDS and 10% β-ME. Boil the beads for 30 min and collect by centrifugation at 1000 rpm for 5 min. Wash twice in 1X resuspension buffer and repeat centrifugation at 10,000 rpm for 5 min. Resuspend in 4 mL of resuspension buffer, aliquot in 100 µL volumes in microcentrifuge tubes and snap freeze.

3.2.2. Preclearing of ChIP Lysates

Preclearing lysates is a good method to eliminate nonspecific background in ChIP assays. ChIP assay protocols using Staph A cells require an initial preblocking step to reduce nonspecific binding *(31,46)*.

3.2.2.1. PREBLOCKING STAPH A BEADS

For every tube (100 µL) of Staph A cells add 10 µL herring sperm DNA (Sigma-Aldrich, cat. no. D7290) (10 mg/mL) and 10 µL BSA (Sigma-Aldrich, cat. no. A3059-50G) (10 mg/mL). Incubate on a nutator at 4°C for at least 3 h or more (overnight incubation can also be done). Centrifuge at 8000g (10,000 rpm) for 3 min. Remove supernatant and wash twice with 1X resuspension buffer. Resuspend the cells in equal volume of resuspension buffer and use 10–15 µL for each preclearing step.

3.2.2.2. PRECLEARING PROTOCOLS WITH PROTEIN A, PROTEIN G, OR PROTEIN A/G

Add 120 µL of 1:1 Protein G slurry (or 400 µL of protein A) to the ChIP lysate (obtained from 2×10^7 cells as mentioned previously in **Subheading 3.1.1.**) in a microcentrifuge tube. Add 15 µL of salmon sperm DNA (Sigma-Aldrich, cat. no. D7696) (10 mg/mL) and 10 µL of BSA (10 mg/mL). Rotate on a nutator for 30 min at 4°C. Thereafter spin the tubes at 16,000*g* (14,000 rpm) for 30 s in a table-top centrifuge at 4°C. Collect supernatant fraction and discard the beads. This supernatant fraction is used for the immunoprecipitation reaction described in **Subheading 3.1.2.**

3.2.2.3. PreClearing Protocols Using Staph A Cells

Use the preblocked Staph A cells as described in **Subheading 3.2.2.1.** to preclear the ChIP lysate. 10–15 µL of preblocked Staph A beads are added to the ChIP lysate and incubated on a nutator for 15 min at 4°C. Centrifuge the tube at 14,000 rpm for 5 min and transfer the supernatant to a fresh tube. Use this precleared lysate for the subsequent immunoprecipitation reaction (**Subheading 3.1.2.**).

3.2.3. DNA Extraction From Immunoprecipitates Using Staph A Beads

1. The immunoprecipitation reaction is set up as detailed in **Subheading 3.1.2.**
2. Centrifuge the samples at 16,000*g* (14,000 rpm) for 3 min at room temperature.
3. Wash the pellets twice with 1.4 mL of 1X resuspension buffer followed by four washes with IP buffer. The washes are done by first dissolving the Staph A pellet in 200 µL of the appropriate buffer and adding 1.2 mL of the same buffer. The tubes are incubated with the wash buffer on a nutator for 3 min at room temperature. Thereafter, the tubes are centrifuged at 16,000*g* (14,000 rpm) for 3 min. The wash buffer is carefully aspirated off. After the last wash, the entire buffer is removed using a gel-loading tip.
4. Elute the immuecomplex by adding 150 µL of IP clution buffer. Mix very well for about 15 min. Microcentrifuge the tubes at 16,000*g* (14,000 rpm) for 3 min. Transfer the supernatants to fresh tubes.
5. Repeat the elution step. Both the elution fractions are pooled. Spin the tubes at 16,000*g* (14,000 rpm) for 5 min at room temperature. Transfer the supernatants to fresh tubes. Add 1 µL of 10 mg/mL RNaseA (Fisher Biotech, cat. no. BP2539-250) and NaCl to a final concentration of 0.3 *M*. Incubate samples at 67°C for 4–5 h (or overnight) to reverse crosslinking between protein and DNA. Take the input DNA sample and incubate at 65°C for 4–5 h (or overnight) to reverse crosslinking.
6. Precipitate the DNA by adding 2.5 vol of ice-cold ethanol to each of the tubes and incubate at –20°C overnight.
7. Centrifuge the DNA at 16,000*g* (14,000 rpm) for 15 min at 4°C. Remove the residual ethanol and air-dry the DNA pellet.
8. Dissolve the pellet in 100 µL TE. The sample may be viscous and may have to be dissolved in a larger volume. Add 25 µL of 5X PK buffer and 1.5 µL of proteinase K to each sample. Incubate for 1–2 h at 45°C.
9. Add 175 µL of TE to each sample. Extract with 300 µL of phenol–chloroform–isoamyl alcohol.

10. Add 30 μL of 5 *M* NaCl, 5 μg of tRNA (Sigma-Aldrich) and 5 μg of glycogen (Roche, cat. no. 901393) to each sample. Mix well and then add 750 μL of ice-cold ethanol. Incubate at −20°C overnight.

11. Next day, centrifuge the tubes at 14,000 rpm for 20 min at 4°C. Aspirate the supernatant carefully and air-dry the DNA pellet. Resuspend the DNA pellet in 30 μL of TE. Use 2–3 μL for PCR analysis. For the total input sample, dilute the sample 300-fold and then perform the PCR reaction *(31,32)*.

3.3. Alternate Protocols for ChIP Assay

3.3.1. Preparation of ChIP Lysate

1. Perform **steps 1–3** as detailed in **Subheading 3.1.1.**
2. Resuspend the pellet in equal volume 200 μL of ChIP lysis buffer.
3. Add an equal volume of glass beads.
4. Vortex at the maximum setting six times for 30 s at 4°C each with 3 min pauses between each vortex.
5. Pierce the bottom of the microcentrifuge tube with a needle and place the tube carefully in another microcentrifuge tube. Spin at 700 g for 2 min. The liquid should move through the first microcetrifuge tube into the bottom tube.
6. Add 200 μL of ChIP lysis buffer and sonicate the samples as described in **Subheading 3.1.1., steps 5–6**.

3.3.2. Fragmentation of Chromatin Using Micrococcal Nuclease (MNase)

1. Prepare nuclear pellet from cells as detailed in **refs. 9** and **36**.
2. Resuspend the nuclear pellet in 1 mL MNase digestion buffer and place on ice.
3. Aliquot two 1.5-mL Eppendorf tubes with 500 μL of resuspended nuclei.
4. Add 1 μL of micrococcal nuclease (Mnase) enzyme (Amersham Bioscience, cat. no. E70196Y) to each tube and mix gently.
5. Incubate the tubes in a 37°C water bath for 10 min. The incubation times as well as MNase concentrations have to be optimized to generate mainly tri-, di-, and mononucleosomes (*see* **Subheading 4., Note 5**).
6. Add 20 μL of stop solution. Chill on ice.
7. The suspension was centrifuged at 13,000*g* (12,000 rpm) for 20 s at 4°C in a refrigerated microfuge. The supernatant is made up of the second soluble fraction S2 (containing larger fragments of chromatin), whereas, the pellet contained the smaller fragment S1 fraction. The pellet was resolubilized in lysis buffer and pooled with the S2 fraction in a fresh clean tube.
8. Continue the ChIP assay as described in **Subheading 1** from **step 6** onward.

3.4. ChIP Assay on Tissues

1. Chop the tissues with a razor blade or scalpel into small pieces and resuspend them in serum free media. Add 1% formaldehyde (some studies have used 0.4% formaldehyde *[47]*) and rotate the tubes for 10 min at room temperature.

2. Stop the crosslinking reaction by adding glycine to a final concentration of 0.125 M. Rotate at room temperature for 5 min.

3. Centrifuge the samples at 250g (400 rpm) for 5 min and discard the supernatant. Wash the sample twice in PBS. Resuspend the chopped tissue in 1–2 mL PBS. Homogenize the tissue on ice using a Dounce homogenizer or Polytron Homogenizer. Centrifuge the sample at 250g to collect all the cells; discard the supernatant.

4. Follow the standard ChIP protocol as detailed in **Subheading 3.1.1., step 5** onward *(45,47,48)*.

Chromatin immunoprecipitation is a powerful approach that allows the elucidation of interactions between endogenous proteins and their native chromatin sites, thereby providing insights into physiological transcription *(2,7,33,49–51)* and transactivation mechanisms *(4,52,53)*. ChIP assays have been extensively used to study histone acetylation and methylation patterns at a variety of genomic loci *(3,18,46,52,54,55)*. In addition, ChIP methodology has been coupled to microarrays (ChIP-chip) containing genomic regions to facilitate genome wide binding patterns of transcription factors to specific promoters *(3,8,56)*. Procedures by Ren et al. *(39,57)* involve immunoprecipitation of the DNA, which is then blunt-ended by T4 DNA polymerase to allow ligation of a universal linker (LM-PCR). The ChIP-enriched DNA and input DNA are then labeled with different fluorescent dyes and hybridized to a single DNA microarray containing promoter sequences (for detailed description of the method refer to **refs.** *38–40,54,56,* and *57*).

4. Notes

1. Despite the many advantages of the ChIP assay, it can be fraught with many obstacles. The first important thing to keep in mind is to make the lysates using large number of cells (at least 2×10^7) to achieve a good signal to noise ratio.

2. Efficient lysis of cells is another important parameter for a successful ChIP assay. It may be a good idea to check the efficacy of cell lysis using Trypan blue exclusion method. In case cells are not being efficiently lysed, use a different lysis buffer (for a list of commonly used lysis buffers refer to **Subheadings 2., items 7–9**). Moreover, cells can be homogenized on ice after addition of cell lysis buffer to facilitate the release of nuclei. In case of heterogeneous samples, it is essential that the cells be sorted to enrich the cell population of interest before performing the ChIP assay. To analyze histone acetylation, add 10 mM sodium butyrate to the lysis buffer, dilution, and elution buffers. Sodium butyrate prevents the loss of histone acetylation during sample preparation by inhibition of histone deacetylase activity.

3. It is essential to perform all steps of cell lysis and immunoprecipitation on ice or at 4°C, wherever indicated to prevent chromatin degradation by endogenous nucleases. Similarly, protease inhibitors should be added wherever mentioned to preserve the integrity of proteins bound to the chromatin.

4. Another crucial step in the ChIP assay is the sonication of DNA. It is essential that sonication should cause efficient shearing of the genomic DNA to about 1 kb in

size. The efficiency of sonication should be checked by agarose gel elecrophoresis when the ChIP assays are being standardized with new cell lines or sonicators.

5. It must be remembered that sonication of crosslinked chromatin gives rise to randomly sheared DNA fragments, which may not produce small enough chromatin fragments at the region of interest. Furthermore, crosslinking the DNA to the protein by formaldehyde may mask critical protein epitopes resulting in poor efficiency of immunoprecipitation. On the other hand, a disadvantage of native ChIP is that certain protein–DNA interactions are transient or intrinsically unstable, precluding detection unless the chromatin–DNA complexes are trapped by crosslinking. The selection of "conventional ChIP" assay vs "native ChIP" assay depends upon the nature of the experimental system. Whereas the "native ChIP" is particularly suited for proteins like core histones, which are tightly bound to chromosomes, the "conventional ChIP" assay works better for studying linker histones, transcription factors and proteins which may relocalize during sample preparation.

6. Although, the use of PCR ensures a high degree of signal amplification, a low efficiency in the immunoprecipitation reaction can give rise to nonspecific signals or no signal. Hence, it is essential to use 2–4 µg of a high-affinity, high specificity antibody for the immunoprecipitation reaction. Preclearing the lysates before immunoprecipitation is a good way to improve specificity and minimize nonspecific background in the ChIP assay **(Subheading 3.2.2.).** However, we have found that even in the absence of preclearing, the ChIP assay works provided all other safeguards are carefully followed. Furthermore, one may have to try different antibodies to optimize the ChIP assay because certain antibodies may efficiently immunoprecipitate the native protein but may not be effective in immunoprecipitating proteins crosslinked to chromatin. The exact conditions of the immunoprecipitation may have to be further optimized to ensure a high recovery of target DNA during the IP. It is crucial to incorporate a positive control as well an irrelevant antibody in the ChIP assay to facilitate critical interpretation of data.

7. Recent studies involving ChIP assays have used real-time PCR for detecting as little as twofold changes in protein–DNA interactions. Real-time PCR allows data to be collected during cycles when amplification is occurring exponentially *(58)*. It increases the precision of ChIP measurements allowing accurate detection of small variations in protein–DNA interactions. Another possibility is to perform duplex amplification PCR *(59,60)*, which involves coamplification of a fragment from the region of interest along with a control fragment. Applications, in particular for allelic studies mechanisms, or on genomic imprinting in mammals, one can apply SSCP *(61,62)* or hot-stop PCR *(63)*, especially to differentiate silent alleles from the active alleles. The different PCR-based approaches have been reviewed in *(4,9,28)*.

8. After the immunoprecipitation reaction, efficient and stringent washing is critical to reduce nonspecific background. A list of wash buffers used by different laboratories has been provided in **Subheading 2., item 11**. The number of washes can be increased if necessary. The wash buffer can be varied according to nature of

antibody or conditions of immunoprecipitation. Some laboratories use 2X PBS as the wash buffer, others wash the immunecomplexes sequentially with 1 mL of the following buffers:

a. Low salt wash buffer.
b. High salt wash buffer.
c. LiCl wash buffer.
d. 1X TE.

9. It is important to be extremely careful and meticulous during this assay. The amplification by PCR is a highly sensitive; it will amplify even minimal non-specific signals. All possible precautions were recommend to prevent contamination from other DNA sources such as amplification in a dedicated space, use of fresh gloves, sterile filter tips, clean microfuge tubes for the assay.

Acknowledgments

Work in the Chellappan lab is supported by NIH/NCI grants CA63136 and CA77301.

References

1. Hecht, A. and Grunstein, M. (1999) Mapping DNA interaction sites of chromosomal proteins using immunoprecipitation and polymerase chain reaction. *Methods Enzymol.* **304,** 399–414.
2. Luo, R. X. and Dean, D. C. (1999) Chromatin remodeling and transcriptional regulation. *J. Natl. Cancer Inst.* **91,** 1288–1294.
3. Kirmizis, A. and Farnham, P. J. (2004) Genomic approaches that aid in the identification of transcription factor target genes. *Exp. Biol. Med. (Maywood)* **229,** 705–721.
4. Johnson, K. D. and Bresnick, E. H. (2002) Dissecting long-range transcriptional mechanisms by chromatin immunoprecipitation. *Methods* **26,** 27–36.
5. Grunstein, M. (1997) Histone acetylation in chromatin structure and transcription. *Nature* **389,** 349–352.
6. Grunstein, M. (1998) Yeast heterochromatin: regulation of its assembly and inheritance by histones. *Cell* **93,** 325–328.
7. Harbour, J. W. and Dean, D. C. (2001) Corepressors and retinoblastoma protein function. *Curr. Top. Microbiol. Immunol.* **254,** 137–144.
8. Bernstein, B. E., Humphrey, E. L., Liu, C. L., and Schreiber, S. L. (2004) The use of chromatin immunoprecipitation assays in genome-wide analyses of histone modifications. *Methods Enzymol.* **376,** 349–360.
9. Umlauf, D., Goto, Y., and Feil, R. (2004) Site-specific analysis of histone methylation and acetylation. *Methods Mol. Biol.* **287,** 99–120.
10. Spencer, V. A., Sun, J. M., Li, L., and Davie, J. R. (2003) Chromatin immunoprecipitation: a tool for studying histone acetylation and transcription factor binding. *Methods* **31,** 67–75.
11. Stallcup, M. R. (2001) Role of protein methylation in chromatin remodeling and transcriptional regulation. *Oncogene* **20,** 3014–3020.

12. Kuo, M. H. and Allis, C. D. (1998) Roles of histone acetyltransferases and deacetylases in gene regulation. *BioEssays* **20,** 615–626.

13. Jenuwein, T. and Allis, C. D. (2001) Translating the histone code. *Science* **293,** 1074–1080.

14. Wang, Y., Fischle, W., Cheung, W., Jacobs, S., Khorasanizadeh, S., and Allis, C. D. (2004) Beyond the double helix: writing and reading the histone code. *Novartis Found Symp.* **259,** 3–17; discussion 17–21, 163–169.

15. Agalioti, T., Chen, G., and Thanos, D. (2002) Deciphering the transcriptional histone acetylation code for a human gene. *Cell* **111,** 381–392.

16. Turner, B. M. (2000) Histone acetylation and an epigenetic code. *BioEssays* **22,** 836–845.

17. Strahl, B. D. and Allis, C. D. (2000) The language of covalent histone modifications. *Nature* **403,** 41–45.

18. Kouzarides, T. (2002) Histone methylation in transcriptional control. *Curr. Opin. Genet. Dev.* **12,** 198–209.

19. Litt, M. D., Simpson, M., Recillas-Targa, F., Prioleau, M. N., and Felsenfeld, G. (2001) Transitions in histone acetylation reveal boundaries of three separately regulated neighboring loci. *EMBO J.* **20,** 2224–2235.

20. Litt, M. D., Simpson, M., Gaszner, M., Allis, C. D., and Felsenfeld, G. (2001) Correlation between histone lysine methylation and developmental changes at the chicken β-globin locus. *Science* **293,** 2453–2455.

21. Weinmann, A. S. and Farnham, P. J. (2002) Identification of unknown target genes of human transcription factors using chromatin immunoprecipitation. *Methods* **26,** 37–47.

22. Orlando, V. (2000) Mapping chromosomal proteins in vivo by formaldehyde-crosslinked-chromatin immunoprecipitation. *Trends Biochem. Sci.* **25,** 99–104.

23. Hebbes, T. R., Clayton, A. L., Thorne, A. W., and Crane-Robinson, C. (1994) Core histone hyperacetylation co-maps with generalized DNase I sensitivity in the chicken β-globin chromosomal domain. *EMBO J.* **13,** 1823–1830.

24. Dasgupta, P., Sun, J., Wang, S., et al. (2004) Disruption of the Rb-Raf-1 interaction inhibits tumor growth and angiogenesis. *Mol. Cell Biol.* **24,** 9527–9541.

25. Dasgupta, P., Betts, V., Rastogi, S., et al. (2004) Direct binding of apoptosis signal-regulating kinase 1 to retinoblastoma protein: novel links between apoptotic signaling and cell cycle machinery. *J. Biol. Chem.* **279,** 38,762–38,769.

26. Fusaro, G., Dasgupta, P., Rastogi, S., Joshi, B., and Chellappan, S. P. (2003) Prohibitin induces the transcriptional activity of p53 and is exported from the nucleus upon apoptotic signaling. *J. Biol. Chem.* **278,** 47,853–47,861.

27. Joshi, B., Ordonez-Ercan, D., Dasgupta, P., and Chellappan, S. (2004) Induction of human metallothionein 1g promoter by VEGF and heavy metals: differential involvement of E2F and MTF transcription factors. *Oncogene* **24,** 2204–2217.

28. Weinmann, A. S., Bartley, S. M., Zhang, T., Zhang, M. Q., and Farnham, P. J. (2001) Use of chromatin immunoprecipitation to clone novel E2F target promoters. *Mol. Cell. Biol.* **21,** 6820–6832.

29. Wells, J., Graveel, C. R., Bartley, S. M., Madore, S. J., and Farnham, P. J. (2002) The identification of E2F1-specific target genes. *Proc. Natl. Acad. Sci. USA* **99,** 3890–3895.

30. Wells, J. and Farnham, P. J. (2002) Characterizing transcription factor binding sites using formaldehyde crosslinking and immunoprecipitation. *Methods* **26,** 48–56.

31. Boyd, K. E., Wells, J., Gutman, J., Bartley, S. M., and Farnham, P. J. (1998) c-Myc target gene specificity is determined by a post-DNAbinding mechanism. *Proc. Natl. Acad. Sci. USA* **95,** 13,887–13,892.

32. Boyd, K. E. and Farnham, P. J. (1999) Coexamination of site-specific transcription factor binding and promoter activity in living cells. *Mol. Cell Biol.* **19,** 8393–8399.

33. Harbour, J. W. and Dean, D. C. (2000) Chromatin remodeling and Rb activity. *Curr. Opin. Cell Biol.* **12,** 685–689.

34. Dorbic, T. and Wittig, B. (1986) Isolation of oligonucleosomes from active chromatin using HMG17-specific monoclonal antibodies. *Nucleic Acids Res.* **14,** 3363–3376.

35. Dorbic, T. and Wittig, B. (1987) Chromatin from transcribed genes contains HMG17 only downstream from the starting point of transcription. *EMBO J.* **6,** 2393–2399.

36. Thorne, A. W., Myers, F. A., and Hebbes, T. R. (2004) Native chromatin immunoprecipitation. *Methods Mol. Biol.* **287,** 21–44.

37. Buck, M. J. and Lieb, J. D. (2004) ChIP-chip: considerations for the design, analysis, and application of genome-wide chromatin immunoprecipitation experiments. *Genomics* **83,** 349–360.

38. Robyr, D. and Grunstein, M. (2003) Genomewide histone acetylation microarrays. *Methods* **31,** 83–89.

39. Ren, B. and Dynlacht, B. D. (2004) Use of chromatin immunoprecipitation assays in genome-wide location analysis of mammalian transcription factors. *Methods Enzymol.* **376,** 304–315.

40. Roh, T. Y., Ngau, W. C., Cui, K., Landsman, D., and Zhao, K. (2004) High-resolution genome-wide mapping of histone modifications. *Nat. Biotechnol.* **22,** 1013–1016.

41. Strahl-Bolsinger, S., Hecht, A., Luo, K., and Grunstein, M. (1997) SIR2 and SIR4 interactions differ in core and extended telomeric heterochromatin in yeast. *Genes Dev.* **11,** 83–93.

42. Morimoto, R. I. (2002) Dynamic remodeling of transcription complexes by molecular chaperones. *Cell* **110,** 281–284.

43. Shang, Y., Hu, X., DiRenzo, J., Lazar, M. A., and Brown, M. (2000). Cofactor dynamics and sufficiency in estrogen receptor-regulated transcription. *Cell* **103,** 843–852.

44. Harlow, E. and Lane, D. (1988) *Antibodies: A Laboratory Manual.* Cold Spring Harbor, New York: 617 pp.

45. Farnham, P. J. (2002) In vivo assays to examine transcription factor localization and target gene specificity. *Methods* **26,** 1–2.

46. Kirmizis, A., Bartley, S. M., Kuzmichev, A., et al. (2004) Silencing of human polycomb target genes is associated with methylation of histone H3 Lys 27. *Genes Dev.* **18,** 1592–1605.
47. Forsberg, E. C., Downs, K. M., and Bresnick, E. H. (2000) Direct interaction of NF-E2 with hypersensitive site 2 of the β-globin locus control region in living cells. *Blood* **96,** 334–339.
48. Chaya, D. and Zaret, K. S. (2004) Sequential chromatin immunoprecipitation from animal tissues. *Methods Enzymol.* **376,** 361–372.
49. Im, H., Grass, J. A., Johnson, K. D., Boyer, M. E., Wu, J., and Bresnick, E. H. (2004) Measurement of protein-DNA interactions in vivo by chromatin immunoprecipitation. *Methods Mol. Biol.* **284,** 129–146.
50. Blais, A. and Dynlacht, B. D. (2004) Hitting their targets: an emerging picture of E2F and cell cycle control. *Curr. Opin. Genet. Dev.* **14,** 527–532.
51. Skowronska-Krawczyk, D., Ballivet, M., Dynlacht, B. D., and Matter, J. M. (2004) Highly specific interactions between bHLH transcription factors and chromatin during retina development. *Development* **131,** 4447–4454.
52. Elefant, F., Cooke, N. E., and Liebhaber, S. A. (2000) Targeted recruitment of histone acetyltransferase activity to a locus control region. *J. Biol. Chem.* **275,** 13,827–13,834.
53. Johnson, K. D., Christensen, H. M., Zhao, B., and Bresnick, E. H. (2001) Distinct mechanisms control RNA polymerase II recruitment to a tissue-specific locus control region and a downstream promoter. *Mol. Cell* **8,** 465–471.
54. Kondo, Y., Shen, L., Yan, P. S., Huang, T. H., and Issa, J. P. (2004) Chromatin immunoprecipitation microarrays for identification of genes silenced by histone H3 lysine 9 methylation. *Proc. Natl. Acad. Sci. USA* **101,** 7398–7403.
55. Nielsen, S. J., Schneider, R., Bauer, U. M., et al. (2001) Rb targets histone H3 methylation and HP1 to promoters. *Nature* **412,** 561–565.
56. Oberley, M. J., Tsao, J., Yau, P., and Farnham, P. J. (2004) High-throughput screening of chromatin immunoprecipitates using CpG-island microarrays. *Methods Enzymol.* **376,** 315–334.
57. Ren, B., Cam, H., Takahashi, Y., et al. (2002) E2F integrates cell cycle progression with DNA repair, replication, and G(2)/ M checkpoints. *Genes Dev.* **16,** 245–256.
58. Tse, C., Sera, T., Wolffe, A. P., and Hansen, J. C. (1998) Disruption of higher-order folding by core histone acetylation dramatically enhances transcription of nucleosomal arrays by RNA polymerase III. *Mol. Cell Biol.* **18,** 4629–4638.
59. Noma, K., Allis, C. D., and Grewal, S. I. (2001) Transitions in distinct histone H3 methylation patterns at the heterochromatin domain boundaries. *Science* **293,** 1150–1155.
60. Gregory, R. I., Randall, T. E., Johnson, C. A., et al. (2001) DNA methylation is linked to deacetylation of histone H3, but not H4, on the imprinted genes Snrpn and U2af1-rs1. *Mol. Cell Biol.* **21,** 5426–5436.
61. Gregory, R. I. and Feil, R. (1999) Analysis of chromatin in limited numbers of cells: a PCR-SSCP based assay of allele-specific nuclease sensitivity. *Nucleic Acids Res.* **27,** E32.

62. Orita, M., Iwahana, H., Kanazawa, H., Hayashi, K., and Sekiya, T. (1989) Detection of polymorphisms of human DNA by gel electrophoresis as single-strand conformation polymorphisms. *Proc. Natl. Acad. Sci. USA* **86,** 2766–2770.
63. Uejima, H., Lee, M. P., Cui, H., and Feinberg, A. P. (2000) Hot-stop PCR: a simple and general assay for linear quantitation of allele ratios. *Nat. Genet.* **25,** 375–376.

10

Manipulating Genes and Gene Copy Number by Bacterial Artificial Chromosomes Transfection

Stephanie F. Phelps, Sharon Illenye, and Nicholas H. Heintz

Summary

Bacterial artificial chromosomes (BACs) provide a well-characterized resource for studying the organization and activity of entire genes, replicons, and other large genomic loci. Protocols and parameters that influence the efficient transfection of these large DNA molecules into cells in culture were described here. By carefully optimizing the conditions for the formation of compact transfection complexes, BACs can be introduced into a variety of mammalian cells with reasonable efficiency. In addition, by cotransfection with a dihydrofolate reductase or hypoxanthine guanine phosphoribosyl transferase BAC, stable cell lines can be generated that carry 2–15 tandem chromosomal copies of the BAC of interest, thus providing a new avenue for studying gene dosage effects.

Key Words: BAC transfection; bacterial artificial chromosomes; BACs; functional genomics; gene amplification; gene dosage.

1. Introduction

Gene expression in higher eukaryotes is a complex process that is influenced by a multitude of factors, including cell and tissue type, stage of development, and environmental stimuli. To understand how diverse signals influence gene expression through *cis*-acting elements, many of which act over long distances, it has become increasingly important to study gene regulation in the context of whole genomic loci. Using this approach, one may develop a better understanding of how the nuclear localization of a gene, its chromatin configuration, and the organization of endogenous regulatory elements influence control of transcription and RNA processing. Because the average gene spans approx 60 kb, most genomic loci cannot be propagated in plasmid or λ vectors owing to cloning capacity limitations. Bacterial artificial chromosomes (BACs), which contain inserts ranging

From: *Methods in Molecular Biology, vol. 383: Cancer Genomics and Proteomics: Methods and Protocols*
Edited by: P. B. Fisher © Humana Press Inc., Totowa, NJ

in size from 100 to 300 kb, offer a means to study those genomic loci that fall within this size range *(1)*. Using homologous recombination to introduce deletions, point mutations, or other modifications in BACs, the interaction of regulatory elements (such as enhancers and promoters) at a distance can be examined in both animal models and cells in culture. BAC transfection, therefore, can be used to study gene regulation in much the same manner as plasmid transfection, but provides the added value of examining dispersed regulatory elements in their native organization. Because BACs may be stably integrated into chromosomes, BAC transfection also can be used to study other aspects of the genome, such as the organization of chromosomal loops or replicons.

There are a multitude of factors involved in achieving high transfection efficiency and long-term expression of transfected DNA in mammalian cell culture. Although methods for achieving high transfection efficiency with plasmids have been refined over many years, these protocols are rarely effective with BACs owing to their large size. BAC transfection protocols require extensive optimization for each cell line under investigation. In addition to cell type, the three most important parameters have been identified that influence the efficiency of BAC transfection are DNA purity, the method used to prepare the transfection complexes, particularly regarding to the ratio of DNA to transfection reagent, and the length of time required for gene expression after transfection *(2)*. In many instances condensation of BACs through DNA looping by a protein cofactor has proven useful in the preparation of transfection complexes *(3)*.

BAC DNA is maintained as a circular supercoiled plasmid at one to two copies in *Escherichia coli*, and therefore, can be purified using equilibrium sedimentation in cesium chloride *(1)*. In our experience, commercial purification methodologies using chromatography columns must be modified in order to obtain yields of clean BAC DNA that approach 30–80 µg for 250 mL of saturated culture. Because the quality of the BAC DNA greatly influences the formation of transfection complexes, and therefore, transfection efficiency in cultured cells, using high-quality BAC DNA to optimize transfection protocols for each cell line of interest were recommended. BAC DNA quality can be examined by pulse field or regular electrophoresis *(4)*, atomic force microscopy *(3)*, or spectrometry, although A260/A280 ratios often are not a reliable measure of purity.

The ratio of BAC DNA to transfection reagent (usually a cationic lipid) is critical in achieving reasonable transfection efficiencies. In the initial experiments with CHOC 400 cells, the quantities of BAC DNA and cationic lipid were simply scaled up from amounts suggested for plasmid transfections, and this approach resulted in transfection efficiencies of 30–50%. When these same conditions were applied to osteosarcoma Saos-2 cells, massive cell death was observed. Upon microscopic examination of these and other transfected cells stained with 4′, 6-diamidino-2-phenylindole (DAPI), observed that extensive aggregation of transfection complexes induces cytotoxicity and reduces transfection efficiencies in some cell types, in agreement with previous observations *(5)*.

Contrary to the conditions used with CHOC 400 cells, Saos-2 cells required use of a different cationic lipid and substantially reduced amounts of BAC DNA *(2)*. A series of titration experiments were performed using BAC DNA with varying ratios of cationic lipid. Transfection of 3×10^5 cells in a 12-well dish with 1.8 μg B16-25 DNA and 1.8–2.7 μL Lipofectamine™ 2000 (1:1 and 1:1.5 ratio DNA: Lipofectamine 2000) resulted in the highest transfection efficiency and low-levels of aggregates. The amount of Lipofectamine 2000 required for BAC transfection was much lower than that recommended by the manufacturer for plasmid transfection. Atomic force microscopy indicated that as the amount of BAC DNA and lipid increased, higher order aggregates formed owing to inter-molecular linking of the lipoplexes *(2)*, which inversely affected the transfection efficiency and resulted in cell death. Based on the previous observations of Oberle et al. *(6)* and Almofti et al. *(7)* excess DNA could nucleate aggregate formation, or excess Lipofectamine 2000 may encourage aggregation of lipoplexes by charge neutralization. Although it may be time-consuming to determine the optimum ratio of DNA to lipid for each cell line, once conditions are established to find the efficiency of BAC transfection is highly reproducible from experiment to experiment.

Although cell type clearly influences the efficiency of BAC transfection; whether variability is owing to inefficient uptake of lipoplexes, sensitivity to aggregation of lipoplexes over time, or if certain cell lines possess mechanisms to eliminate foreign DNA from the nucleus, as suggested by Coonrod et al. *(8)* was not determined. Despite the problems of low transfection efficiency experienced with certain cell lines (such as Saos-2), cell lines from human, mouse, and hamster can be transfected with BACs, so long as the formation of the lipoplexes has been optimized.

Another consideration when using BACs to transfect mammalian cells for the analysis of gene expression is the time required for expression from the transfected BAC. Compared with cDNA constructs driven by viral promoters, the detection of mRNA or protein expression from transfected BACs appears to require significantly more time. Analysis of expression of green fluorescent protein in Saos-2 cells transfected with B16-25, a BAC that contained a CMV–EGFP marker and the human *TP53* gene, showed varying intensities of fluorescence after 48 h, whereas the expression of p53 could only be detected after 5 d. No doubt expression levels are influenced by the strength of the gene promoter (i.e., cytomegalovirus [CMV] vs the endogenous *TP53* promoter) and the number of copies of BAC DNA per transfected cell, which in stable cell lines ranges from 2 to 15 copies *(9)*. Because previous studies have shown that the efficiency of replication of plasmid DNA in transfected cells increases as a function of size *(10)*, the large size of BACs may encourage their replication and segregation, thereby increasing levels of gene expression with cell division. This notion is supported by the observations of Tseng et al. *(11)* reported that efficient entry of plasmid DNA into the nucleus is only achieved after breakdown of the nuclear membrane during cell proliferation.

When cotransfected with marker BACs that provide for selection, such as dihydrofolate reductase (*Dhfr*) or hypoxanthine guanine phosphoribosyl transferase (*Hprt*), BACs of interest become cointegrated with the marker BAC into chromosomes and therefore are very stable over time *(9)*. Moreover, the recent studies show that when cells are cotransfected with *Dhfr* and BACs of interest, selection in methotrexate can be used to amplify the cointegrated BACs, allowing one to increase gene copy number substantially and study gene dosage effects.

2. Materials

2.1. BAC DNA Preparation

1. Chloramphenicol: dissolve chloramphenicol to a final concentration of 12.5 mg/mL in 100% ethanol and store at −20°C.
2. Luria bertanii (LB): 10 g bacto-tryptone, 5 g bacto-yeast extract, and 10 g NaCl. Autoclave at 121°C for 20 min. Add chloramphenicol to a final concentration of 12.5 µg/mL before use.
3. LB plates: 10 g bacto-tryptone, 5 g bacto-yeast extract, 10 g NaCl, and 15 g bacto-agar/1 L. Autoclave at 121°C for 20 min. Add chloramphenicol to a final concentration of 12.5 µg/mL after medium has cooled to 65°C and then pour plates. Once plates have solidified, store at 4°C.
4. 1000-mL Triple-baffled Erlenmeyer shaker flask (*see* **Note 2**) (Bellco Glass, Vineland, NJ).
5. Bacterial shaker at 37°C.
6. High purity plasmid purification columns and solutions (*see* **Note 3**) (Marligen Biosciences, Inc., Ijamsville, MD).
7. TE: 10 m*M* Tris-HCl pH 8.0 and 0.1 m*M* EDTA.
8. Water bath at 65°C.
9. Spectrophotometer or fluorometer.

2.2. Transient BAC Transfection

1. Dulbecco's modified Eagle's medium (DMEM; Invitrogen, Carlsbad, CA).
2. Fetal bovine serum (FBS; Hyclone, Logan, UT).

3. Antibiotics.
4. BAC DNA at 0.5–1 µg/µL.
5. GST-Z2: a glutathione-*S*-transferase (GST) fusion protein containing three zinc fingers and the proline region of RIP60 (*see* **ref. 12**, available from N. H. Heintz).
6. 10X GST-Z2 binding buffer: 200 mM Tris-HCl pH 7.5, 50 mM KCl, 50 mM MgCl$_2$, 10 mM β–mercaptoethanol, 20 mM ZnCl$_2$.
7. LipofectamineJ, PlusJ Reagent, LipofectamineJ 2000 and Opti-MEM[7] I medium (Invitrogen).
8. 37°C incubator with 5% CO$_2$.

2.3. BAC Cotransfection

1. α-Modification of Eagle's medium (αMEM) without ribosides and deoxyribosides (Sigma Chemicals, St. Louis, MO).
2. FBS (Hyclone).
3. Antibiotics.
4. Growth medium: αMEM, 10% FBS, 2 mM glutamine, 1 µg/mL adenosine, 1 µg/mL thymidine.
5. BAC DNA at 0.5–1 µg/µL.
6. GST-Z2: GST fusion protein containing three zinc fingers and the proline region of RIP60 *(12)*.
7. 10X GST-Z2 binding buffer: 200 mM Tris-HCl pH 7.5, 50 mM KCl, 50 mM MgCl$_2$, 10 mM β-mercaptoethanol, 20 mM ZnCl$_2$.
8. LipofectamineJ, PlusJ Reagent, and Opti-MEM[7] I medium (Invitrogen).
9. 37°C Incubator with 5% CO$_2$.

2.4. Generation of Stable Cell Lines by BAC Transfection

1. α-Modification of Eagle's medium (αMEM) without ribosides and deoxyribosides (Sigma Chemicals).
2. FBS (Hyclone).
3. Antibiotics.
4. Growth medium: αMEM, 10% FBS, 2 mM glutamine, 1 µg/mL adenosine, 1 µg/mL thymidine.
5. *Dhfr* BAC clone 163L20 from the RPCI-23 mouse BAC library (http://bacpac.chori.org).
6. BAC DNA at 0.5–1 µg/µL.
7. GST-Z2: GST fusion protein containing three zinc fingers and the proline region of RIP60 *(12)*.
8. 10X GST-Z2 binding buffer: 200 mM Tris-HCl pH 7.5, 50 mM KCl, 50 mM MgCl$_2$, 10 mM β-mercaptoethanol, 20 mM ZnCl$_2$. Stable at 4°C for 1 mo.
9. LipofectamineJ, PlusJ Reagent, and Opti-MEM[7] I medium (Invitrogen).
10. 10% Dialyzed FBS (Invitrogen).
11. *Dhfr* selection medium: αMEM, 10% dialyzed FBS, 2 mM glutamine.
12. 37°C Incubator with 5% CO$_2$.

2.5. Amplification of Gene Copy Number by Methotrexate

1. The reagents used in the generation of stable cell lines by BAC transfection.
2. Methotrexate (+)-Amethopterin (Sigma Chemicals). A 2-mM stock solution of methotrexate is made in α-modification of Eagle's medium (α-MEM) without serum and stored at −20°C. Dilutions are made in α-MEM and stored at 4°C for 1 mo.

3. Methods

3.1. BAC DNA Preparation

1. Inoculate an isolated bacterial colony containing the BAC of interest into 5 mL LB plus 12.5 μg/mL chloramphenicol (*see* **Note 1**) and incubate at 37°C with shaking at 225 rpm for 4–6 h. Use this culture to inoculate 250 mL LB plus 12.5 μg/mL chloramphenicol medium contained in a triple-baffled Erlenmeyer flask (*see* **Note 2**). Incubate the flask at 37°C with shaking at 225 rpm for 16–20 h.
2. BAC DNA is isolated using a Marligen High Purity plasmid purification maxiprep column and the high purity BAC protocol suggested by the manufacturer (*see* **Notes 3–7**).
3. The BAC DNA pellet is air-dried for 7–10 min and 40–100 μL prewarmed (65°C) TE is added. The DNA is allowed to begin resuspending by incubation at 65°C for 10 min. The DNA is then gently mixed by pipetting, and placed at 4°C overnight to several days to fully resuspend (*see* **Note 8**).
4. To determine BAC DNA concentration, the DNA is removed from 4°C, gently mixed, and centrifuged for 2 min at maximum speed to remove resin or other debris. The supernatant is transferred to a new tube and concentration is determined by absorbance spectrophotometry at 260 and 280 nm or fluorometry. The BAC DNA is diluted with TE to a final concentration of 0.5–1 μg/μL and stored at 4°C (*see* **Note 9**). A typical BAC DNA yield from a 250 mL culture is 30–80 μg.

3.2. Transient BAC Transfections

3.2.1. BAC Transfection Using GST-Z2, LipofectamineJ, and PlusJ Reagent

All amounts are per 35-mm dish and can be scaled up or down proportionally based on culture dish surface area.

1. Plate cells in a 35-mm dish such that they are approx 50–75% confluent on the following day (*see* **Notes 10–12**).
2. On the day of transfection, combine up to 1 μg BAC DNA and up to 1 μg GST-Z2 in 1X GST-Z2 binding buffer to a final volume of 40 μL, mix gently. Incubate at room temperature for 15–20 min (*see* **Notes 13** and **14**).
3. Dilute the DNA/GST-Z2 mixture with 100 μL Opti-MEM[7] I medium. Add 2.5 μL PlusJ reagent, mix, and incubate at room temperature for 10 min.
4. In another tube dilute 4 μL LipofectamineJ in 100 μL Opti-MEM[7] I medium.
5. Combine the diluted DNA/GST-Z2/PlusJ and LipofectamineJ, mix gently and incubate at room temperature for 20 min.

6. Remove the culture media from the cells and replace with 800 μL Opti-MEM[7] I medium.

7. Add the 200 μL of transfection complexes from **step 5** to the cells in a dropwise fashion and rock the dish gently to mix. Incubate the dish at 37°C in a CO_2 incubator for 3 h (*see* **Note 15**).

8. After 3 h incubation, add 1 mL of DMEM, 10% FBS and +/− antibiotics medium and return the dish to the 37°C CO_2 incubator for 1–5 d (*see* **Notes 16–18**).

3.2.2. BAC Transfection Using LipofectamineJ and PlusJ Reagent

All amounts are per 35-mm dish and can be scaled up or down proportionally based on culture dish surface area.

1. The day before the experiment, seed the cells in a 35-mm dish in antibiotic-free medium so that they are approx 50–80% confluent on the day of transfection (*see* **Notes 10–12**).

2. On the day of the transfection resuspend 0.2–5 μg BAC DNA, 6–15 μL PlusJ Reagent in DMEM without serum to a final volume of 100 μL (*see* **Note 13**). Incubate the reaction at room temperature for 15 min.

3. In a separate tube dilute 4–12 μL LipofectamineJ in a final volume of 100 μL DMEM.

4. Gently mix the diluted LipofectamineJ and the DNA/PlusJ together and incubate at room temperature for an additional 15 min.

5. During the incubation remove the medium from the cells and replace with 1 mL serum-free DMEM.

6. Add the transfection complexes to the cells and mix by moving the dish back and forth and side-to-side. Return the dish to the incubator and allow the complexes to attach for 5–8 h (*see* **Note 15**).

7. At the end of the incubation period, remove the transfection medium and replace with DMEM supplemented with 10% FBS with or without antibiotics. Return the dish to a 37°C CO_2 incubator (*see* **Notes 16** and **17**).

8. Culture the cells for 1–5 d before analysis (*see* **Note 18**).

3.2.3. BAC Transfection Using LipofectamineJ 2000

All amounts are per 12-well plate and can be scaled up or down proportionally based on culture dish surface area.

1. Plate cells in antibiotic-free medium in a 12-well plate so that they are approx 80–90% confluent on the day of transfection (*see* **Notes 10–12**).

2. On the day of transfection in a 1.5-mL tube, dilute 1.8–3 μg BAC DNA in Opti-MEM[7] I medium to a final volume of 100 μL (*see* **Note 13**). Incubate at room temperature for at least 5 min.

3. Varying concentration of LipofectamineJ 2000 were diluted with Opti-MEM[7] I to a final volume of 100 μL to achieve a ratio of DNA weight to LipofectamineJ 2000 volume that ranged from 1:1 to 1:3 (*see* **Note 17**).

4. The diluted BAC DNA and LipofectamineJ 2000 were gently mixed together and incubated at room temperature for 20 min.
5. During the incubation remove the medium from the cells and replaced it with 1 mL serum-free DMEM.
6. Transfection complexes were added to the cells and the plate was returned to the 37°C CO_2 incubator for 4 h (*see* **Note 15**).
7. At the end of 4 h, the transfection medium was replaced with DMEM, 10% FBS and antibiotics and the cells were maintained under standard conditions until analysis (*see* **Notes 16** and **18**).

3.3. BAC Cotransfection

All amounts are per six-well plate and can be scaled up or down proportionally based on culture dish surface area.

1. The day before the experiment seed the cells in a six-well plate in antibiotic-free medium so that they are approx 50% confluent on the day of transfection (*see* **Notes 10–12**).
2. On the day of the transfection combine the following reagents in order in a 1.5-mL tube: dH_2O, 10X GST-Z2 binding buffer, BAC DNA (0.1–1 μg) and GST-Z2 (1 μg for every 1 μg of BAC DNA) to a final volume of 25 μL (*see* **Notes 13** and **14**). Incubate at room temperature for 15 min.
3. After the incubation add 100 μL Opti-MEM[7] I to the BAC DNA/GST-Z2 mixture. Before use, mix PlusJ reagent. Add 2.5 μL of PlusJ reagent to the diluted DNA, mix and incubate at room temperature for 10 min.
4. In another 1.5-mL tube combine 100 μL Opti-MEM[7] I and 3.9 μL LipofectamineJ.
5. To prepare the transfection complexes combine the diluted LipofectamineJ (from **step 4**) and the BAC DNA/GST-Z2/ PlusJ mix (from **step 3**). Pipet up and down several times to gently mix. Incubate at room temperature for 20 min.
6. During the incubation, remove the medium from the cells and replace with 800 μL Opti-MEM[7] I per well.
7. Add the transfection complexes (from **step 5**) to the cells in a dropwise fashion. Mix the plate back and fourth 10 times, right to left 10 times and repeat. Do not swirl to mix. Return the plate to 37°C and incubate for 3 h (*see* **Note 15**).
8. After 3 h add 1 mL of complete medium with serum and return the plate to 37°C CO_2 incubator until analysis (*see* **Notes 16** and **18**).

3.4. Generation of Stable Cell Lines by BAC Transfection

1. Choose a cell line that is *Dhfr⁻* or *Hprt⁻* (*see* **Note 19**).
2. Follow the protocol for BAC Cotransfection with the following exceptions: use 0.1 μg *Dhfr* BAC DNA and up to 0.9 μg BAC DNA of interest (*see* **Note 20**).
3. To select for *Dhfr⁺* cells, the day after the transfection change the medium to selection medium.
4. Continue to incubate the cells in selection medium for 1–2 mo changing the medium on the cells twice a week. Cells that are *Dhfr⁺* will form colonies that can be picked or pooled and cells that lack *Dhfr* will die (*see* **Notes 21** and **22**).

3.5. Amplification of Gene Copy Number Using Methotrexate

1. Create a stable cell line by cotransfection of a *Dhfr* BAC and a BAC of interest using the protocol above. Select for *Dhfr* positive cells.
2. To amplify the *Dhfr* gene incubate the cells in selection medium with 50 nM methotrexate.
3. Each time the cells are split increase the methotrexate concentration (*see* **Notes 23** and **24**).
4. Maintain the cells in methotrexate to retain the increased copy number (*see* **Notes 23** and **25**).

4. Notes

1. Do not use an enriched medium, such as SOB or TB.
2. The use of a triple-baffled flask for the 250 mL overnight grow allows for optimum aeration and increased BAC DNA yields.
3. The RNase A concentration in E1 is increased to 400 µg/mL, the E2 solution is prepared immediately before use from 10 N NaOH and 10% SDS stocks, and the NaCl concentration of E5 is increased to 0.9 M. For 250 mL overnight growth, use 40 mL each of E1, E2, and E3, respectively.
4. On the addition of E2, it is crucial to mix the solution thoroughly too adequately release the BAC DNA. Generally this requires a more pronounced arm movement (a 45°C arch) that is repeated four to six times. The solution is then swirled another three to four times during the incubation period to ensure a homogenous mixture.
5. The centrifuge bottle is swirled on the addition of E3 and then the bottle is closed and mixed by inversion several time until a fine "snowflake" precipitate is visible.
6. To increase BAC DNA yields, store isopropanol precipitations at 4°C overnight before proceeding to the 30 min centrifugation step.
7. Because BAC DNA pellets can be quite small after the isopropanol precipitation and centrifugation, the supernatant is poured off the pellet and 5 mL 70% ethanol is added and the Oakridge tube is swirled to dislodge the pellet. 1.5 mL of the resuspended pellet/ethanol is transferred to a 1.5-mL microfuged tube and centrifuged at room temperature, maximum speed for 5 min. The ethanol is poured off and another 1.5 mL of the resuspended pellet/ethanol is added and the centrifugation is repeated. These steps are repeated until all the contents of the bottle are transferred to the 1.5-mL tube. The bottle is rinsed with an additional 1 mL of 70% ethanol, transferred to the 1.5-mL tube and centrifuged. This final wash is poured off, the tube is quickly spun and the remaining supernatant is removed with a pipet.
8. Because BAC DNA is such a large molecule it is extremely important to allow sufficient time for the DNA to resuspend completely.
9. BAC DNA concentrations of less than 1 µg/µL store better at 4°C than higher concentrations, which tend to become viscous and unusable after several months.
10. The cells used for BAC transfection experiments must be in log phase, meaning that they have been split within the last 24 h or new media was added the day before. Do not use cells from a confluent culture for BAC transfection experiments.
11. The percentage of confluency on the day of transfection must be empirically determined for each cell line to achieve the highest efficiency.

12. To achieve high transfection efficiency, the cells must be spread evenly across the surface of the culture vessel.

13. All reagents used during the transfection must be at room temperature (not just taken out of 4°C). All mixing is done by flicking the bottom of the tube, no vortexing.

14. No precipitate should be visible after mixing the DNA and GST-Z2. A total of 0.1–1 μg of BAC DNA can be used, but at higher DNA concentrations each batch of GST-Z2 needs to be optimized.

15. When transporting the cells to the incubator it is important to make sure that the vessels are not swirled.

16. After the initial transfection period and before the complete media is added to the cells, examine the transfection dishes/wells microscopically. There should be no visible precipitate and the cells should appear healthy.

17. All transfection reagents must be optimized for each cell type. Use the ranges suggested by the manufacturer as a starting point. With transfection of BAC DNA, higher DNA concentrations result in higher cell toxicity and lower transfection efficiency are noted.

18. Because the copy number of the transfected BAC in cultured cells is much lower than for a plasmid transfection, analysis of the transfection should be delayed so that there is adequate expression from the BAC.

19. CHO-DUKX are a *Dhfr*$^{-/-}$ derivative of CHO cells *(13)* and A9 mouse fibroblasts are *Hprt*$^{-/-}$ (ATCC, cat no. CCL-1.4).

20. By varying the ratio of the BAC of interest to the *Dhfr* BAC during cotransfection, one can obtain a range of copy number in the stable cell lines, which provides a means of examining gene dosage effects.

21. CHO-DUKX cells positive for *Dhfr* expression initially appear needle-like and form an aster-shaped colony; whereas parental *Dhfr*$^-$ cells appear viable and flat, but will not grow on replating. Stable *Dhfr*$^+$ cell lines generally grow faster than the parental cells.

22. Examination of stable cell lines by fluorescent *in situ* hybridization indicates that the *Dhfr* BAC and BAC of interest integrate into a single site in the genome.

23. The methotrexate concentration should be increased so that only 10–50% of the cells survive. Maintain two sets of plates; one on a high dosage of methotrexate and the other on a lower dose as a backup. Include a control plate of cells that do not contain the *Dhfr* BAC to confirm that the methotrexate selection is working. In CHO-DUKX cells containing *Dhfr*, gene amplification is slow and stepwise for several weeks and then proceeds more rapidly.

24. The BAC of interest may not coamplify with the *Dhfr* gene on exposure to methotrexate, suggesting gene dosage effects. One important consideration is that amplification of the BAC of interest may be toxic to cells, or cellular mechanisms may limit the degree of gene amplification.

25. CHOC 400 cells were selected for resistance to methotrexate by increasing doses of the drug gradually over 3–6 mo, reaching a final concentration of 400 μ*M*. At this level of drug the *Dhfr* gene copy number has been estimated at 500–1000 copies.

References

1. Shizuya, H., Birren, B., Ung-Jin, K., et al. (1992) Cloning and stable maintenance of 300-kilobase-pair fragments of human DNA in *Escherichia coli* using an F-factor-based vector. *Proc. Natl. Acad. Sci. USA* **89**, 8794–8797.
2. Montigny, W. J., Phelps, S. F., Illenye, S., and Heintz, N. H. (2003) Parameters influencing high-efficiency transfection of bacterial artificial chromosomes into cultured mammalian cells. *BioTechniques* **35**, 796–807.
3. Montigny, W. J., Houchens, C. R., Illenye, S., et al. (2001) Condensation by DNA looping facilitates transfer of large DNA molecules into mammalian cells. *Nucleic Acids Res.* **29**, 1982–1988.
4. Gama, S. M., De Gasperi, R., Wen, P. H., et al. (2002) BAC and PAC DNA for the generation of transgenic animals. *BioTechniques* **33**, 51–53.
5. Lawrence, M. J. (1994) Surfactant systems: microemulsions and vesicles as vehicles for drug delivery. *Eur. J. Drug Metab. Pharmacokinet.* **19**, 257–269.
6. Oberle, V., Bakowsky, U., Zuhorn, I. S., and Hoeskstra, D. (2000) Lipoplex formation under equilibrium conditions reveals a three-step mechanism. *Biophys. J.* **79**, 1447–1454.
7. Almofti, M. R., Harashima, H., Shinohara, Y., Almofti, A., Baba, Y., and Kiwada, H. (2003) Cationic liposome-mediated gene delivery: biophysical study and mechanism of internalization. *Arch. Biochem. Biophys.* **410**, 246–253.
8. Coonrod, A., Li, F. Q., and Horwitz, M. (1997) On the mechanism of DNA transfection: efficient gene transfer without viruses. *Gene Ther.* **4**, 1313–1321.
9. Illenye, S. and Heintz, N. H. (2004) Functional analysis of bacterial artificial chromosomes in mammalian cells: mouse Cdc6 is associated with the mitotic spindle apparatus. *Genomics* **83**, 66–75.
10. Heinzel, S. S., Krysan, P. J., Tran, C. T., and Calos, M. P. (1991) Autonomous DNA replication in human cells is affected by the size and the source of the DNA. *Mol. Cell Biol.* **11**, 2263–2272.
11. Tseng, W. C., Haselton, F. R., and Giorgio, T. D. (1999) Mitosis enhances transgene expression of plasmid delivered by cationic liposomes. *Biochim. Biophys. Acta* **1445**, 53–64.
12. Houchens, C. R., Montigney, W. J., Zeltser, L., Dailey, L., Gilbert, J. M., and Heintz, N. H. (2000) The *dfhr* oriβ-binding protein RIP60 contains 15 zinc fingers: DNA binding and looping by the central three fingers and an associated proline-rich region. *Nucleic Acids Res.* **28**, 570–581.
13. Urlaub, G. and Chasin, L. A. (1980) Isolation of Chinese hamster cell mutants deficient in dihydrofolate reductase activity. *Proc. Natl. Acad. Sci. USA.* **77**, 4216–4220.

11

Molecular Cytogenetic Applications in Analysis of the Cancer Genome

Pulivarthi H. Rao, Subhadra V. Nandula, and Vundavalli V. Murty

Summary

Cancer cells exhibit nonrandom and complex chromosome abnormalities. The role of genomic changes in cancer is well established. However, the identification of complex and cryptic chromosomal changes is beyond the resolution of conventional banding methods. The fluorescence microscopy afforded by imaging technologies, developed recently, facilitates a precise identification of these chromosome alterations in cancer. The three most commonly utilized molecular cytogenetics methods comparative genomic hybridization, spectral karyotype, and fluorescence *in situ* hybridization, that have already become benchmark tools in cancer cytogenetics, are described in this chapter. Comparative genomic hybridization is a powerful tool for screening copy-number changes in tumor genomes without the need for preparation of metaphases from tumor cells. Multicolor spectral karyotype permits visualization of all chromosomes in one experiment permitting identification of precise chromosomal changes on metaphases derived from tumor cells. The uses of fluorescence *in situ* hybridization are diverse, including mapping of alteration in single copy genes, chromosomal regions, or entire chromosomes. The opportunities to detect genetic alterations in cancer cells continue to evolve with the use of these methodologies both in diagnosis and research.

Key Words: Cancer; chromosome aberrations; comparative genomic hybridization; fluorescence *in situ* hybridization; spectral karyotyping; molecular cytogenetics.

1. Introduction

The cancer cell accumulates a number of genetic changes during the process of tumor initiation and progression. In order to understand the fundamental biology of cancer, it is essential to identify as many of these genetic alterations during cancer development as possible. Theodor Boveri's hypothesis in 1914 that chromosome abnormalities play a central role in transformation of normal cell to neoplastic cell laid the foundation for the

From: *Methods in Molecular Biology, vol. 383: Cancer Genomics and Proteomics: Methods and Protocols*
Edited by: P. B. Fisher © Humana Press Inc., Totowa, NJ

current understanding of chromosomal basis of cancer. Therefore, it is critical to identify chromosomal changes in the tumor genome to accrue knowledge of molecular mechanisms of the neoplastic process.

Conventional cytogenetic analysis is one of the most widely accepted DNA genome-screening tools for identification of chromosomal aberrations in cancer cells. However, there are two built-in limitations to conventional cytogenetic analysis. One limitation is representation. The success rate of detection of clonally abnormal karyotypes in short-term cultures of most types of tumors is low. Such a low cytogenetic success rate introduces a bias in representation of true genetic aberration in the tumor genome. The other limitation is complexity. In addition to recognizable chromosomal aberrations, most tumors exhibit marker chromosomes whose derivation is difficult to determine by conventional cytogenetic analysis, which introduces another level of bias. To overcome these limitations a host of molecular cytogenetic techniques such as fluorescence *in situ* hybridization (FISH), comparative genomic hybridization (CGH), multicolor spectral karyotype (SKY)/multicolor FISH (m-FISH), and array-based CGH have recently been developed *(1–9)*. In this chapter, detailed methodologies and limitations of three of the most commonly used molecular cytogenetic methods in the analysis of the cancer genome are discussed **(Table 1)**.

2. Materials

2.1. Preparation of Human Lymphocyte Metaphase Chromosomes

1. RPMI medium 1640 (GIBCO, Invitrogen, Carlsbad, CA; cat. no. 21870-076).
2. L-Glutamine 200 m*M* (GIBCO, cat. no. 25030-081), 100 mL (make 5-mL aliquots; add 5–500 mL medium).
3. Penicillin/streptomycin (GIBCO, cat. no. 15070-063) (make 5-mL aliquots; add 5–500 mL medium).
4. Fetal bovine serum qualified, heat inactivated (GIBCO, cat. no. 16140-071).
5. Phytohaemagglutinin (PHA) lyophilized (GIBCO, cat. no. 10576-015). (Rehydrate with 10 mL of sterile double-distilled [dd]H_2O and store at [20°C].)
6. Colcemid, KaryoMAX Colcemid (10 µg/mL) (GIBCO, cat. no. 15210-040).

2.2. Preparation of Labeled DNA Probes for CGH

1. 10X Nucleotide mix (*see* **item 31**).
2. 25 µ*M* dTTP (Roche Diagnostics, Indianapolis, IN; cat. no. 105 1482).
3. 25 µ*M* dATP (Roche Diagnostics, cat. no. 105 1440).
4. 25 µ*M* dGTP (Roche Diagnostics, cat. no. 105 1466).
5. 25 µ*M* dCTP (Roche Diagnostics, cat. no. 105 1458).
6. Fluorescein 12-dUTP (Dupont NEN, Boston, MA; cat. no. NEL-413).
7. Texas Red 5-dUTP (Dupont NEN, cat. no. NEL-417).
8. DNA polymerase I (0.5 U/µL/0.4 mU/µL DNase I) (Invitrogen, cat. no. 18162016).
9. DNA polymerase I (10 U/µL) (Invitrogen, cat. no. 18010025).

Table 1
Advantages and Limitations of Various Molecular Cytogenetic Techniques

Fluorescent *in situ* hybridization (FISH)	Comparative genomic hybridization (CGH)	Spectral karyotyping (SKY)
Advantages		
Highly specific and sensitive	Global view of chromosomal losses, gains, and amplifications	Detects complex and cryptic chromosomal translocations
Mapping of genes to chromosomes	No need for chromosomes from clinical samples and requires only a small amount of DNA (~2 µg)	Defines the origin of markers, hsrs and double minute chromosomes. Sensitive to detect deletions 10 to 15 mb
Detects deletions, inversions, numerical changes, and translocations in interphase nuclei and metaphase chromosomes	Sensitive to detect deletions 10–15 mb	Detects the chromosomal boundaries in interphase nuclei
Sensitive to detect deletions >1 kb		
Limitations		
Highly focused, prior knowledge of gene(s) is required	Unable to detect balanced translocations and inversions	Cannot detect inversions and duplications of specific band(s) and subtle deletions
	Detects abnormalities only if the tumor cells are greater than 50% of the specimen	

10. DNase I type II (Sigma, St. Louis, MO; cat. no. D4527).
11. 2-Mercaptoethanol (Sigma, cat. no. M6250).
12. Agarose (Sigma, cat. no. A9539).
13. G-50 Sephadex columns (Roche Diagnostics, cat. no. 1523023).
14. Pepsin (Sigma, cat. no. P6887).
15. Human Cot-1 DNA (Invitrogen, cat. no. 15279-011).
16. Lambda DNA/Hind III marker (Promega, Madison, WI; cat. no. G1711).
17. Formamide (Fisher, cat. no. F84-1).
18. 4′,6-diamidino-2-phenylindole (DAPI) (Sigma, cat. no. D-9542).
19. VECTASHIELD mounting medium (Vector Laboratories, Burlingame, CA; cat. no. H-1000).
20. Diamond pencil.
21. Humidifier boxes.

22. Rubber sealant.
23. Water baths (37, 45, and 75°C).
24. Circulating water bath (15°C) or PCR thermal cycler.
25. Glass slides, 22-mm^2 and 24 × 50-mm^2 cover slips.
26. Vertical Coplin jars.
27. DAPI Stock solution (100 μg/mL):
    a. Dissolve 1 mg DAPI in 10 mL of ddH$_2$O. Add few drops of methanol to dissolve DAPI before adding ddH$_2$O. Store at −20°C.
    b. Working solution: add 1–2 μL of DAPI stock solution to 1 mL of antifade mounting medium (VECTASHIELD mounting medium). Aliquot into small brown tubes and store at −20°C.
28. Denaturation solution (pH 7.0):
    a. 350 mL Formamide.
    b. 50 mL 20X SSC.
    c. 100 mL ddH$_2$O.
    d. Stir well and adjust pH to 7.0 if necessary with 1 *N* HCl. Store at 4°C.
29. DNase working solution:
    a. Stock solution: dissolve 3 mg of DNase in 1 mL of 0.15 *M* NaCl and 50% glycerol. Aliquot 100-μL volumes into small tubes and store at −20°C.
    b. Working solution: add 1 μL stock solution to ice cold 500 μL ddH$_2$O immediately before use, discard after each use (*see* **Note 1**).
30. Master hybridization buffer (pH 7.0):
    a. 5 mL Formamide.
    b. 1 mL 20X SSC.
    c. 1 gm Dextran sulfate.
    d. Make up to 10 mL with ddII$_2$O.
    e. Filter through 0.22-μ filter and store at −20°C.
31. 10X Nucleotide mix:
    a. 100 μL 10 m*M* dATP.
    b. 100 μL 10 m*M* dGTP.
    c. 100 μL 10 m*M* dCTP.
    d. 2.5 mL Tris-HCl (pH 7.2).
    e. 250 μL 1 *M* MgCl$_2$.
    f. 34 μL 14.7 *M* 2-Mercaptoethanol.
    g. 50 μL 10 mg/mL Bovine serum albumin.
    h. 1.866 mL ddH$_2$O. Store at −20°C.
32. Pepsin:
    a. Stock solution (10%) = 100 mg/mL, dissolve in sterile water, make 50-μL aliquots, store at −20°C. Mix 20 μL pepsin in 100 mL prewarmed 0.01 *N* HCl and adjust pH to 2.0. Incubate slides at 37°C for 5–10 min and wash slides in 1X PBS and 1X PBS/MgCl$_2$ Store at −20°C.
33. Washing solution (pH 7.0):
    a. 250 mL Formamide.
    b. 50 mL 20X Saline sodium citrate (SSC).

   c. 200 mL ddH$_2$O.

   d. Stir well and adjust pH to 7.0 if necessary with 0.1 N HCl. Store at 4°C.

34. Common stock solutions (*10*):

   a. 3 M Sodium acetate pH 5.2. Dissolve 408.1 g of sodium acetate·3H$_2$O in 800 mL of H$_2$O. Adjust the pH to 5.2 with glacial acetic acid and sterilize by autoclaving.

   b. PBS. Add 8 g NaCl, 200 mg KCl, 1.44 g Na$_2$HPO$_4$ (dibasic, anhydrous), and 240 mg KH$_2$PO$_4$ (monobasic, anhydrous) to 1 L H$_2$O and adjust pH using HCl. Sterilize by autoclaving.

   c. 5 M Sodium chloride. Dissolve 292.2 g of NaCl in 800 mL of H$_2$O; adjust volume to 1 L with H$_2$O. Sterilize by autoclaving.

   d. 20X SCC. Dissolve 175.3 g of NaCl and 88.2 g of sodium citrate in 800 mL H$_2$O. Adjust the pH to 7.0 with NaOH and make up to 1 L with H$_2$O. Sterilize by autoclaving.

   e. 1 M Tris-HCl, pH 7.2. Add 121 g of Tris base to 900 mL H$_2$O. Adjust pH to 7.2 by adding 76 mL of HCl and adjust the volume to 1 L with H$_2$O. Sterilize by autoclaving.

   f. Nonidet P-40-sodium bicarbonate buffer (PN) buffer (pH 8.0): add 1 g sodium bicarbonate and 5 mL of Nonidet P-40 to 900 mL-dH$_2$O. Stir well and store at 4°C.

2.3. Tumor Chromosome Preparations for SKY and FISH

1. RPMI medium 1640 (GIBCO, cat. no. 21870-076).
2. L-Glutamine 200 mM (GIBCO, cat. no. 25030-081), 100 mL (make 5-mL aliquots; add 5–500 mL medium).
3. Penicillin/streptomycin (GIBCO, cat. no. 15070-063) (make 5-mL aliquots; add 5–500 mL medium).
4. Fetal bovine serum qualified, heat inactivate (GIBCO, cat. no. 16140-071).
5. Colcemid, KaryoMAX Colcemid (10 µg/mL) (GIBCO, cat. no. 15210-040).

2.4. Spectral Karyotyping

1. Formamide (Fisher, Pittsburg, PA; cat. no. F84-1).
2. DAPI (Sigma, cat. no. D-9542).
3. VECTASHIELD mounting medium (Vector Laboratories, H-1000).
4. Diamond pencil.
5. Humidifier boxes.
6. Rubber sealant.
7. Water baths (37, 45, and 75°C).
8. Formaldehyde (Sigma, cat. no. F1268).
9. Glass slides, 22-mm^2 and 24 × 50-mm^2 cover slips.
10. Pepsin (Sigma, cat. no. P6887).
11. 1X PBS (Invitrogen, cat. no. 20012-027).
12. 1 M MgCl$_2$ (Sigma, cat. no. M-1028).
13. PBD (PBS + Tween-20).
14. Human spectral karyotyping reagent (Applied Spectral Imaging, Visa, CA; cat. no. SKY 000029).

15. Mouse spectral karyotyping reagent (Applied Spectral Imaging, cat. no. SKY 000031).
16. Concentrated antibodies detection kit (CAD) (Applied Spectral Imaging, cat. no. SKY 000133).
17. Tween-20 (Sigma, cat. no. P 5927).
18. 0.01 M HCl: 1 mL 1 M HCl in 99 mL dH$_2$O. Prewarm to 37°C.
19. 4X SSC/Tween-20: add 100 mL 20X SSC and 0.5 mL Tween-20 in 400 mL dH$_2$O and heat for 30 min at 45°C.
20. PBS/MgCl$_2$: add 50 mL 1 M MgCl$_2$ to 950 mL dH$_2$O.
21. 1% Formaldehyde: add 2.7 mL of 37% formaldehyde to100 mL PBS/MgCl$_2$.
22. Pepsin stock:
    a. *See* for details **Subheading 2.2.**, **item 32**.
23. Denaturation solution (pH 7.0):
    a. *See* for details **Subheading 2.2.**, **item 28**.
24. Washing solution (pH 7.0):
    a. *See* for details **Subheading 2.2.**, **item 33**.

2.5. Preparation of Labeled DNA Probes for FISH

1. Spectrum Green-dUTP (50 nM) (Vysis, Des Plaines, IL; cat. no. 30-803200).
2. Spectrum Red-dUTP (50 nM) (Vysis, cat. no. 30-803400).
3. For other materials please refer to **Subheading 2.2.**

3. Methods

3.1. Comparative Genomic Hybridization

1. CGH is used in delineating chromosomal gains, losses, and amplifications of a tumor genome by mapping these changes to a normal metaphase. CGH is based on the competitive hybridization of differentially labeled normal (reference) and tumor (test) DNA *(1,11)*. The standard direct labeling protocol uses normal and tumor DNA that has been labeled by incorporation of flurochrome-conjugated nucleotides. An alternative method involves a slightly different procedure using normal and tumor DNA that have been labeled with biotin or dioxigenin. This latter method requires additional steps for detecting these labels with fluorochrome-conjugated secondary agents. The repetitive sequences present in both genomes are suppressed by adding an excess of unlabeled Cot-1 DNA to the hybridization mixture or copre-cipitating with the differentially labeled tumor and normal genomic DNA. Hybridized probes are detected with different colored fluorochromes (e.g., red and green). The ratio of green to red fluorescence intensities along the length of the chromosome reflects the relative amounts of DNA sequences in the test DNA. Measuring the ratio of color intensities along the length of the chromosomes will provide DNA copy number changes in the tumor genome. The labeled normal DNA included in the hybridization process serves as a control for regional variations in the ability to hybridize to the target chromosomes.
2. Although high-level amplifications are visible directly in a fluorescence microscope, a quantitative measurement of fluorescence intensity values based on digital image

analysis is crucial for a precise estimate of low copy-number changes. This analysis includes image acquisition of a green and red fluorescence with a charge-coupled device (CCD) camera. Using custom computer software, the painted chromosomes are then segmented and the florescence values determined perpendicular to the axis of the chromosome on a pixel-to-pixel basis. The result of measurement of the fluorescence values can now be visualized by means of a value table where certain colors refer to gains or losses in the tumor genome. The final step in a quantitative fluorescence measurement includes the calculation of average ratio profiles along the chromosomal axis based on data from a minimum of 10 chromosomes (5 metaphase spreads). The average ratio of 1.0 indicates equal copy numbers of the respective chromosomes, a ratio of 0.5 indicates a deletion of one homologue and a ratio of 1.5 reflects a trisomy in the genome. Gene amplification can be visualized as a peak fluorescence ratio of >2.0 and a localized signal to specific chromosome bands.

3.2. Preparation of Human Lymphocyte Metaphase Chromosomes

1. Prewarm 40 mL RPMI 1640 medium containing 20% fetal bovine serum (FBS), 1% L-glutamine, and 1% penicillin/streptomycin. Add 400 μL of PHA to the medium.
2. Collect 10 mL of whole blood in a heparin tube (green top).
3. Spin 10 mL of blood at 250g (1500 rpm) for 10 min or let sit at room temperature for 2–3 h.
4. Collect 2 mL of supernatant (lymphocyte layer/buffy coat), distribute to two to three T25 flasks or 15-mL tubes and culture for 72 h at 37°C (shake flasks once a day).
5. Add two drops of colcemid 2 h before harvesting using a 24-gauze needle (stock: 10 μg/mL).
6. Transfer to 15-mL tubes and centrifuge for 5 min at 250g (1500 rpm).
7. Remove supernatant and add 5 mL of prewarmed (37°C) 0.56% hypotonic solutions (KCI) to each tube drop-by-drop, tapping tubes, and add more KCI to a total volume of 10 mL.
8. Incubate at 37°C water bath for 15–20 min.
9. Centrifuge for 5 min at 1500 rpm (250g) and remove supernatant.
10. Add 2 mL of freshly prepared fixative (3 methanol: 1 acetic acid) per tube drop-by-drop by tapping tubes.
11. Add more fixative to total 10 mL/tube and let sit at room temperature for 1–2 h.
12. Centrifuge for 5 min at 1500 rpm (250g) and remove supernatant. Repeat **steps 11–12** for three times.
13. Resuspend in 1–2 mL of fixative each tube (adjust the cell concentration depending on the size of the pellet).
14. Drop suspension onto clean glass slides (dip slides in ethanol and wipe with a Kleenex tissue).
15. Check each batch of chromosome preparations for CGH and store at 4°C. (*see* **Note 2**).

3.3. Preparation of Labeled DNA Probes for CGH

High-molecular weight genomic DNA from a normal donor (reference DNA) and a tumor (test DNA) is required for CGH. Several DNA extraction kits are available commercially for genomic DNA preparation (QIAGEN, Promega). Degraded DNA should be avoided because it will yield probes that are too small upon nick translation, thereby resulting in poor-quality CGH (*see* **Note 3**).

The DNA extracted from microdissected tumor cells can be used for CGH. However, the DNA isolated from these cells might be insufficient for CGH analysis. Several whole genome amplification methods are currently available to generate the large quantities of DNA that are required for CGH and other genome screening methods *(12)*.

3.3.1. Probe Labeling

1. To label test or reference DNA by nick translation add the reagents in the following order:
   a. 1 µg Test DNA or reference DNA in 38 µL dd H_2O.
   b. 5 µL 10X dNTP mix.
   c. 1 µL dTTP mix.
   d. 1 µL Fluorescein isothiocynate (FITC) or Texas Red.
   e. 5 µL Enzyme mix containing DNA polymerase I/DNase I.
   f. 1 µL DNA polymerase I.
2. Mix contents of the tube well and incubate at 15°C for 1 h and 45 min. (The nick translation time needs to be adjusted depending on the size of the genomic DNA.)
3. Place tubes on ice, while keeping them protected from light.
4. Check the DNA fragment size by running a 1% agarose gel, using a 3-µL aliquot of the reaction.
5. Remove unincorporated nucleotides by running through a Sephadex G-50 column.
6. If fragment size is in the appropriate range (500- to 3000-bp), heat the tubes at 75°C for 10 min, to inactivate the enzymes (*see* **Note 4**).

3.3.2. DNA Purification

1. Allow the G-50 Sephadex column to warm up to room temperature. Gently invert it several times to resuspend the medium, while flicking it to remove any air bubbles.
2. Remove the top cap from the column, followed by the bottom tip. This sequence is absolutely necessary to avoid creating a vacuum, and uneven flow of buffer. Place on a collection tube. Allow the buffer to drain by gravity and then discard the elute.
3. Place the column and collection tube in a 15-mL centrifuge tube, and centrifuge for 3 min at 3000 rpm in a swinging bucket rotor. Discard the eluted buffer.
4. Keeping the column in an upright position, carefully apply the DNA sample to the center of the column bed.
5. Place the column on a second collection tube while keeping it in an upright position. Centrifuge for 3 min at 3000 rpm. Please note that speed and length of centrifugation should be calibrated for individual centrifuges.
6. Collect the elute from the collection tube. This contains the purified DNA sample.

3.3.3. Precipitation of Probe DNA for Hybridization

1. Add 40 µL FITC-labeled tumor DNA, 40 µL Texas Red-labeled normal reference DNA, 20 µL Human Cot-1 DNA, 10 µL 3 *M* sodium acetate, and 400 µL cold absolute ethanol.
2. Mix contents well and leave at −70°C for at least 1 h.
3. Centrifuge at 14,000 rpm for 30 min at 4°C.
4. Remove the supernatant. A good size pellet should be visible at the bottom of the tube.
5. Wash the pellet with 500 µL cold 70% ethanol.
6. Centrifuge at 14,000 rpm for 30 min at 4°C.
7. Remove the supernatant as thoroughly as possible. Air-dry pellet in the dark for at least 2 h, until completely dry.
8. Resuspend the pellet in 10 µL hybridization mixture. Mix thoroughly by tapping the bottom of the tube several times. Allow the pellet to dissolve over 2–3 h.

3.3.4. Slide Pretreatment

1. Pretreat all slides before use, for 1 h in 2X SSC at 37°C.
2. Rinse slides in distilled water.
3. Treat the slides in series of ethanol (70, 80, and 100% absolute ethanol) for 2 min each. Air-dry slides (*see* **Note 5**).

3.3.5. Slide Denaturation

1. Place the slides in denaturation solution (*see* **Subheading 2.2.**, **item 28**) for 2 min in a 74°C water bath in a Coplin jar. Denature only two slides at a time.
2. After denaturation, immediately place the slides in ice cold 70% ethanol. Wash the slides in 70, 80, and 100% ethanol for 2 min each. Air-dry the slides.

3.3.6. Probe Denaturation and Hybridization to Target Chromosomes

1. Denature the probe in a 74°C water bath for 6 min.
2. Upon removal from the water bath, immediately apply the probe onto the slide and place a 22-mm^2 cover slip over the probe.
3. Seal the edges of the cover slip with rubber cement. Up to two different hybridizations can be done on one slide, using half of the slide for each probe. It is useful to note that, sealing the middle of the slide first will prevent the cover slip from moving when applying the rubber cement.
4. Place the slides in a humidified chamber (Tupperware with paper towels moistened with formamide solution). Incubate the slides at 37°C for 48 h.

3.3.7. Posthybridization Washes

1. Prewarm 50% formamide/2X SSC, 2X SSC, and 0.1% SSC in a 45°C water bath.
2. Remove slides from the humid chamber and carefully remove rubber cement and cover slips from the slide.
3. Wash in 50% formamide/2X SSC for 10 min shaking intermittently.

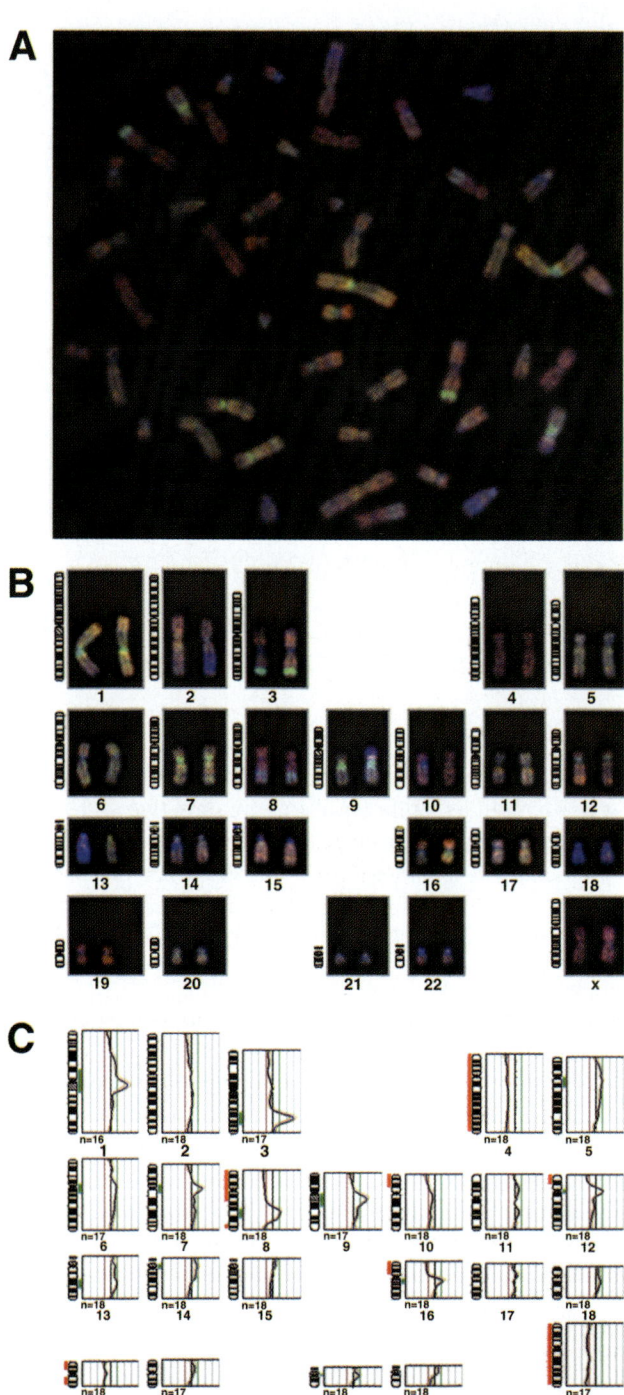

4. Wash twice in 2X SSC for 5 min each.
5. Wash once in 0.1X SSC for 5 min.
6. Wash slides once in 1X PN buffer for 5 min at room temperature.
7. Rinse in distilled water. Air-dry slides in the dark.
8. Apply 20 µL DAPI/antifade across the slide.
9. Cover the slide with a 24 × 50-mm^2 cover slip.

3.3.8. Microscopy and Image Analysis for CGH

1. A fluorescence microscope equipped with a CCD camera and appropriate filters (DAPI, fluorescein, and Texas Red or rhodamine) is required to visualize metaphase spreads hybridized with differentially labeled test and reference DNA **(Fig. 1A)**. The differences in the relative intensities of the two fluorochromes can provide a color ratio profile indicative of copy-number changes in the test DNA relative to the reference DNA. The following criteria is essential for high-quality CGH, (1) well spread metaphase chromosomes with adequate length, (2) uniform red and green hybridization (no granularity), (3) the red and green fluorescence distribution should be similar between two sister chromatids, two homologous chromosomes and the same chromosome in all the metaphases, (4) good DAPI banding for the identification of chromosomes, and (5) background fluorescence level surrounding the chromosomes should be low and uniform.
2. After direct visual inspection, the metaphase chromosomes are subjected to quantitative analysis. This type of analysis can only be derived from digital imaging. Several digital imaging systems are available commercially for CGH analysis (Applied Imaging [San Jose, CA], Applied Spectral Imaging, Leica Microsystems Inc. [Bannockburn, IN], and Metasystems [Watertown, MA]). By digital imaging, first individual chromosomes will be segmented, local background subtracted, and the median axises of the chromosome defined *(13,14)*. These chromosomes will then be normalized to standard length and combined,

---

Fig. 1. *(Opposite page)* Comparative genomic hybridization analysis. **(A)** A metaphase spread after simultaneous hybridization with differentially labeled normal (red) and tumor (medulloblastoma in this case) DNA (green) by CGH. Chromosomal regions that were over-represented in the tumor are visualized in green, whereas regions that were lost or deleted from the tumor are seen as red color. **(B and C)**. Quantitative analysis of CGH. **(B)** Chromosomes from **Fig. 1A** identified and karyotyped using quantitative imaging processing software (QUIPS, Applied Imaging). **(C)** Quantitative digital image analysis of fluorescence intensity ratios. Green to red fluorescence ratio profiles is shown for all chromosomes. The mean ratio (blue line) and ±1 SD (black lines) of 13–20 measurements for each chromosome are shown. The ratio profiles for each chromosome are shown from pter to qter. The average value (1) representing the mean green to red ratio for the entire case (6–10 metaphases) and red and green lines indicate threshold values of 0.8 and 1.2 for loss and gain, respectively.

a minimum of 10 chromosomes to show statistically the mean and 95 or 99% confidence intervals of the red: green signal ratio. Control experiments (normal vs normal DNAs) are very helpful for interpretation of CGH results from tumor samples with a particular batch of slides and reagents. Only ratio changes that exceed the fluctuation seen in the control experiments are interpreted as evidence for real loss or gain in the tumor compliment **(Fig. 1B,C)**. The normal variation in a given CGH experiment should not exceed ratios of 0.80–1.20 (±1 SD). In most cases, the telomeric, peri-centromeric or heterochromatic regions fall outside this range due to low signal intensities. Therefore, these regions should be excluded from the analysis. In addition, caution should also be exercised in interpreting ratio changes at 1p32-p36, 16p, 19, and 22 because of the high abnormal ratios in these regions. Chromosomal amplifications can be detected as strong localized FITC signal at the chromosomal site. For the precise assignment of amplification to chromosomal bands, the peak of the ratio profile should be compared with corresponding DAPI band of the same chromosome.

3.4. Tumor Chromosome Preparations for SKY and FISH

The investigation of karyotypic changes in tumor cells requires ability to arrest cells at the metaphase stage, achieve spreading of the chromosomes to finally study them under a microscope using an array of conventional and molecular cytogenetics methods. Essentially, it is feasible to prepare chromosomes from any actively dividing tumor cells. Conventional karyotypic examination requires preparation of metaphases and staining by banding methods to enable identification of individual chromosomes. Although there are a number of techniques used for the recognition of specific chromosomal structures or regions, the G-banding method has been the most accepted technique. Owing to limitation of space it is recommended that the reader refer to standard methods *(15,16)*.

3.5. Spectral Karyotyping

Conventional karyotype analysis is one of the widely accepted genome-screening tools for identification of chromosomal aberrations in cancer cells. However, metaphases with poor morphology and complexly rearranged marker chromosomes, a common feature of tumor cells, are often difficult to interpret unambiguously. The advent of multicolor FISH methods provides new ways to identify precisely the nature of complex chromosome changes *(4,5)*. This method is based on the hybridization of combinatorially labeled chromosome-specific painting probes to metaphase spreads, allowing simultaneous visualization of each chromosome pair by a unique color, in a single experiment. The principle of multicolor FISH lies in labeling individual whole chromosome probe pools in different fluorochromes or different ratios of fluorochromes such that each chromosome ultimately displays a distinct color. Two distinctly

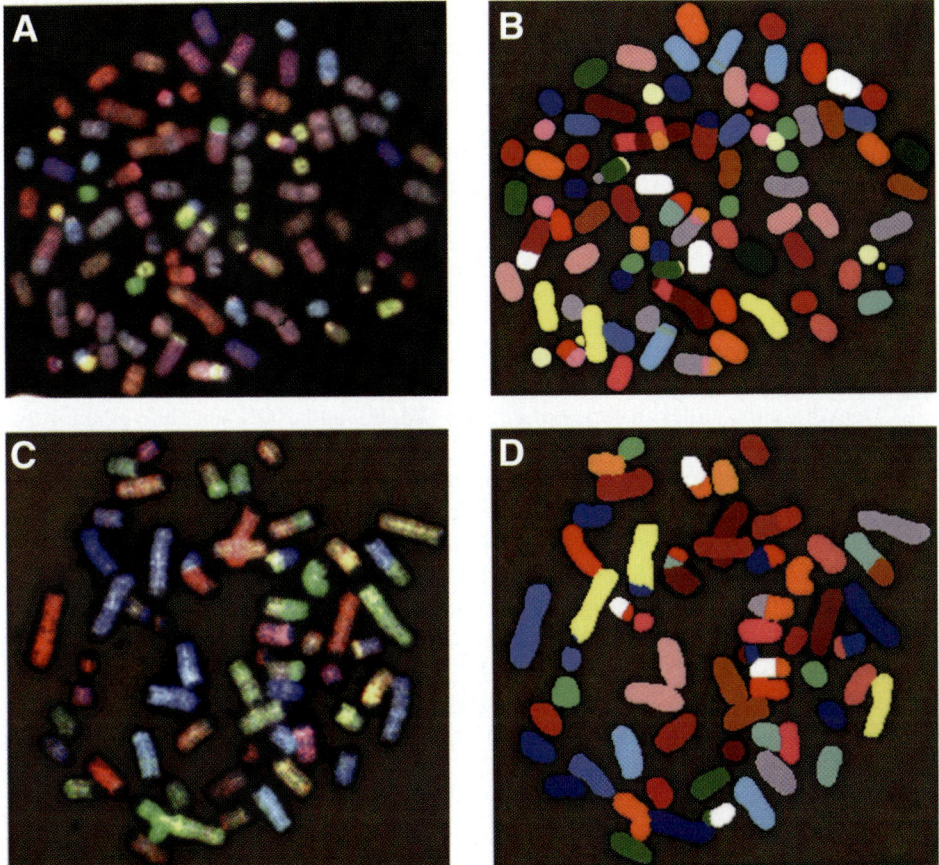

Fig. 2. SKY images of metaphase chromosomes from human (**A** and **B**) and mouse (**C** and **D**) tumors with complex chromosomal rearrangements. (**A** and **C**) Spectra-based display colors, (**B** and **D**) spectral classification from the same metaphase spreads.

different protocols have been developed for further analysis. One, termed multiplex FISH (m-FISH) *(4)*, makes use of digital images acquired separately for each of five fluorochromes employed in labeling probes. The images are acquired by use of appropriate filters and a CCD camera and analyzed by dedicated software that generates a composite image in which each chromosome is pseudo-colored on the basis of the composition of fluorochrome(s) used. The other procedure is SKY *(5)*, which makes use of an entirely different system of image acquisition from the probe-hybridized metaphase, which combines CCD imaging with Fourier spectroscopy. The images are analyzed by dedicated software, which generates composite images as in the case of m-FISH (**Fig. 2**). The power of these new labeling and image acquisition/analysis techniques in the

definition of chromosomal changes in cancer cells are manifold. Thus, chromosomal breakpoints can be mapped and the composition of marker chromosomes
determined with high precision. However, the breakpoint determination on
painted chromosomes/inverted DAPI is often difficult mainly owing to the poor
chromosomal morphology after hybridization. Therefore, G-banded karyotypes
from the same tumor will be useful in determining breakpoints.

3.5.1. Slide Pretreatment

1. Prepare chromosome spreads and mark the area of hybridization target with a diamond pencil.
2. Equilibrate the slides with 2X SSC for 30 min at RT.
3. Rinse the slides in distilled water.
4. Dehydrate the slides in series of ethanol (70, 80, and 100% absolute ethanol) for
   2 min each. Air-dry slides.
5. Slides that appear particularly cytoplasmic should undergo pepsin treatment.
6. Prewarm 50 mL of 0.01 M HCl in a Coplin jar at 37°C. Add 5–25 µL of pepsin
   stock solution and mix well. Incubate slides at 37°C in a Coplin jar for 5–10 min.
7. Wash slides in 1X PBS for 5 min at RT. Repeat the wash a second time in PBS.
8. Wash slides in 1X PBS/MgCl$_2$ for 5 min at RT.
9. Incubate slides in 1% formaldehyde in 1X PBS/MgCl$_2$ for 10 min at RT.
10. Wash slides in 1X PBS for 5 min at RT.
11. Dehydrate the slides in a series of 70, 80, and 100% absolute ethanol for 2 min each.
    Air-dry slides (*see* **Notes 7** and **8**).

3.5.2. Slide Denaturation

1. Place the slides in denaturation solution (70% formamide/2X SSC) for 2 min in a
   74°C water bath. Denature not more than two slides at a time.
2. After denaturation, immediately place the slides in ice cold 70% ethanol. Pass slides
   through 70, 80, and 100% ethanol for 2 min each. Air-dry the slides.

3.5.3. Probe Denaturation and Hybridization to Target Chromosomes

1. Denature the sky paint mixture in a 78–80°C water bath for 7 min and incubate in
   a water bath at 37°C for 1 h.
2. Upon removal from the water bath, immediately apply the probe onto the slide and
   place a 22-mm^2 cover slip over the probe.
3. Seal the edges of the cover slip with rubber cement.
4. Place the slides in a humidified chamber (Tupperware with paper towels moistened with formamide solution).
5. Incubate the slides at 37°C for 48 h.

3.5.4. Posthybridization Washes

1. Prewarm 50% formamide/2X SSC, 1X SSC, and 4X SSC/0.1% Tween-20 in a
   45°C water bath.

2. Remove the slides from humid chamber and carefully remove rubber cement and cover slips from the slide.
3. Wash twice in 50% formamide/2X SSC for 10 min each at 45°C.
4. Wash twice in 1X SSC for 5 min each at 45°C.
5. Wash once in 4X SSC/0.1% Tween-20 for 5 min at 45°C.
6. Rinse in distilled water at room temperature.

3.5.5. Detection

1. Take the slides out of the distilled water and drain well. Add 60 μL of vial no. 3 (supplied by ASI) over the specimen area and add a plastic cover slip. Place slides in a humidified chamber and incubate at 37°C for 30 min. Adding blocking reagent prior to this step is optional.
2. Wash slides three times in 4X SSC/0.1% Tween-20 for 5 min each.
3. Repeat **step 1** with vial no. 4 (supplied by ASI) and incubate for 30 min in a humidified chamber at 37°C.
4. Wash the slides three times in 4X SSC/0.1% Tween-20 for 5 min each.
5. Rinse the slides with distilled water for 2–3 min and air-dry the slides.
6. Place 15–20 μL of DAPI in antifade over the hybridized area and add a cover slip of appropriate size. Now the slides are ready for capturing and analysis.

3.5.6. Microscopy and Image Analysis for SKY

SKY is based on spectral imaging, which is a combination of spectroscopy and imaging. In contrast to conventional epifluorescence microscopy in which flurochrome discrimination is based on the measurement of a single intensity through fluorochromes with a specific optical filter, spectral imaging allows one to measure and analyze the full spectrum of light at all given pixels of the image. These images can be captured and analyzed using hardware and software developed by applied spectral imaging (ASI).

3.6. Fluorescence In Situ Hybridization

FISH is a powerful molecular cytogenetic method based on the hybridization of specific nucleic acid sequences to the target genome *(8)*. The DNA or RNA sequences are first labeled with reporter molecules and later the probe and the target, for example, chromosomes or nuclei are denatured. Complimentary sequences in the probe and target are then allowed to reanneal. After washing and incubation in fluorescently labeled affinity reagents, a discrete fluorescent signal is visible at the site of probe hybridization. An alternate method for labeling is the use of direct fluorochromes that incorporate into the DNA. This method does not require additional steps for detecting labels with fluorochrome-conjugated secondary agents. Several reporter molecules, for example, red (Texas Red, Spectrum Red, Cy 3) and green (FITC, Spectrum Green), are available for direct labeling of DNA. Multicolor *in situ* hybridization is relatively new method, in

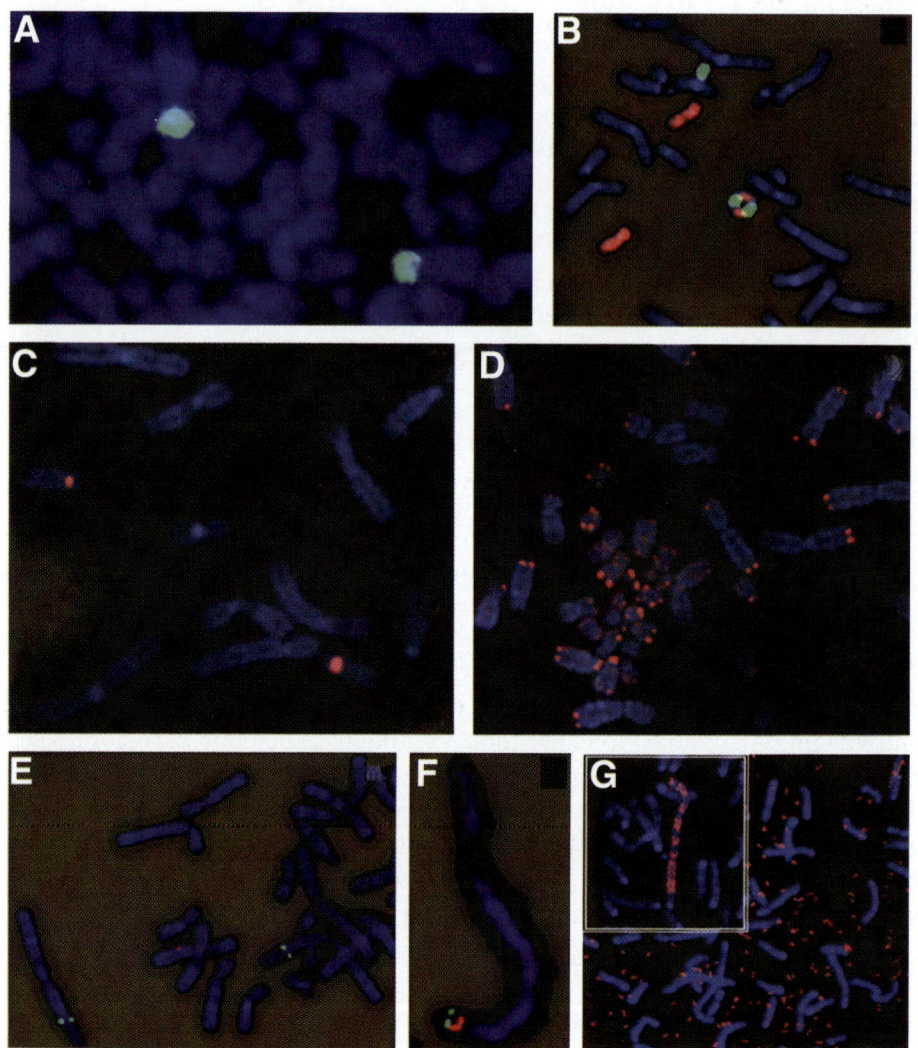

Fig. 3. Partial metaphase spreads showing examples of hybridization of various types of FISH probes. **(A)** A chromosomal arm probe (12p) labeled with Biotin and hybridized to normal human metaphase chromosomes, **(B)** A complex ring chromosome in a case of malignant fibrous histiocytoma. Whole chromosome painting probes—WCP 17 (SpectrumOrange) and WCP 22 (SpectrumGreen) hybridized to metaphase spreads derived from a malignant fibrous histiocytoma, **(C)** Alpha satellite probe for chromosome 18 labeled with SpectrumRed hybridized to normal human metaphase chromosomes, **(D)** Cy 3-labeled peptide nucleic acid (PNA) telomeric FISH probe hybridized to human tumor chromosomes, **(E)** SpectrumGreen labeled single copy DNA probe hybridized to human chromosome 1p31 region, **(F)** Biotin and dioxigenin labeled single copy DNA probes hybridized to human pachytene chromosome 6,

Table 2
Comparison of Sensitivity of Various FISH Mapping Techniques

Source	Resolution
Metaphase chromosomes	>1 mb
Mechanically stretched chromosomes	>200 kb
Pachytene chromosomes	~380–500 kb
Interphase nuclei	~50-kb–1 mb
DNA fibers	~1–500 kb

which several DNA probes can be labeled using different fluorochromes. This method is based on the availability of fluorochromes, optical filters, and image acquisition and registration software. Nick translation is the most common method for the labeling of DNA probes. However, it is useful to label small cDNA probes using the random priming method. Several DNA probes are commercially available that include chromosome-specific repeats (centromeric and telomeric), whole chromosome painting probes (WCP), chromosome arm probes, and single-copy sequences (human DNA sequences cloned in plasmid, phage, cosmid, PAC, bacterial artificial chromosome [BAC], yeast artificial chromosome vectors [YAC]) (**Fig. 3**). The relative position and orientation of clones along the chromosomal axis can be visualized and quantified by FISH. The wide array of currently available FISH techniques extends the resolution of visual mapping from a few megabases to only a few kilobases *(9,17)* (**Table 2**).

3.6.1. Preparation of Plasmid, Phage, Cosmid, PAC, and BAC DNA for FISH

The DNA from any of the cloning vectors can be extracted using standard alkaline lysis protocols or a Qiagen (Valencia, CA) plasmid purification kit.

3.6.2. Preparation of Labeled DNA Probes for FISH (Probe Labeling)

1. To label Plasmid/Phage/Cosmid/PAC/BAC DNA by nick translation add the reagents in the following order:
   a. 1 μg DNA in ddH$_2$O (make up to 50 μL of total reaction volume).
   b. 5 μL 10X dNTP mix.
   c. 1 μL dTTP mix.
   d. 1 μL Spectrum Green or Spectrum Red.
   e. 5 μL Enzyme mix containing DNA polymerase I/DNase I.

Fig. 3. *(Continued)* (**G**) Identification of gene amplification in mouse breast cancer model. A BAC clone containing the *Met* gene labeled with SpectrumOrange and hybridized to metaphase chromosomes derived from mouse breast cancer cells. Multiple copies of double min are shown. Inset shows homogenously stained region with multiple copies of *Met* gene.

2. Mix contents of the tube well and incubate at 15°C for 1 h and 45 min. (The nick translation time can be adjusted depending upon the size of the genomic DNA. The bigger the probe the longer the requirement of incubation time.)
3. Place tubes on ice, while keeping them protected from light.
4. Check the DNA fragment size by running a 1% agarose gel, using a 2- to 3-µL aliquot of the reaction.
5. Heat the tubes at 75°C for 10 min, to inactivate the enzymes (*see* **Note 9**).

3.6.3. Precipitation of Probe DNA for Hybridization

1. Add the following contents to the nick translation reaction mix before precipitation:
   a. 5 mL Human Cot-1 DNA (for YACs, BACs, and PACs to suppress the repetitive sequences).
   b. 5 µL 3 *M* Sodium acetate.
   c. 300 µL Cold absolute ethanol.
2. Mix contents well and leave at –70°C for at least 1 h.
3. Centrifuge at 14,000 rpm for 30 min at 4°C.
4. Remove supernatant. A good size pellet should be visible at the bottom of the tube.
5. Wash the pellet with 500 µL cold 70% ethanol.
6. Centrifuge at 14,000 rpm for 30 min at 4°C.
7. Remove the supernatant as thoroughly as possible. Air-dry the pellet in the dark for at least 2 h, or until completely dried.
8. Resuspend the pellet in 10 µL ddH$_2$O. Mix thoroughly by tapping the bottom of the tube several times. Allow the pellet to dissolve over 1 h. When ready to carry out hybridization, prepare probe mixture by dissolving 2–3 µL of labeled DNA in 10 µL of hybridization mixture.

3.6.4. Slide Pretreatment

1. Pretreat all slides before use, for 1 h in 2X SSC at 37°C.
2. Rinse the slides in distilled water.
3. Treat the slides in 70, 80, and 100% absolute ethanol series for 2 min each.
4. Air-dry the slides. Please note that, metaphase spreads that appear particularly cytoplasmic should undergo pepsin treatment.

3.6.5. Slide Denaturation

1. Place the slides in denaturation solution (70% formamide/2X SSC) for 2 min in a 74°C water bath. Denature a maximum of two slides at a time.
2. After denaturation, immediately place the slides in ice cold 70% ethanol. Wash the slides in 70, 80, and 100% ethanol for 2 min each.
3. Air-dry the slides.

3.6.6. Probe Denaturation and Hybridization to Target Chromosomes

1. Denature the probe in a 74°C water bath for 6 min.
2. Upon removal from the water bath, immediately apply the probe onto the slide and place a 22-mm^2 cover slip over the probe.

3. Seal the edges of the cover slip with rubber cement.
4. Place the slides in a humidified chamber (Tupperware with paper towels moistened with formamide solution). Incubate the slides at 37°C for 24 h.

3.6.7. Posthybridization Washes

1. Prewarm 50% formamide/2X SSC, 2X SSC, and 0.1% SSC in a 45°C water bath.
2. Remove the slides from humid chamber and carefully remove rubber cement and cover slips from the slide.
3. Wash in 50% formamide/2X SSC for 10 min.
4. Wash twice in 2X SSC for 5 min each.
5. Wash once in 0.1X SSC for 5 min.
6. Wash the slides once in 1X PN buffer for 5 min at room temperature.
7. Rinse in distilled water.
8. Air-dry the slides in the dark.
9. Apply 20 µL DAPI/antifade across the slide.
10. Cover the slide with a 24 × 50-mm^2 cover slip.

3.6.8. Microscopy and Image Analysis for FISH

A laboratory that focuses mainly on single copy gene mapping on interphase, metaphase or prophase chromosomes, a fluorescence microscope equipped with dual band pass filter or triple band pass filter for the simultaneous visualization of multiple probes or paints is required. Several digital imaging systems are available commercially for FISH/CGH, for example, Applied Imaging, Metasystems, Applied Spectral Imaging, and Leica.

3.7. Future Perspective

Our current awareness of chromosome abnormalities in cancers has permitted simultaneous detection of multiple abnormalities using multicolor-multiplex FISH-type tests using the technologies described in this chapter. The combination of conventional and molecular cytogenetic methods of analyses will remain methods of choice in diagnosis and classification of malignant diseases into clinically and biologically relevant classes. With the availability of genome sequences of various organisms, including human, refined genome-wide screening approaches to define the cytogenetic changes such as microarray technologies will gain importance in both clinical and research scenarios. These technologies will ultimately provide personalized genetic diagnosis and treatment a reality.

4. Notes

1. Check the activity of DNase on normal DNA.
2. The quality of metaphase spreads is crucial for CGH. The criteria for superior quality metaphase preparations include medium density, well spread, slight or no

cytoplasm, and little debris on the slide. It is better to avoid chromosomes that appear bright or hollow. Sometimes synchronized chromosome preparations will give granularity on the chromosomes and this can adversely affect the fluorescence ratios.

3. The percentage of tumor cells in the specimen is critical for detecting chromosomal copy-number changes in cancer. If the tumor cell content in the tissue is less than 40%, the resulting CGH karyotype will be normal. In those cases, the tumors' cells can be microdissected using laser capture microscope (Palm Laser-MicroBeam System, Wolfratshausen, Germany or Pix Cell II, Arcturus Biosciences, Inc., Mountain View, CA).

4. Optimum DNA fragment size is in the range of 500 to 3000 bp. If fragment size is larger than 3000 bp, continue the incubation for another 15–20 min after adding 1–2 µL of a DNase working solution, if necessary.

5. Metaphase spreads that appear particularly cytoplasmic should be subjected to pepsin treatment.

6. CGH was the first high-throughput whole genome analysis methods for estimating DNA copy-number changes in the entire genome in a single experiment. Particular advantage of this method is that DNA from tumor specimens can be used to circumvent the need for tumor metaphases. CGH can be performed on good quality DNA isolated from any form of tissue, fresh, frozen, or archival.

7. The quality of metaphase spreads is crucial for SKY. The criteria for superior quality metaphase preparations include well spread, slight or no cytoplasm and little debris on the slide. It is better to avoid chromosomes that appear bright or hollow. The timing of pepsin treatment varies from sample to sample and therefore, it is better to evaluate pepsin digest placing a $24 \times 60\text{-mm}^2$ cover slip under phase-contrast microscope. If excessive cytoplasm remain around the metaphase chromosomes, additional treatment with pepsin is necessary.

8. SKY on previously G-banded metaphase preparations is possible and the results obtained can be with similar hybridization intensities to slides that were not subjected to G-banding. Before doing SKY, the slides should go through xylene and methanol treatment. Previously G-banded chromosomes need a shorter denaturation time (15–30 s).

9. Optimum DNA fragment size is in the range of 100 to 500 bp. If fragment size is larger than 500 bp, continue the incubation for another 15–20 min after adding 1–2 µL of a DNase working solution, if necessary.

References

1. Kallioniemi, A., Kallioniemi, O. –P., Sudar, D., et al. (1992) Comparative genomic hybridization for molecular cytogenetic analysis of solid tumors. *Science* **258,** 818–821.
2. Rao, P. H., Cigudosa, J. C., Ning, Y., et al. (1998) Multicolor spectral karyotyping identifies new recurring breakpoints and translocations in multiple myeloma. *Blood* **92,** 1743–1748.

3. Singh, B., Gogineni, S. K., Sacks, P. G., et al. (2001) Molecular cytogenetic characterization of head and neck squamous cell carcinoma and refinement of 3q amplification. *Cancer Res.* **61,** 4506–4513.

4. Speicher, M. R., Ballard, S. G., and Ward, D. C. (1996) Karyotyping human chromosomes by combinatorial multi-fluor FISH. *Nat. Genet.* **14,** 312.

5. Schrock, E., du Manoir, S., Veldman, T., et al. (1996) Multicolor spectral karyotyping of human chromosomes. *Science* **273,** 494–497.

6. Pinkel, D., Segraves, R., Sudar, D., et al. (1998) High resolution analysis of DNA copy number variation using comparative genomic hybridization to microarrays. *Nat. Genet.* **20,** 207–211.

7. Cai, W. W., Mao, J. H., Chow, C. W., Damani, S., Balmain, A., and Bradley, A. (2002) Genome-wide detection of chromosomal imbalances in tumors using BAC microarrays. *Nat. Biotechnol.* **20,** 393–396.

8. Trask, J. B. (1991) Fluorescence *in situ* hybridization: applications in cytogenetics and gene mapping. *TIG* **7,** 149–154.

9. Heiskanen, M., Peltonen, L., and Paloticm, A. (1996) Visual mapping by high resolution FISH. *TIG* **12,** 379–382.

10. Sambrook, J., Fritsch, E. F., and Maniatis, T. (1989) *Molecular Cloning, A Laboratory Manual, 2nd ed.* Cold Spring Harbor Laboratory Press, New York.

11. Kallioniemi, O. P., Kallioniemi, A., Piper, J., et al. (1994) Optimizing comparative genomic hybridization for analysis of DNA sequence copy number changes in solid tumors. *Genes Chromosomes Cancer* **10,** 231–243.

12. Lasken, R. S. and Egholm, M. (2003) Whole genome amplification: abundant supplies of DNA from precious samples or clinical specimens. *Trends Biotechnol.* **21,** 531–535.

13. du Manoir, S., Kallioniemi, O. P., Lichter, P., et al. (1995) Hardware and software requirements for quantitative analysis of comparative genomic hybridization. *Cytometry* **19,** 4–9.

14. Piper, J., Rutovitz, D., Sudar, D., et al. (1995) Computer image analysis of comparative genomic hybridization. *Cytometry* **19,** 10–26.

15. Rooney, D. E. (ed.) (2001) *Human Cytogenetics: Malignancy and Acquired Abnormalities.* Oxford University Press, New York.

16. Verma, R. S. and Babu, A. (eds) (1995) *Human Chromosomes: Principles and Techniques.* McGraw-Hill, New York.

17. Hauptschein, R., Gamberi, B., Rao, P. H., et al. (1998) Cloning and mapping of human chromosome 6q26-27 deleted in B-cell non-Hodgkin's lymphoma and multiple tumor types. *Genomics* **50,** 170–186.

12

Monitoring Methylation and Gene Expression in Cancer

Hetty Carraway and James Herman

Summary

Identification of patterns of DNA methylation in higher order eukaryotes has become necessary through recognition that disease processes such as cancer can result from abnormal methylation patterns. Abnormal DNA methylation affecting the promoter region of genes can halt their expression, making DNA methylation a marker of gene inactivation. The explosion of studies involving changes in DNA methylation in the last 10 yr, particularly in the study of cancer, are largely the result of two factors: the increasing awareness of the importance of epigenetic silencing in cancer and the improvements in the techniques used to determine changes in DNA methylation. In this chapter, focus is made on the specific methods of how to perform bisulfite modification of DNA and polymerase chain reaction as well as nested methylation specific polymerase chain reaction, and discuss other techniques to evaluate DNA methylation along with the limitations inherent in each process.

Key Words: Bisulfite sequencing; CpG island ("C" nucleotide immediately followed by a "G" nucleotide. The "p" in CpG refers to the phosphate group linking the two bases), cytosine; epigenetics; guanine; methylated allele; methylation specific PCR (msp); nested methylation; primer; specific PCR (NMSP), restriction enzyme; unmethylated allele.

1. Introduction

The increasing awareness that epigenetic changes could play an important role in cancer was initially focused on the silencing of bone fide tumor suppressor genes in cancer. These observations were fueled by the discovery of genes that were genetically altered in both familial and sporadic forms of cancer, the classic tumor suppressor genes such as retinoblastoma, von Hippel-Lindau, INK4a, BRCA1, and others that could also be inactivated by promoter region methylation. Studies in the early 1990s were largely accomplished by using restriction enzymes to preferentially cleave unmethylated sequences in the promoter regions of these genes, while leaving methylated sequences uncut. Southern blot analysis provided the final readout of the methylation status of these CpG islands (*1*), von

From: *Methods in Molecular Biology, vol. 383: Cancer Genomics and Proteomics: Methods and Protocols*
Edited by: P. B. Fisher © Humana Press Inc., Totowa, NJ

Hippel-Lindau *(2)*, INK4a *(3)*. This technique provided an assessment of the overall methylation status of the CpG islands, but required large amounts of high-molecular weight DNA (generally >5 µg). Another limitation of Southern hybridization is that it detects methylation only if present in several percent of the alleles, and provided information about CpG sites only within those sequences recognized by methylation-sensitive restriction enzymes.

Combining the use of methylation sensitive enzymes and the PCR resulted in a more sensitive method of methylation detection. Like Southern hybridization, this method was only able to monitor CpG methylation in methylation-sensitive restriction sites. Because of the false-positive rate of methylation detection, this approach was felt to be unreliable for detection of hypermethylation in small samples, or in samples where methylated alleles represent a small fraction of the population.

The breakthrough in this area was the recognition that 5-methylcytosine information could be fixed in the DNA by selective deamination of cytosine to uracil with the protection of 5-methylcytosine from the same deamination, initially forming the basis of genomic bisulfite sequencing, but also forming the basis of all other bisulfite based techniques. This deamination allows the epigenetic modification of cytosine to be effectively converted into genetic information, which can then be subjected to the same molecular techniques that are used for genetic analysis. Following modification, a specific region or gene of interest can be amplified using the PCR. The nature of this amplification divides bisulfite PCR-based techniques into two categories: nonselective amplification or methylation selective amplification.

Most modern methods for determining gene or locus specific methylation rely on this pivotal observation that bisulfite modification allows a conversion of the methylation signal in DNA to a sequence change that can be directly determined *(4)*. Methylation-selective amplification includes MSP (methylation specific PCR *[5]*) and NMSP (nested methylation specific PCR). Nonselective methods of amplification are all similar in that these techniques seek to amplify all sequences regardless of methylation status following bisulfite modification. Amplification is intended to multiply both methylated and unmethylated sequences in an unbiased fashion to a higher number. This nonselective process then requires a second step, which will analyze and determine the actual methylation changes or patterns, and these readouts of the methylation changes are the ways in which these techniques are unique.

Genomic bisulfite sequencing, the original bisulfite technique for determining methylation patterns in DNA, remains the most comprehensive method for methylation analysis *(4)*. Following PCR amplification with nonbiased primers, the methylation status at each CpG site can be determined using either gel based or automated sequencing *(4)*. Two alternatives provide slightly different types of information: direct sequencing from PCR products or sequencing of individual plasmid clones. The advantage of direct sequencing of a purified PCR product is that, the resulting sequence is representative of the sample that was bisulfite-treated and amplified. Specifically, it is a pooled average of the methylation of each CpG site over the amplified region. This approach saves effort, time, and expense of the isolation of multiple plasmid preparations and the added expense of multiple sequence analyses and may better represent the overall methylation status of a sample. However, this convenience comes at a cost of a loss of much of the information in the methylation patterns contained in the original sequence: all information concerning the methylation status of individual copies of DNA is lost in the average methylation result obtained. In contrast, the genomic sequencing of individual plasmid clones provides a detailed analysis of the methylation status of each CpG site on each copy of DNA *(6)*. This allows the evaluation of complex patterns of methylation and a better evaluation of mixed patterns of methylation. For example, cloned sequencing can determine whether a segment of DNA with 50% overall methylation has resulted from heterogeneous methylation averaging 50% methylation on all copies of DNA from that resulting from a mixture of complete methylation on some alleles with completely unmethylated DNA on the remaining DNA copies.

Combining bisulfite sequence analysis with restriction analysis allows for a simpler and less laborious approach, although it does not give the detailed analysis that sequencing offers. This type of technique does offer examination of more samples or more genes for each sample with a given amount of resources (time, expense, and labor). In practice, it is often helpful to combine the use of genomic bisulfite sequencing on some well-studied samples with a simpler analysis on a larger panel of tumors or samples. The simplest approach of these methods is bisulfite restriction analysis. This approach uses the predicted changes in restriction enzyme sites created by the conversion of only unmethylated cytosine to uracil by bisulfite treatment. This technique uses primers that are identical to those used for bisulfite sequencing and genomic DNA is bisulfite treated and amplified by the PCR before restriction analysis. The restriction products are run on gels to determine the level of methylation, comparing uncut to cut fragments. A more quantitative approach of this restriction analysis is accomplished by the use of radioactive probes from the fragmented DNA *(7)*. Limitations of this approach are the limited number of sites appropriate for restriction analysis, the inability to choose the specific sites analyzed for methylation analysis, the need for added steps (restriction, gel analysis) and the use of radioactivity for accurate quantitation. However, this approach is relatively simple and can provide some degree of methylation quantitation.

Methylation-sensitive single nucleotide primer extension is based on bisulfite treatment of DNA followed by single nucleotide primer extension *(8)* and is a more precise approach for quantitatively determining methylation status at individual CpG sites. Advantages of this approach include the avoidance of restriction enzymes and the precise quantitation for the methylation status at individual CpG sites. Potential disadvantages include the need for radioactivity and the need for separate amplifications and evaluations for each CpG site examined.

Another method to assess CpG methylation density of a DNA region has been called ERMA (enzymatic regional methylation assay). This method determines an exact measurement of the methylation density of the region studied, essentially averaging the methylation on all copies of DNA amplified by PCR over this region *(9)*. However, the method requires radioactivity and does not provide the methylation status of individual CpG sites that MS-SnuPE would provide or the detailed allelic methylation patterns obtained by genomic bisulfite sequencing. It may be best suited for examination of methylation changes induced by demethylation agents.

Methylation-specific PCR represents a simple, rapid, and inexpensive method for determination of methylation patterns from very small samples of DNA, including those obtained from paraffin embedded samples. Methylation specific PCR allows for evaluation of single-stranded alleles of a gene, specifically evaluating for the presence of methylated cytosines. Treatment of the single-stranded DNA with sodium bisulfite converts unmethylated cytosines to uracil, whereas methylated cytosines remain unchanged. This modified DNA is then amplified with primers specific for the newly generated sequence, and the products analyzed by gel electrophoresis.

Before proceeding with the MSP protocol on precious samples, there are a few hints to ensure the success of this protocol. One essential step is to ensure that primer designs of the genes of interest are optimal. First, is the gene of interest a CpG island *(10)*? The accepted definition of CpG island is regions of DNA greater than 200 bp, with guanine/cytosine content greater than 0.5 and an observed or an expected presence of CpG more than 0.6. Primers should be designed such that they are smaller than 200 bp, and be located near the transcription start site of the gene of interest. During the design, try to have similar annealing temperatures for the unmethylated primer and methylated primer. Inherent in all of the earlier-mentioned methods of methylation analysis following bisulfite modification of DNA is the assumption that these sequences are amplified in an unbiased manner during PCR. Primer design plays an essential role in this approach, and the ideal primer for any of the unbiased PR approaches incorporates or overlies a number of cytosines at non-CpG sites whose modification from cytosine to uracil allows the distinction of unmodified from bisulfite modified DNA. Additionally, it is important that the primers cover no CpG sites or at least a minimum number of CpG sites that would bias toward either methylated or unmethylated sequences. Alternatively, intentional mismatches at these CpG sites can be incorporated to avoid such bias. If either mixed bases or intentional mismatches are used, these should be located in the 5′ portion of the primer to minimize effects on primer annealing and extension. Further details on primer design are annotated in the discussion section, as it is one of the most important aspects of MSP. Once the primers are obtained, significant time should be spent up front optimizing the annealing temperature for those specific primers using a gradient temperature on the PCR block. This can be done using bisulfite-treated cell lines as a positive control for the gene of interest.

There are other ways in which to evaluate DNA for methylation. One additional way to look for methylation in a highly sensitive manner is by a procedure called NMSP or nested-MSP *(11)*. This technique utilizes a two-step technique with a first step utilizing a multiplex of flanking primer sequences of multiple genes to amplify the specific areas of interest and a second step utilizing gene specific MSP primers to identify the unmethylated and methylated sequences. Detailed protocol algorithm for this technique is outlined in **Subheadings 2.** and **3.**

As stated previously, bisulfite sequencing can be used to confirm the presence of methylation, and is often performed for verification of the MSP as the gold standard. Methods for quantification of methylation include real time PCR. This approach monitors the production of PCR product with florescent detection, using a variety of methods for measurement including simple double-strand DNA dyes, molecular beacons, or Taqman technology. This type of quantitation must be distinguished from that of the unbiased approaches previously described, because what is being quantified here is the amount of DNA having a particular methylation pattern, recognized by the PCR primers and/or probe, and not the overall level of methylation. Quantitative results of this and all above quantitative methods might also be greatly affected by the relative amounts of tumor and normal cells included in the issue or sample used for DNA extraction. However, the advantage of this technique is that many samples can be tested quantitatively at one time (96-well plate) but the trade-off is that experiments are completed in triplicate and might be time-consuming. Other approaches include evaluation of global methylation such as ALU methylation or HPLC and techniques, which utilize restriction enzymes to identify methylated cytosines such as COBRA. Finally, microarray analyses exist which use CpG islands for analyses.

1.1. Determination of DNA Methylation Patterns by MSP

In methylation specific PCR (MSP), DNA is modified by sodium bisulfite treatment, converting unmethylated (but not methylated) cytosines to uracil. Following removal of bisulfite and completion of the chemical conversion, this modified DNA is used as a template for PCR. Two PCR reactions are performed for each DNA sample, one specific for DNA originally methylated for the gene of interest, and one specific for DNA originally unmethylated. PCR products are separated in nondenaturing polyacrilamide gels and the bands are visualized by staining with ethidium bromide. The presence of a band of the appropriate molecular weight indicates the presence in the original sample of unmethylated alleles, methylated alleles, or both. Methylation-specific PCR is a very specific, sensitive, rapid, and economical method to determine methylation patterns of CpG islands.

The protocols for DNA extraction from paraffin embedded tissues and unstained slides are available on request. Tissue embedded on slides and stained with hematoxylin and eosin should not be used, as the staining process will alter the methylation evaluation results based on the crosslinking of DNA in the staining procedure (*see* **Note 1**).

2. Materials

1. Sample DNA.
2. Positive control DNA, previously determined to be methylated.
3. Negative control DNA, previously determined to be unmethylated.
4. 2 and 3 *M* NaOH.
5. 10 m*M* Hydroquinone (prepare fresh before use, *see* protocol).
6. 3 *M* Sodium bisulfite (prepare fresh before use, *see* protocol).
7. Mineral oil.
8. DNA Wizard cleanup kit (Promega) or equivalent.
9. 80% Isopropanol, room temperature.
10. Autoclaved distilled water, 60–70°C and room temperature.
11. Electrophoresis buffer (preferably Tris borate EDTA [TBE] buffer or sodium borate).
12. 10 mg/mL Glycogen.
13. 10 *M* Ammonium acetate.
14. 100 and 70% Ethanol, ice cold.
15. 10X PCR amplification buffer with 15 m*M* MgCl$_2$.
16. 25 m*M* 4 dNTP mix.
17. 300 ng/μL Sense and antisense primers.
18. Taq DNA polymerase.
19. Vacuum manifold.
20. Dedicated micropipetors for setting up PCR reactions.
21. 0.5-μL PCR tubes or strips.
22. Thermal cycler with tube-controlled temperature monitoring.
23. Additional reagents and equipment for nondenaturing polyacrlamide gel electrophoresis and gel staining and photography/digital camera as well as nanospectophotometer.

3. Methods

3.1. Sixteen Hour Bisulfite Treatment

1. Dilute DNA (up to 1 μg) into 50 μL with distilled water in a 1.5-mL Eppendorf tube. Prepare separate dilutions of samples, positive control, and negative control DNA and process in parallel.

2. Add 5.5 µL of 2 *M* NaOH. Incubate 10 min at 37°C. This step creates single-stranded DNA, which is sensitive to reaction with sodium bisulfite.
3. Add 30 µL of freshly prepared 10 m*M* hydroquinone (Sigma) to each tube. Make fresh by adding 55 mg (0.055 g) of hydroquinone to 50 mL of water. Hydroquinone is an antioxidant, which prevents oxidation of intermediates formed during bisulfite treatment.
4. Add 520 µL of freshly prepared sodium bisulfite (Sigma) to each tube. Make fresh by adding 1.88 g of sodium bisulfite per 5 mL of water, and adjusting pH to 5.0 with concentrated NaOH (usually three to five drops).
   (One 5-mL aliquot makes enough for about 10 samples.)

Na Bisulfite (g)	1.88	3.76	5.64	7.52
ddH$_2$O (mL)	5	10	15	20
No. of reactions	9	18	27	36

5. Ensure the reagents are mixed with DNA.
6. Layer with 200 µL of mineral oil.
7. Incubate at 50°C for 16–20 h. Also include two Eppendorf tubes of distilled water for DNA wizard clean up after the 16 h are completed.
8. Next day (or after completion of 16–20 h incubation).
9. Make sure manifold is clean and washed out.
10. Take out sample from Eppendorf tube and place into equivalently labeled tube/column set up on manifold. Leave oil in Eppendorf tube and discard.
11. Add 1 mL of DNA wizard cleanup (Promega A7280) to each tube in miniprep column in kit. Mix gently. Other DNA recovery systems, such as glass milk, are also suitable; separation of the DNA from the high concentration of sodium bisulfite is all that is necessary.
12. Apply vacuum, the manifold makes this convenient.
13. Wash with 2 mL of 80% isopropanol.
14. Remove and discard the tube. Immediately place the column in a clean, labeled 1.5-mL Eppendorf tube and add 50 µL of heated water (60–70°C).
15. Spin the Eppendorf and column in microcentrifuge for 1 min. The warmed water elutes the modified DNA from the resin.
16. Add 5.5 µL of 3 *M* NaOH to each tube, and incubate 5 min at room temperature. This completes the chemical conversion of the cytosine to uracil.
17. Add glycogen as a carrier (1 µL of 10 mg/mL glycogen per sample).
18. Add 17 µL of 10 *M* ammonium acetate (NH$_4$Ac) and 3 vol of ice cold 100% ethanol (150 µL).
19. Precipitate DNA at –20°C (the longer the better, at least 1 h).
20. Place tubes in cold room centrifuge and spin at maximum speed for 25 min.
21. Take off 100% ethanol.
22. Add ice cold 70% (150 µL) ethanol and spin at maximum for 10 min in cold room. Discard supernatant.
23. Spin vacuum dry or air-dry.
24. Add 20 µL autoclaved distilled water to resuspend pellet (*see* **Note 2**).

3.2. Methylation Specific PCR

3.2.1. Prepare the PCR Mixes for MSP

1. Thaw 10X PCR buffer, four dNTP mix, and primers. Determine the number of samples to be analyzed, including a positive control for both the unmethylated and methylated reactions, and a water control. Prepare separate master mixes for both the methylated and unmethylated PCR reactions, each containing the following amounts of components per 23 µL PCR reaction.
   a. 2.5 µL 10X PCR buffer.
   b. 1.5 µL 100 m*M* 4dNTP mix.
   c. 1 µL 300 ng/µL sense primer.
   d. 1 µL 300 ng/µL antisense primer.
   e. 1 µL Red-taq.
   f. 16 µL Sterile distilled water.
   g. Mix well.
2. Place 23-µL aliquots of each PCR master mix into separate PCR tubes labeled for each sample. Label tubes in order as follows: number 1 unmethylated (1U), number 1 methylated (1M), number 2 unmethylated (2U), number 2 methylated (2M), so each eight tube strip will have four samples with each U and M (i.e., 1U, 1M, 2U, 2M, 3U, 3M, 4U, 4M).
3. Add 2 µL of bisulfite modified DNA template to each tube to bring the final volume for each tube to 25 µL. Make sure to include the positive cell line controls for methylated and unmethylated reactions and a water control.
4. Add two drops of mineral oil (25–50 µL) to each tube. Be sure the mineral oil completely covers the surface of the reaction mixture to prevent evaporation. Cover the tube strips with cover for tubes. Place the strips into thermal cycler. If the thermal cycler has a heated lid to prevent condensation, mineral oil might not be necessary, but longer run times are typical.

3.3. PCR Program for Thermocycler

3.3.1. Amplify PCR Products

3.3.1.1. STAGE 1: 1 CYCLE

1. Step 1: 95°C × 5 min × 1 cycle.
2. Step 2: nothing.
3. Hold temperature none.

3.3.1.2. STAGE 2: 35 CYCLES

1. Step 1: 95°C × 30 s (denaturation).
2. Step 2: 60°C × 30 s (annealing) (*see* **Note 3**).
3. Step 3: 72°C × 30 s (extension).
4. Step 4: none.
5. Hold temperature none.

3.3.1.3. STAGE 3: 1 CYCLE

1. Step 1: 72°C × 5 min × 1 cycle (extension).
2. Step 2: none.
3. Hold temperature none.

3.3.1.4. STAGE 4

1. Step 1: none.

3.4. Analyze PCR Products by Gel Electrophoresis

3.4.1. Prepare 6% Nondenaturing Polyacrylamide Gel With 1X TBE Buffer

The size range of the products typically generated by MSP is 80–200 bp, making acrylamide gels optimal for size resolution. High percentage horizontal agarose gels can be used as an alternative. Additionally, there are multichannel pipets which will also decrease time in gel loading and yield equivalent results.

1. Put glass and white strips together and place into dry plastic bag.
2. Place glass and bag contraption into the holding apparatus and screw bolts to hold plates upright.
3. Select appropriate size comb.
4. Using a 100-mL conical, mix the following.
    a. 30 mL Water.
    b. 6 mL Acrylamide: BIS 19:1 from refrigerator.
    c. 4 mL 10X TBE.
    d. 300 µL Ammonium persulfate (APS).
    e. 60 µL TEMED.
    f. Swirl to mix.
5. Pour solution into bag and place comb into top of plates.
6. Wait 10 min for polymerization.
7. Remove glass plate from plastic bag and take off excess gel and wash.
8. Place glass plates in gel box.
9. Load each sample into adjacent lanes to allow for direct comparison between unmethylated and methylated reactions. Include positive and negative controls.
10. Run gel.
11. Stain gel with ethidium bromide and visualize the photograph under UV illumination. Digital technology exists where it is possible to quantitate the amount of methylation, although this is not as accurate as other methods (*see* **Note 4**).

3.5. Determination of DNA Methylation Patterns by NMSP

This nested technique is designed to improve the sensitivity of MSP. It enables a very small amount of DNA to be used and to be evaluated with a panel of genes. The DNA will be bisulfite modified exactly the same as described in **Subheading 3.1.**, only the PCR technique will be altered.

As previously described, the DNA will be amplified with flanking primers to the area of interest (usually flanks the gene of interest). This occurs during the first step PCR amplification. It may be important to note that during this first round of PCR, multiple flanks can be used in this reaction so as to amplify areas of multiple genes of interest. This technique allows for amplification of specifically those areas of the genes of interest, and the resulting product will be evaluated separately with the one gene of interest for methylation status. The products from this first round of PCR are then taken and run on a gel to verify resultant product (*see* **Notes 5** and **6**). The first round products are taken and diluted, where upon the diluted PCR sample is used for the sec round of PCR with the methylated and unmethylated primers for the gene of interest.

3.6. Prepare the PCR Mixes for Nested PCR: Step One

1. Determine how many genes you want to have in your multiplex.
2. We will use this example of four genes in the multiplex (p16, RAR-beta, MGMT, ECAD) flanking PCR. It is recommended that a minimum of 2 and a maximum of 5 genes be used, although people have been successful in using up to 8–10 genes in the multiplex.
3. It is recommended to use 0.25 µL of each flank up and flank down per reaction. As the number of flanks increase, the total volume of the flank up and flank down will increase in the master mix. This increase in volume of the flank volume can be subtracted from the water amount to keep the total volume of the master mix 21 µL. If greater than 0.25 µL is used for each flank up/down reaction, there will not be enough volume of water to counteract the increase in volume for flanks, and the final volume of the master mix will be greater than 21 µL per reaction.
4. Thaw 10X PCR buffer, 4dNTP mix, and flanks. Determine the number of samples to be analyzed, including a positive control for both the unmethylated and methylated reactions, and a water control. Prepare one master mix for all the PCR reactions, which will contain the following amounts of components per 21 µL PCR reaction.
   a. 2.5 µL 10X PCR buffer.
   b. 1.5 µL 100 m*M* 4dNTP mix.
   c. 1 µL 300 ng/µL Flank primers up (4 genes × 0.25 = 1 µL).
   d. 1 µL 300 ng/µL Flank primers down (4 genes × 0.25 = 1 µL).
   e. 1 µL Red-taq.
   f. 14 µL Sterile distilled water.
5. Mix well.
6. Place 21-µL aliquots of PCR master mix into separate PCR tubes labeled for each sample.
7. Add 4 µL of bisulfite modified DNA template to each tube to bring the final volume for each tube to 25 µL. Make sure to include the positive and negative cell line controls as well as a water control. It is important to note that for cell lines use only 2 µL of DNA.

8. Add two drops of mineral oil (25–50 μL) to each tube. Be sure the mineral oil completely covers the surface of the reaction mixture to prevent evaporation. Cover the tube strips with cover for tubes. Place the strips into thermal cycler.

3.7. PCR Program for Thermocycler for First Step PCR

3.7.1. Stage 1: 1 Cycle

1. Step 1: 95 C × 5 min × 1 cycle.
2. Step 2: nothing.
3. Hold temperature none.

3.7.2. Stage 2: 35 Cycles

1. Step 1: 95°C × 30 s (denaturation).
2. Step 2: 56°C × 30 s (annealing).
3. Step 3: 72° × 30 s (extension).
4. Step 4: none.
5. Hold temperature none.

3.7.3. Stage 3: 1 Cycle

1. Step 1: 72°C × 5 min × 1 cycle (extension).
2. Step 2: none.
3. Hold temperature none.

3.7.4. Stage 4

1. Step 1: none.
2. Remove products from Thermal cycler.
3. Use 7.5 μL of product to run on the polyacriamide gel and confirm for presence of product.
4. Then use a designated pipet (usually a 2- to 20-μL pipet) to dilute the product. Add 998 μL of autoclaved distilled water to a 1.5-mL Eppendorf tube and add 2 μL of each sample into their respectively labeled Eppendorf tube to reach a dilution of 1:500. The designated pipet is important in order to avoid contamination of your other PCR products when setting up MSP or even when extracting your DNA. Careful attention to avoid any contamination is essential, especially with the multiplex technique.
5. This Eppendorf tube of diluted product is what is used for the second round of PCR. This product can be used for analysis of methylation status for each gene that was placed in the multiplex. Thus, in this example, we can use these samples to move to evaluate MSP separately for p16, RAR-β, MGMT, and ECAD). Again, you are ready to proceed with Step Two Nested PCR.

3.8. Step Two for Nested PCR

3.8.1. Prepare the PCR Mixes

1. Thaw 10X PCR buffer, 4dNTP mix, and primers. Determine the number of samples to be analyzed, including a positive control for both the unmethylated and

methylated reactions, and a water control. Prepare separate master mixes for both the methylated and unmethylated PCR reactions, each containing the following amounts of components per 23 µL PCR reaction.

    a. 2.5 µL 10X PCR buffer.

    b. 1.5 µL 100 m*M* 4dNTP mix.

    c. 1 µL 300 ng/µL Sense primer.

    d. 1 µL 300 ng/µL Antisense primer.

    e. 1 µL Red-taq.

    f. 16 µL Sterile distilled water.

    g. Mix well.

2. Place 23-µL aliquots of each PCR master mix into separate PCR tubes labeled for each sample. Label tubes in order as follows: number 1 unmethylated (1U), number 1 methylated (1M), number 2 unmethylated (2U), number 2 methylated (2M), so each eight tube strip will have four samples with each U and M (1U, 1M, 2U, 2M, 3U, 3M, 4U, 4M).

3. Add 2 µL of diluted (1:500) first round PCR product to each tube to bring the final volume for each tube to 25 µL. Make sure to include the positive cell line controls for methylated and unmethylated reactions and a water control.

4. Add two drops of mineral oil (25–50 µL) to each tube. Be sure the mineral oil completely covers the surface of the reaction mixture to prevent evaporation. Cover the tube strips with cover for tubes. Place the strips into thermal cycler.

3.9. PCR Program for Thermo-Cycler for Second Step PCR

3.9.1. Stage 1: 1 Cycle

1. Step 1: 95°C × 5 min × 1 cycle.
2. Step 2: nothing.
3. Hold temperature none.

3.9.2. Stage 2: 30 Cycles

1. Step 1: 95°C × 30 s (denaturation).
2. Step 2: 60°C × 30 s (annealing).
3. Step 3: 72°C × 30 s (extension).
4. Step 4: none.
5. Hold temperature none.

3.9.3. Stage 3: 1 Cycle

1. Step 1: 72°C × 5 min × 1 cycle (extension).
2. Step 2: none.
3. Hold temperature none.

3.9.4. Stage 4

1. Step 1: none.

3.10. Analyze PCR Products by Gel Electrophoresis

As described in **Subheading 3.4.**

4. Notes

1. **General Caution:** as with all PCR methodology, exercise great care to ensure lack of contamination of the preparations with adventitious DNA: ideally, reserve a separate area for setting up PCR such as a PCR hood.
2. Because the DNA is single-stranded, treat the modified DNA like RNA (keep it cold, minimize freeze–thaw cycles, and store at −70°C if possible).
3. When optimizing the annealing temperature for each primer set for gene of interest, the T_m (melting temperature) is equal to 2 (A + T) + 4 (C + G).
   a. Annealing temperature = T_m − 2 × 30 s.
   b. Extension time: 1 kb = 1 min, 500 bp = 30 s.
   c. Need to take into account time for temperature transmission through the tube.
4. Some laboratories have now started using sodium borate instead of TBE for gel preparation and running of gels. Sodium borate is less costly and the gels can run more quickly. The results are also equivalent.
5. Critical parameters of MSP: the most critical parameter affecting the specificity of methylation specific PCR is determined by the primer design. After the modification of DNA by bisulfite, the two daughter strands of any given gene are no longer complementary after treatment; either strand can serve as the template for subsequent PCR amplification, and the methylation pattern of each strand could then be determined. In practice, it is often easiest to deal with only one strand, most commonly the sense strand. CpG dinucleotides are almost always symmetrically methylated, meaning that detection of methylation on one strand implies similar methylation patterns on the other strand.

   MSP uses the sequence differences resulting from bisulfite treatment as the basis for primer design, placing the primers precisely over the mismatches in order to identify the unmethylated strand separately from the methylated strand. Methylation induced gene silencing is associated with methylation changes in the proximal promoter region of a specific gene. Although each region may differ in CpG content and length, 5' CpG islands should be the starting point for the examination of methylation changes leading to gene silencing. Primers should be designed to amplify a region that is 80–200 bp in length, and should incorporate enough cytosines in the original sequence to assure that unmodified DNA will not serve as a template for the primers. The intentional design of shorter primers allows for a more robust amplification, allows for shorter cycle times, and is needed for Taqman based approaches owing to system constraints. In addition, the number and position of cytosines within the CpG dinucleotide determines the specificity of the primers for methylated or unmethylated templates. Typically, 1–3 CpG sites are included in each primer, and concentrated in the 3' region of each primer. This provides optimal specificity and minimizes false-positives because of mispriming.

In general, each (sense or antisense) primer should be at least 20 bp long. To facilitate simultaneous analysis of the unmethylated and methylated reactions of a given gene in the same thermocycler, the length of the primers is adjusted to give nearly equal melting/annealing temperatures. Specifically, the melting temperature should not be >70°C. One trick to make this possible is to avoid cytosines on the methylated primer set in order to keep the melting temperature lower. This design also usually results in the unmethylated product being a few basepairs larger than the methylated product, which provides a convenient way to recognize each lane after electrophoresis.

Because methylation specific PCR utilizes specific primer recognition to discriminate between methylated and unmethylated alleles, stringent amplification conditions must be maintained. This means that annealing temperatures should be at the maximum temperature, which allows annealing and subsequent amplification. In practice, newly designed primers are typically tested with an initial annealing temperature 5–8°C below the calculated melting temperature. Nonspecificity can be remedied by slight increases in annealing temperature, while faint, or weak PCR product bands might be improved by a drop in temperature of 1–3°C. As with all PCR protocols, great care must be taken to ensure that the template DNAs and reagents do not become contaminated with exogenous DNAs or PCR products.

A note should be made about primer design for bisulfite sequencing, where it is preferred to have longer primers as this would allow longer sequencing runs yielding more information in genomic bisulfite sequencing, or for bisulfite restriction analysis which might incorporate more potential restriction sites within the PCR amplicon. However, practical limitations produced by the bisulfite fragmentation of the DNA limit the size of DNA easily amplified during PCR. Realistic PCR products should be <500 bp, although with large amounts of good quality DNA, this might be increased somewhat.

Added sensitivity to simple MSP has been accomplished through use of florescent primers or nested PCR approaches (11). The preferential amplification of MSP, however, largely eliminates any useful degree of quantitation, although much like RT-PCR approaches with limiting cycle number, one might gain gross quantitative information using this approach.

Approaches which restore a degree of quantitation to MSP take advantage of the recent advances in quantitative real-time PCR. These approaches monitor the production of PCR product with florescent detection, using a variety of methods for measurement including simple double-stranded DNA dyes, molecular beacons or Taqman technology. This type of quantitation must be distinguished from that of the unbiased approaches in **Subheading 1.**, since what is being quantified here is the amount of DNA having no particular methylation pattern, recognized by the PCR primers and/or probe, and not the overall methylation. Quantitative results of this and all above quantitative methods might also be greatly affected by the relative amounts of tumor and normal cells included in the tissue or sample used for DNA extraction.

6. Top 10 list of things to remember when doing MSP:
   a. Design of primers for the assay is critical. No amount of technical improvements can fix a problem with nonspecific or poorly designed primers. Because this assay, as well as genomic bisulfite sequencing relies on complicated changes and specific location of primers, this step should be carefully performed and discussion with persons experienced with these techniques encouraged.
   b. Bisulfite protocol can be shortened to a 4 h instead of 16 h protocol, although the shortened protocol is not favored at this time.
   c. If, actively using bisulfite-treated DNA keep at −20°C, if not using it store at −80°C. Minimize freeze/thaw since the bisulfite-treated DNA is single-stranded.
   d. Optimize annealing temperatures for each gene.
   e. Keep PCR strips cold while aliquoting master mix and DNA into tubes by having an "ice block" that fits the PCR strips. This ice block is kept in the −20°C freezer when not in use.
   f. Use positive and negative controls in experiments, as well as a water control.
   g. Remember to put in mineral oil in the tubes before you place strips into thermal cycler.
   h. Make sure you have appropriate cycle number and annealing temperature for MSP vs nested MSP.
   i. Do everything possible to decrease the risk of contamination. Keep PCR products away from DNA extraction.
   j. If things are not working: go back to your primer design and evaluate it, make sure it is appropriate and has correct design.

References

1. Ohtani-Fujita, N., Fujita, T., Aoike, A., Osifchin, N. E., Robbins, P. D., and Sakai, T. (1993) CpG methylation inactivates the promoter activity of the human retinoblastoma tumor-suppressor gene. *Oncogene* **8,** 1063–1067.
2. Herman, J. G., Latif, F., Weng, Y., et al. (1994) Silencing of the VHL tumor suppressor gene by DNA methylation in renal carcinoma. *Proc. Natl. Acad. Sci. USA* **91,** 9700–9704.
3. Merlo, A., Herman, J. G., Mao, L., et al. (1995) 5′ CpG island methylation is associated with transcriptional silencing of the tumor suppressor p16/CDKN2/MTS1 in human cancers. *Nat. Med.* **1,** 686–692.
4. Frommer, M., McDonald, L. E., Millar, D. S., et al. (1992) A genomic sequencing protocol that yields a positive display of 5-methylcytosine residues in individual DNA strands. *Proc. Natl. Acad. Sci. USA* **89,** 1827–1831.
5. Herman, J. G., Graff, J. R., Myohanen, S., Nelkin, B. D., and Baylin, S. B. (1996) Methylation-specific PCR: A novel PCR assay for methylation status of CpG islands. *Proc. Natl. Acad. Sci. USA* **93,** 9821–9826.
6. Cameron, E. E., Baylin, S. B., and Herman, J. G. (1999) p15 CpG island methylation in primary acute leukemia is heterogeneous and suggests density as a critical factor for transcriptional silencing. *Blood* **94,** 2445–2451.

7. Xiong, Z. and Laird, P. W. (1997) COBRA: a sensitive and quantitative DNA methylation assay. *Nucleic Acids Res.* **25,** 2532–2534.

8. Gonzalgo, M. L. and Jones, P. A. (1997) Rapid quantitation of methylation differences at specific sites using methylation-sensitive single nucleotide primer extension (Ms-SNuPE) *Nucleic Acids Res.* **25,** 2529–2531.

9. Galm, O., Rountree, M. R., Bachman, K. E., Jair, K. W., Baylin, S. B., and Herman, J. G. (2002) Enzymatic regional methylation assay: a novel method to quantify regional CpG methylation density. *Genome Res.* **12,** 153–157.

10. Bird, A. P. (1986) CpG-rich islands and the function of DNA methylation. *Nature* **321,** 209–213.

11. House, M. G., Herman, J. G., Guo, M. Z., et al. (2003) Aberrant hypermethylation of tumor suppressor genes in pancreatic endocrine neoplasms. *Ann. Surg.* **238,** 423–432.

13

The Use of Phage Display Peptide Libraries for Basic and Translational Research

Renee Brissette and Neil I. Goldstein

Summary

Phage display is a molecular technique, whereby genes are displayed in a functional form on the outer surfaces of bacteriophages by fusion to viral coat proteins. The gene product is encoded by a plasmid contained within the virus, which can be recovered and sequenced, linking the genetic information to the function of the protein. Phage display offers a powerful tool for the identification of short peptides or single chain antibodies that can bind and regulate the function of target proteins. One major advantage of phage display lies in its ability to rapidly identify target-specific peptides with pharmacological activity as agonists or antagonists.

Key Words: Agonists; antagonists; bacteriophage; peptides; phage display; protein function; single-chain antibodies.

1. Introduction

Display technologies can be divided into biological display systems that employ a biological host and/or biological reactions and nonbiological systems. Li *(1)* described four major display modules including protein/peptide on RNA/DNA, viral/phage display, cell-based display, and nonbiological display. Protein-DNA/RNA systems have the distinct advantage that they are not restricted by selection pressures from the biological host. Phage display systems offer powerful tools for the identification of short peptides or single-chain antibodies that can bind and regulate the function of target proteins. One of the obvious advantages of phage display is the ease of introduction into and amplification within the host bacterial cells. Cell-based display of cDNA libraries requires efficient transfer into mammalian cells, which is more problematic than introduction of libraries into bacterial or yeast cells. Nonbiological systems offer the obvious advantage of creating different types of building blocks and are not limited to naturally occurring amino acids.

From: *Methods in Molecular Biology, vol. 383: Cancer Genomics and Proteomics: Methods and Protocols*
Edited by: P. B. Fisher © Humana Press Inc., Totowa, NJ

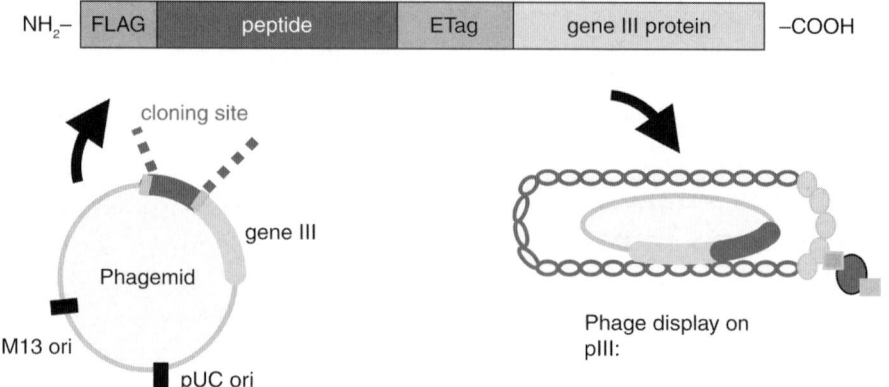

Fig. 1. Cartoon of the M13 phage showing orientation of a peptide within its genome. Two protein tags, FLAG and Etag are included in the vector for identification purposes.

Phage display is a molecular technique, whereby genes are displayed in a functional form on the outer surface of a bacteriophage by fusion to viral coat proteins *(1,2)*. The gene product is encoded by a plasmid contained within the virus, which can be recovered and sequenced, linking the genetic information to the function of the protein. The bacterial viruses are usually members of the Ff filamentous phage family are M13, f1, and fd. Foreign genes have been fused to three coat proteins; pIII, pVIII, and pVI **(Fig. 1)**. In general, there are three different phage display systems according to the arrangement of the coat protein genes *(1,2)*. In the first type there is a single-phage chromosome (or genome) bearing a single gene III, VIII, or VI into which is fused the foreign DNA. The resulting fusion protein is theoretically displayed on all 5 copies of pIII or pVI and all 2700 copies of pVIII. The phage genome encodes two copies of the coat protein in the second system where one copy is fused to the foreign gene and the other wild type. The resulting phage is a mosaic because its coat contains both recombinant and wild-type coat protein. The third system differs from the second in that the DNA encoding the two copies of the coat protein is on different genomes. The wild-type copy is encoded by the "helper" phage, while the recombinant copy is on a plasmid called a "phagemid."

George Smith *(3)* constructed the first gene fusion between gene III and the gene encoding Eco RI endonuclease. The displayed protein was demonstrated which could be affinity purified from a library of random inserts using antibody to Eco RI endonuclease. Cwirla et al. *(4)* created a phage library of random residues displayed on the virion from synthetic oligonucleotides. This phage library was "panned" or affinity purified using a monoclonal antibody against L-enkephalin. The isolated phage as amplified by *Escherichia coli* infection and then used in two more rounds of panning. The phage clones sequences showed relatedness to the known specificity of the monoclonal antibody, i.e., a Tyr-Gly at the N-termini of the random peptides. Since then this successful screening approach had been exploited by employing various types of phage libraries and various targets.

Phage display peptide libraries can be designed as random or constrained, the later having known amino acids (e.g., cysteines) at specific positions within the peptide sequence. Alternatively whole proteins or their domains can be displayed on phage. In these constructs mutations are introduced into the sequence of the protein or domain with the aim of selecting proteins with enhanced or new function. Additionally, genomic or cDNA phage display libraries have been constructed from normal and disease tissues *(5,6)*.

Recently, lytic phage has been used as selection system for cloning RNA-binding proteins from cDNA libraries. One drawback of using M13 phage is the requirement that all components of the phage particle be synthesized and exported from the bacterial cell. With lytic phage, the particles are assembled and released without the need for particle export. For example, T7 has been used to identify novel RNA-binding proteins, by utilizing a ribonuclease-deficient host to minimize contamination with cellular nucleases *(5,6)*.

Peptides binding to individual targets can be identified by in vitro selection (biopanning *[7,8]*), while those binding to normal or diseased tissues can be found using in vivo selection *(9–15)*. For biopanning (**Fig. 2**), a display library is incubated with an immobilized target followed by extensive washing to remove nonreacting phage. Binders are usually eluted using acid or high salt and are enriched by amplification in the appropriate host cell. Biopanning usually involves three to five rounds in order to obtain high affinity binders. The primary structure of the peptide or antibody can then be determined by sequencing the DNA of individual clones. For the identification of in vivo binders, an initial ex vivo selection is followed by several rounds of enrichment in the target animal. Using this approach, it is easy to identify peptides that bind specifically to target tissues in different stages of development.

Biopanning allows the selection of target-specific peptides with affinity constants in the millimolar to nanomolar range (*see* **Note 1**). Affinity maturation of desired clones can be done through the generation of secondary libraries and subsequent binder selection under stringent conditions *(7,8)* (*see* **Note 2**). In the case of peptides, secondary libraries are designed to define the necessary amino acid residues at each position for optimal binding, specificity, and affinity (**Fig. 3**). In order to achieve these types of results, secondary libraries are generated where an amino acid at any position in the sequence can change to any other amino acid. Panning of secondary libraries will identify substitutions that either do or do not affect binding. At the same time, a comparison of peptide binding to related proteins will help to identify residues that confer specificity

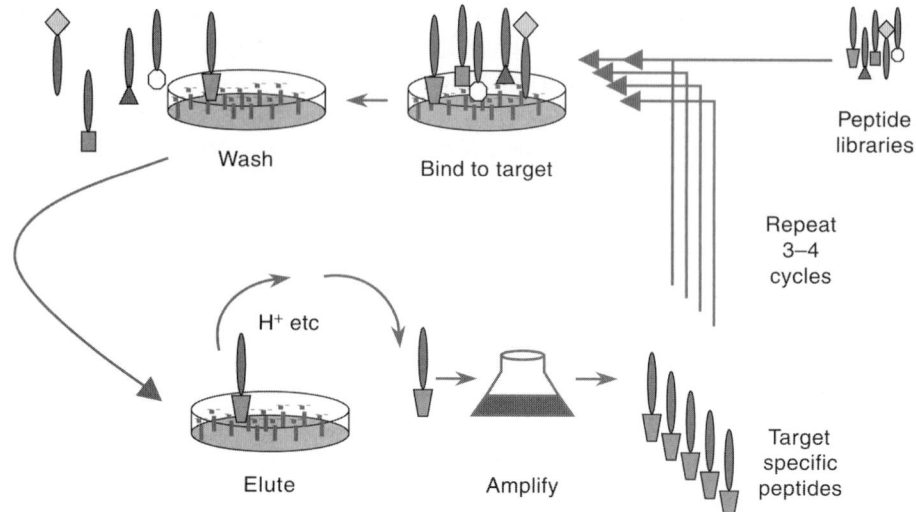

Fig. 2. Biopanning protocol. A target is first coated on an immobilized surface (e.g., microtiter plates). A phage library is added for several hours and nonbinders removed by extensive washing. Target-specific phage are eluted with a low pH buffer and amplified overnight at 37°C in combination with helper phage. Usually three to five rounds are needed to identify target-specific binders and eliminate nonspecific phage.

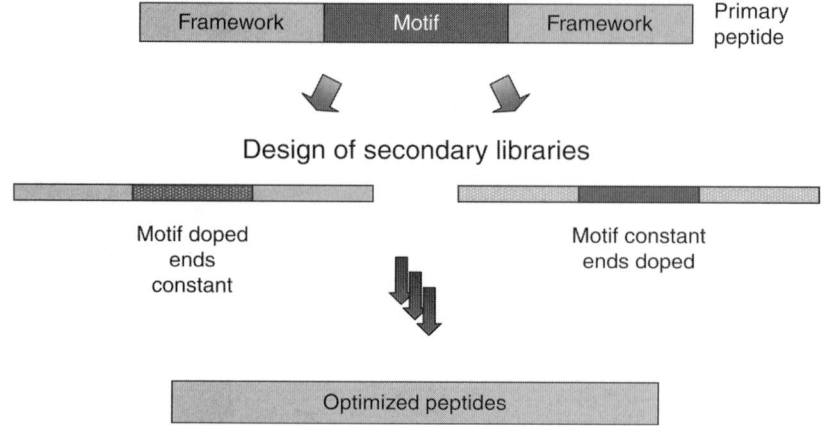

Fig. 3. Procedure for the generation of secondary libraries. Primary binders may, often times, require maturation in terms of affinity, biological potency, and selectivity. This is accomplished by generating secondary libraries based on the sequence of a primary target-specific binder. If a motif is identified, it is important to develop two types of libraries: one, where the motif is held constant and the flanking amino acids are allowed to change and another, where the motif is held constant with amino acid substitutions in the flanking regions.

to the original target. In addition, detailed analysis of secondary binders will enable identification of an "optimal" peptide sequence which, when synthesized, can be tested for improved affinity and biological activity in the appropriate assays. When necessary, tertiary libraries can be built for further optimization. Typically, the strategy is to design tertiary libraries differing from secondary libraries in that the entire sequence is doped, so that on an average, 10–15% of the amino acids in the sequence are altered. Results of panning of tertiary libraries will usually define peptides with significant improvements in binding affinities and potency (*see* **Notes 3** and **4**).

By using large and diverse phage display libraries, peptides can be obtained, which bind to any given target and exert a pharmacological effect on that target. Random peptide libraries have been used to isolate molecular mimics for important polypeptide hormones including insulin, erythropoietin (EPO), interleukin (IL)-1, interferon (IFN)-β, and thrombopoietin (TPO). In the case of insulin, peptides were identified that bound to and subdivided the insulin receptor into "hotspots" or domains of pharmacological activity *(8,16)*. A series of peptides (called Hotspot Pharmacophores) were identified which either bound to a hotspot, designated Site 1, where they acted as either agonists or antagonists or to a second, non-overlapping hotspot, called Site 2, where they could only act as antagonists **(Fig. 4)**. The Sites 1 and 2 peptides were distinguished by different motifs which, in turn, determined pharmacological activity. An insulin mimetic was built by combining specific Sites 1 and 2 "monomers" in the appropriate orientation (N-Site 2–Site 1-C). The resulting molecule was found to have an affinity close to that of the natural ligand and full agonist activity. In addition, the mimetic had the same pharmacological properties as insulin when tested in an animal model *(16)*. Other laboratories have isolated peptides mimicking or inhibiting the biological functions of protein hormones or growth factors using phage display. Sato and Sone *(17)* isolated an IFN-β mimetic by panning and antibody that bound to its cognate receptor. Yanofsky et al. *(18)* generated a peptide antagonist to IL-1 that was active in the nanomolar range. Wrighton et al. *(19)* and Livnah et al. *(20)* have reported on the isolation of peptides binding to the EPO receptor with full agonist activity in vivo. Some of these peptides were found to form cysteine loops via disulfide bonding and, like the IL-1 antagonist peptides mentioned earlier, show no significant sequence homology to the natural ligand. Importantly, X-ray crystallography revealed that one of these peptides spontaneously formed noncovalent homodimers that allowed the dimerization of two EPO receptors. Another group *(21)* isolated two families of small peptides binding to the human TPO receptor.

Another valuable use of phage display is in the area of proteomics and functional genomic research (*see* **Note 5**). The high-throughput format of phage display makes it a leading technology to establish the functional relevance of

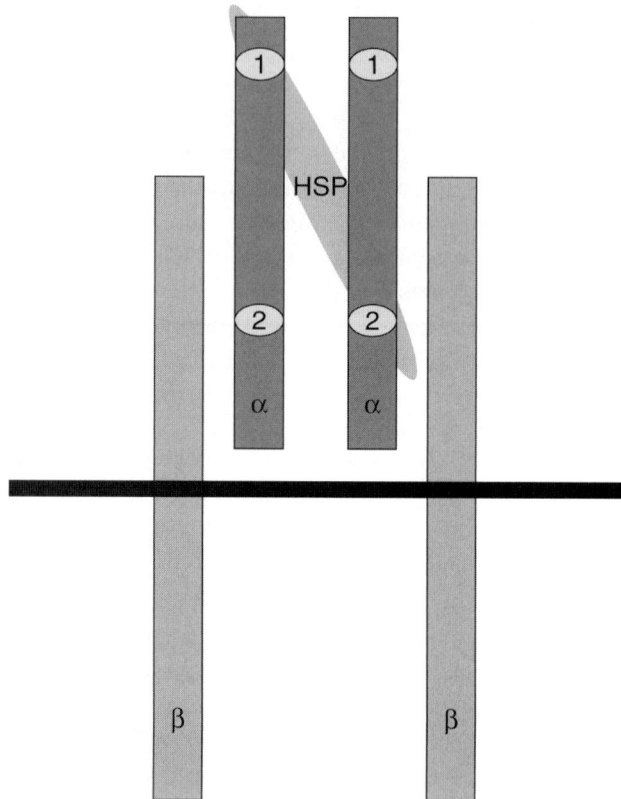

Fig. 4. Generating target-specific peptides using phage display: Identification of "HotSpots" on the IR. A soluble form of the insulin receptor was panned with high-diversity peptide libraries. After several rounds of panning, IR-specific peptides were identified that bound to one of two nonoverlapping sites on the receptor, designated as Sites 1 and 2. Site 1 peptides had pharmacological activity as agonists or antagonists, while Site 2 peptides could only act as antagonists. Reconstitution of insulin activity was accomplished by synthesizing peptides in a specific orientation: N terminus-Site 2 peptide-Site 1 peptide-C terminus. The resulting mimetic had pharmacological properties close to that of the natural ligand *(16)*.

newly identified gene products. There are several reports of using phage displayed peptides and cDNAs as a means of identifying protein–protein interaction partners *(22–26)*.

2. Materials

1. Commercially available phage display peptide libraries can be purchased from vendors such as Strategene (www.strategene.com), Novagen (www.novagen.com), and New England Biolabs (www.neb.com).

2. A primer on generating a phage display peptide library can be found at the following website http://www.biosci.missouri.edu/smithgp/PhageDisplayWebsite/PhageDisplayWebsiteIndex.html.
3. Commercially available phage display cDNA libraries can be obtained from vendors such as Spring Bioscience (http://www.springbio.com).
4. Although not the focus of this article, phage display single chain antibody libraries can be obtained from sources such as the MRC (www.hgmp.mrc.ac.uk/geneservice/reagents/products).
5. Coating buffer: 50 mM NaHCO$_3$, pH 8.5.
6. Nonfat milk-phosphate buffered saline (NFM-PBS): 2% nonfat milk in PBS.
7. 50 mM glycine–HCl containing 0.1% BSA (pH 2.2).
8. 100 μL of 1 M Tris-HCl (pH 8.0).
9. 2X YT-glucose medium: YT medium containing glucose 2%.
10. YT-AK medium: YT medium containing ampicillin (50 μg/mL) and kanamycin (50 μg/mL).
11. YT-AG medium: YT medium with ampicillin (50 μg/mL) and glucose (2%).
12. 30% PEG-8000 in 1.6 M NaCl.
13. Helper phage: final concentration = 5×10^{10}/mL.
14. PBSα-Tween: PBS with 0.05% Tween.
15. SpectraMax Microplate Spectrophotometer (Molecular Devices, Sunnyvale, CA).

3. Methods
3.1. Biopanning

1. Individual wells of a 96-well microtiter plates are coated with antigen (in 50 μL of coating buffer) at concentrations ranging from 50 to 500 ng/well). Plates are incubated overnight at 4°C.
2. The next day, unbound antigen is removed and the coated wells are blocked with 300 μL of NFM-PBS for 1 h at room temperature.
3. The plates are then washed three times with PBS. The phage libraries are thawed and mixed with 0.1 vol of NFM-PBS, 100 μL of each library is added to the antigen-coated wells and the plates are incubated for 1–3 h at room temperature.
4. Each well is washed 13 times with NFM-PBS and the phage eluted with 100 μL of glycine-HCL containing BSA following a 5-min incubation.
5. The eluted phage from each library are pooled, neutralized with 100 μL of Tris-HCl, and added to 10 mL of log phase *E. coli* TG1 and amplified in 2X YT-glucose medium for 1 h at 37°C.
6. Helper phage (M13K07) and ampicillin are then added and the cells are incubated for an additional hour at 37°C.
7. The cells are pelleted at 1000g 10 min, resuspended in 2X YT-AK medium and incubated overnight at 37°C.
8. The next day, the infected bacterial cells are centrifuged at 2500g 10 min and the pellet discarded. The supernatant contains the phage and is precipitated with 1 vol of PEG-8000 in NaCl by incubating on ice for 1 h.

9. The precipitant is centrifuged at 10,000 rpm (8000*g*) 30 min at 4°C for 30 min and the phage pellet resupended in about 1 mL of NFM-PBS. The phage is then used for the next round of panning.

10. Normally, three to four rounds of panning are done for secondary libraries. Usually, 96 random clones are picked from rounds three to four and grown in 96-well cluster plates as a master stock and used for sequencing.

3.2. Assay for Peptide Binders

1. A master stock is prepared by isolating individual colonies and growing them overnight at 37°C in 500 µL of 2X YT-AG.

2. Forty microliters of the master stock is transferred from each master to another set of cluster tubes containing 500 µL of 2X YT-AG and helper phage.

3. The tubes are incubated at 37°C with constant shaking for 2 h. The cultures are centrifuged at 2500*g* at 4°C for 20 min, the supernatant is discarded, the bacterial pellet resuspended in 400 µL of 2X YT-AK and the plate incubated overnight at 37°C.

4. At that time, the cells are removed by centrifugation at 2500*g* and the supernatants transferred to a new set of cluster tubes and used in an enzyme-linked immunosorbent assay.

5. For the enzyme-linked immunosorbent assay protocol, each well of a MaxiSorp plate (Nunc) is coated with 100 µL of target (1 µg/mL) overnight at 4°C. The wells are blocked with NFM-PBS for 1.5 h at room temperature.

6. Phage is added at 100 µL/well and the plates incubated for 3 h at room temperature. After washing three times with PBS-Tween, plates are probed with an anti-M13 antibody conjugated to horse-radish peroxidase (1:3000 in PBS-NFM) for 1 h at room temperature followed by addition of 100 µL of ABTS for 15–30 min at room temperature.

7. The OD is measured at 405 nm after 30–60 min at room temperature.

3.3. In Vivo Selection of Tissue-Specific Phage Binders

1. The screening process involves two ex vivo selection rounds followed by two to three in vivo selection (adapted from **ref. 13**).

2. For the ex vivo rounds, cell suspensions are prepared from the target tissue (e.g., tumor) and incubated overnight at 4°C with your peptide library.

3. Wash cells to remove unbound phage and the bound phage rescued and amplified in *E. coli*.

4. Inject the ex vivo preselected phage pool intravenously into mice carrying the target cells through the tail vein, allowed to circulated for 7 min and hear perfused with PBS to remove unbound intravascular phage.

5. Excise the target tissue and control tissues (brain, kidney, speen, lung, white pancreas, and liver) to allow for comparison of homing efficiencies.

6. Prepare cell suspensions by mechanical disruption of the tissues, washed to remove unbound phage and the bound phage was rescued and amplified by adding *E. coli*.

7. Reinject the phage pool into target mice and the cycle repeated. In each experiment, nonrecombinant control phage was used as a control for relative selectivity.

8. Collect sets of 96 phage clones randomly from each homing phage population and the peptide-encoding DNA inserts amplified by PCR and sequenced.

4. Notes

1. Antyra has developed high diversity phage display libraries that bind to domains called Hotspots.
2. In many cases, it becomes necessary to "improve" the initial peptide. In order to change its affinity, potency, and selectivity. This can be done by generating secondary libraries (**Fig. 3**). In many cases, the primary biopan yields peptides containing a "motif" or consensus region, which is defined as a stretch of amino acid homology found in multiple peptides with differing overall sequences. To determine optimal amino acid requirements within a motif, a secondary library can be generated, where the flanking regions (or framework) are held constant, while the "motif" is allowed to change. It is also important to determine the critical amino acids in the flanking regions; in this case, the motif is held constant and the flanking regions are mutated. The secondary library is prepared from "doped" oligonucleotides so that half of the amino acid residues (on average) in the core or flanking sequences are altered per peptide. Panning of this library will identify substitutions within the core or flanking regions that either affect or have no effect on binding to the target. Amino acids optimal for binding are selected during panning. This includes residues at randomized positions where any amino acid is allowed by the design of the library. A description of the preparation of secondary libraries for the insulin receptor (IR) has been previously published (*8*).
3. Based on detailed analysis of binders obtained from panning of the secondary libraries optimal peptide sequences are identified, synthesized, and tested for binding affinities and agonist or antagonist activity using in vitro and cell-based assays. If needed, tertiary libraries can be designed based on such peptide sequences. Typically our strategy is to design tertiary libraries differing from secondary libraries in that the entire sequence is doped, so that on an average, 10–15% of the amino acids in the sequence are altered. Results of panning of tertiary libraries will usually define peptides with significant improvements in binding affinities and potency.
4. After a target-specific peptide has been identified, it is necessary to synthesize that peptide for use in the appropriate biological assay. For initial assays, PEPscreen custom peptide libraries from Sigma-Genosys (www.sigma-genosys.com) were routinely used, where peptides are supplied in a 96-well format making it easier for high throughput screening.
5. In addition to the obvious utility of peptides for research and therapeutic purposes, they can also be used to identify potential interaction partners. In this case, peptide recognition modules are revealed which are mapped back to the complete genome sequences by computational analysis. For example, peptides have been found to mimic the contact points of natural protein partners and often contained information that could lead to the identification of the natural partner through simple and rapid searching of genomic databases. Several examples of the use of phage display for proteomics are described in the literature (*22,23,26*).

References

1. Li, M. (2000) Applications of display technology in protein analysis. *Nat. Biotechnol.* **18,** 1251–1256.
2. Smith, G. P. (1985) Filamentous fusion phage: expression vectors that display cloned antigens on the virion surface. *Science* **228,** 1315–1317.
3. Smith, G. P., Schultz, D. A., and Ladbury, J. E. (1993) A ribonuclease S-peptide antagonist discovered with a bacteriophage display library. *Gene* **128,** 37–42.
4. Cwirla, S. E., Peters, E. A., Barrett, R. W., and Dower, W. J. (1990) Peptides on phage: a vast library of peptides to identify ligands. *Proc. Natl. Acad. Sci. USA* **87,** 6378–6382.
5. Danner, S. and Belasco, J. G. (2001) T7 phage display: a novel genetic selection system for cloning RNA-binding proteins form cDNA libraries. *Proc. Natl. Acad. Sci. USA* **98,** 12,954–12,959.
6. Crameri, R. and Suter, M. (1993) Display of biologically active proteins on the surface of filamentous phages: a cDNA cloning system for selection of functional gene products linked to the genetic information responsible for their production. *Gene* **137,** 69–75.
7. Ravera, M. W., Carcamo, J., Brissette, R., et al. (1998) Identification of an allosteric binding site on the transcription factor p53 using a phage-displayed peptide library. *Oncogene* **16,** 1993–1998.
8. Pillutla, R. C., Hsiao, K. C., Beasley, J. R., et al. (2002) Peptides identify the critical hotspots involved in the biological activation of the insulin receptor. *J. Biol. Chem.* **277,** 22,590–22,594.
9. Pasqualini, R. and Ruoslahti, E. (1996) Organ targeting in vivo using phage display peptide libraries. *Nature* **380,** 364–366.
10. Nowakowski, G. G., Dooner, M. S., Valinski, H. M., Mihaliak, A. M., Quesenbery, P. J., and Becker, P. S. (2004) A specific heptapeptide from a phage display peptide library homes to bone marrow and binds to primitive hematopoietic stem cells. *Stem Cells* **22,** 1030–1038.
11. Spear, M. A., Breakefield, X. O., Beltzer, J., et al. (2001) Isolation, characterization and recovery of small peptide phage display epitopes selected against viable malignant glioma cells. *Cancer Gene Ther.* **8,** 506–511.
12. Ruoslahti, E. and Rajotte, D. (2000) An address system in the vasculature of normal tissues and tumors. *Annu. Rev. Immunol.* **18,** 813–827.
13. Joyce, J. A., Laakkonen, P., Bernasconi, M., Bergers, G., Ruoslahti, E., and Hanahan, D. (2003) Stage-specific vascular markers revealed by phage display in a mouse model of pancreatic islet tumorigenesis. *Cancer Cell* **4,** 393–403.
14. Rafi, S., Avecilla, S. T., and Jin, D. K. (2003) Tumor vasculature address book: identification of stage-specific tumor vessel zip codes by phage display. *Cancer Cell* **4,** 331–333.
15. Lee, I., Buckley, C., Blades, M. C., Panayi, G., George, A. J., and Pitzalis, C. (2002) Identification of synovium-specific homing peptides by in vivo phage display selection. *Arthritis Rheum.* **46,** 2109–2120.

16. Schaffer, L., Brissette, R. E., Spetzler, J. C., et al. (2003) Assembly of high affinity insulin receptor agonists and antagonists from peptide building blocks. *Proc. Natl. Acad. Sci USA* **100,** 4435–4439.

17. Sato, A. and Sone, S. (2003) A peptide mimetic of human interferon (IFN)-β. *Biochem. J.* **371,** 603–608.

18. Yanofsky, S. D., Baldwin, D. N., Butler, J. H., et al. (1996) High affinity type I interleukin 1 receptor antagonists discovered by screening recombinant peptide libraries. *Proc. Natl. Acad. Sci. USA* **93,** 7381–7386.

19. Wrighton, N. C., Farrell, F. X., Chang, R., et al. (1996) Small peptides as potent mimetics of the protein hormone erythropoietin. *Science* **273,** 458–463.

20. Livnah, O., Stura, E. A., Johnson, D. L., et al. (1996) Functional mimicry of a protein hormone by a peptide agonist: the EPO receptor complex at 2.8A. *Science* **273,** 469–471.

21. Cwirla, S. E., Balasubramanian, P., Duffin, D. J., et al. (1997) Peptide agonist of the thrombopoietin receptor as potent as the natural cytokine. *Science* **276,** 1696–1699.

22. Pillutla, R. C., Hsiao, K., Brissette, R., et al. (2001) A surrogate-based approach for post-genomic partner identification. *BMC Biotechnol.* **1,** 6–14.

23. Blume, A. J., Beasley, J., and Goldstein N. I. (2000) The use of peptides in Diogenesis®: a novel approach to drug discovery and phenomics. *Biopolymers* **55,** 347–356.

24. Zozulya, S., Lioubin, M., Hill, R., Abram, C., and Gishizky, M. (1999) Mapping signal transduction pathways by phage display. *Nat. Biotechnol.* **17,** 1193–1198.

25. Tong, A. H., Drees, B., Nardelli, G., et al. (2002) A combined experimental strategy to define protein interaction networks for peptide recognition modules. *Science* **295,** 321–324.

26. Hsiao, K. C., Brissette, R. E., Wang, P., et al. (2003) Peptides identify multiple hotspots within the ligand binding domain of the TNF receptor 2. *Proteome Sci.* **1,** 1–10.

14

Yeast and Mammalian Two-Hybrid Systems for Studying Protein–Protein Interactions

Shu-ichi Matsuzawa and John C. Reed

Summary

An important step in the analysis of protein function is identification of the interaction partners of each protein. The two-hybrid system has been widely used to identify and explore protein–protein interactions. By using various two-hybrid systems, numerous protein interactions that regulate apoptosis signaling have been discovered that reveal unexpected functions of cancer-relevant proteins. Methods for performing two-hybrid experiments using either yeast or mammalian cells will be described in this chapter.

Key Words: Mammalian two-hybrid system; protein interaction; yeast two-hybrid system; prey; bait; reporter.

1. Introduction

Protein–protein interactions play crucial roles in various cellular events such as cell growth, cell death, protein trafficking, DNA replication, transcription, translation, and metabolism. The two-hybrid system is a useful method for studying protein–protein interactions in vivo cells. The yeast two-hybrid system, originally developed by Song and Fields *(1)*, has been used in many laboratories not only to study known protein–protein interactions but also to identify novel interacting partners for proteins of interest. The yeast two-hybrid method has also been developed and modified for use in bacterial and mammalian systems. These methods can be exploited for high-throughput screens of protein interactions, studies of the effects of mutations on protein interactions, and drug screening, among other uses.

The term "two-hybrid" refers to expression of candidate interacting proteins as chimeric fusion proteins containing complementary protein domains that when brought into close apposition result in reconstitution of function of the separately fused domain to yield a reportable signal. The most commonly

From: *Methods in Molecular Biology, vol. 383: Cancer Genomics and Proteomics: Methods and Protocols*
Edited by: P. B. Fisher © Humana Press Inc., Totowa, NJ

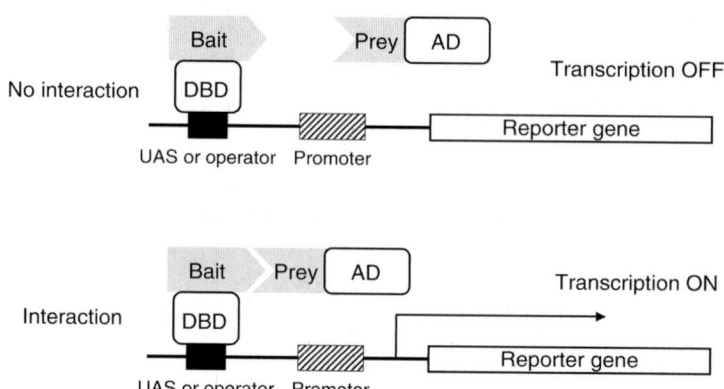

Fig. 1. Diagram of the two-hybrid system. Interaction between "bait" and "prey" brings the activation domain (AD) into close proximity with the DNA binding domain (BD), resulting in transcription of reporter genes.

employed two-hybrid method is based on the modular nature of transcription factors, and uses the transactivation and the DNA-binding domains (DBD) of transcription factors as fusion partners that are appended to test proteins. When a protein expressed as a fusion containing the DBD interacts with another partner expressed as a fusion containing the transcription domain, then a functional transcription factor is reconstituted, this can be measured using a standard reporter gene assay. Many other variations of the two-hybrid method have been reported *(2–5)*, but none has proved as robust and reliable as the method is based upon merger of complementary domains from transcription factors.

The principle of the two-hybrid system that describes here is shown in **Fig. 1**. The system requires at least three plasmids, called "bait," "prey," and "reporter." The bait plasmid encodes a protein of interest that is fused to a DBD, while the prey plasmid encodes a target protein that is fused to the transcriptional activation domain (AD). These plasmids are then co-introduced into appropriate host cells along with the reporter plasmids. Should the bait and prey proteins physically interact, this allows the DBD to recognize a specific DNA sequence upstream of the reporter gene with the AD facilitating assembly of the transcription machinery and transcription of the reporter genes downstream of the DBD. Thus, the interaction between protein of interest and its potential partner can be measured by the level of reporter gene activity. Several of the currently available yeast and mammalian two-hybrid systems are listed in **Table 1**.

2. Materials

2.1. Equipments

1. Incubator (30°C).
2. 5% CO_2 Incubator (37°C).

Table 1
Yeast and Mammalian Two-Hybrid System

Host	DNA binding domain (plasmid, marker)	Activation domain (plasmid, marker)	Host strain	Reporter (plasmid, marker)	Reference	Vender
	Lex A (pGilda, His)	B42 (pJG4-5, Trp)	EGY48 or 191	LEU2[a] or LacZ (pSH18-34, Ura)	*6,7*	Origene Invitrogen
Yeast	Lex A (pBTM116, Trp)	VP16 (pVP16, Leu)	L40	HIS3[a] or LacZ (pSH18-34, Ura)	*8*	
	GAL4 (pAS1, Trp)	GAL4 (pACT, Leu)	Y190	HIS3[a] and LacZ (pSH18-34, Ura)	*9*	Stratagene
	GAL4 (pCMV-BD)	NF-κB (pCMV-AD)	Mammalian cell line[b]	Luciferase (pFR-Luc)	–	Stratagene
Mammalian	GAL4 (PSV40-GAL4)	VP16 (PSV40-VP16)	Mammalian cell line[b]	LacZ (pGAL/lacZ)	*10–12*	Invitrogen
	GAL4 (pM)	VP16 (pVP16)	Mammalian cell line[b]	CAT or SEAP[c] (pG5CAT or pG5SEAP)	*10–12*	BD Biosciences

[a]These genes are integrated in chromosome of host strains.
[b]Any type of mammalian cell lines that have high transfection efficiency (i.e., HEK293, HeLa, CHO, COS, CV-1, PPC1, or NIH3T3).
[c]Secreted alkaline phosphatase.

3. Centrifuge.
4. Microcentrifuge.
5. Shaker (30°C).
6. Luminometer.

2.2. Chemicals

1. Glucose (Sigma, St. Louis, MO).
2. Galactose (Sigma).
3. Raffinose (Sigma).
4. PEG3350 (50% [w/v]) (Sigma). Dissolves in dH_2O and sterilized by autoclave.
5. DMSO (methyl sulfoxide, 99.9% high-performance liquid chromatography grade, Sigma).
6. X-gal (Sigma).
7. Lipofectamine™ 2000 reagent (Invitrogen, Carlsbad, CA).
8. Bacto yeast extracts (BD Biosciences, Spark, MD).
9. Bacto tryptone (BD Biosciences).
10. Bacto agar (BD Biosciences).
11. Lithium acetate (LiOAc) (Sigma).

2.3. Stock Solutions

1. Amino acid stock solutions. Dissolve in dH_2O sterilize with filter membrane (0.22 μm) and store at 4°C:
   a. 10 mg/mL Leucine (Sigma).
   b. 10 mg/mL Tryptophan (Sigma).
   c. 10 mg/mL Histidine (Sigma).
2. 10^4X Metal solutions: dissolve in dH_2O, sterilize with filter membrane (0.22 μm) and store at 4°C:
   a. Metal solution A (50 mL): 30 mg H_3BO_3, 50 mg $MnSO_4 \cdot 7H_2O$.
   b. 150 mg $ZnSO_4 \cdot 7H_2O$, and 20 mg $CuSO_4 \cdot 5H_2O$.
   c. Metal solution B (50 mL): 100 mg $Na_2MoO_4.2H_2O$.
   d. Metal solution C: (50 mL): 125 mg $FeCl_3 \cdot 6H_2O$.
3. 0.05% KI solution: dissolve 0.05 g KI in 200 mL dH_2O, sterilize with filter membrane (0.22 μm), and store at 4°C.
4. 10^3X Vitamin stock solution: dissolve 20 mg each of thiamine, pyridoxine, nicotinic acid, pantothenic acid, 0.2 mg biotin, and 1 g inositol in 100 mL of dH_2O, filter sterilize (0.22 μM), and store at 4°C.
5. 4X Burkholder's minimum media (BMM) stock solution (1 L):
   Dissolve 6 g KH_2PO_4, 2 g $MgSO_4 \cdot 7H_2O$, and 1.32 g $CaCl_2 \cdot 2H_2O$ in 900 mL of dH_2O, add 0.8 mL KI solution (0.05%), 0.2 mL each of metal solutions A, B, and C, and adjust the final volume to 1 L. Autoclaving is not necessary and the solution should be stored at room temperature.
6. 10X TE (pH 7.5): 100 mM Tris-HCl, 10 mM EDTA, pH 7.5. Autoclave at 120°C for 20 min.

7. ssDNA (single-strand carrier DNA). Dissolve Salmon sperm DNA (Sigma) in 1X TE buffer at a concentration of 10 mg/mL. Place in a beaker with a stir-bar and stir at 4°C for overnight. Place solution in a beaker and shear the DNA by sonicating briefly using a sonication bath to an average size of 7 kb but ranging from 2 to 15 kb.
8. 0.1M LiOAc/0.5X TE pH 7.5: mix 5 mL 1 M LiOAc (pH 7.5), 2.5 mL 10X TE (pH 7.5), and 42.5 mL dH$_2$O.
9. 40% PEG3350/0.1 M LiOAc/1XTE pH 7.5: mix 5 mL of 1 M LiOAc (pH 7.5), 5 mL of 10X TE (pH 7.5), and 40 mL 50% PEG3350.
10. Z-buffer: 60 mM Na$_2$HPO$_4$, 40 mM NaH$_2$PO$_4$, 10 mM KCl, and 1 mM MgSO$_4$, pH 7.0.
11. Lysis buffer for mammalian cells: 10 mM Tris-HCl, pH 7.8, 10 mM NaCl, 0.4 mM EDTA, 0.2 mM MgSO$_4$, 1 mM dithiothreitol, and 1% Triton X-100.

2.4. Media

1. YPD medium:

YPD	Media	Plate
Bacto yeast extracts (g)	5	5
Bacto tryptone (g)	10	10
Glucose (g)	10	10
Bacto agar (g)	–	8
Distilled/deionized water (mL)	375	375

Autoclave 120°C for 30 min.

2. Burkholder's minimum media:

BMM	Media	Plate
4X BMM (mL)	125	125
Vitamin stock (mL)	0.5	0.5
Glucose or galactose/raffinose (g)	10	10
L-Asparagine (g)	1	1
Amino acid stock solution (mL)	1	1
Difco bacto agar (g)	–	8
Distilled/deionized water (mL)	375	375

Autoclave at 120°C for 30 min. Vitamin stock and amino acids solution must be added after autoclaving. Other sugars such as 2% galactose and 1% raffinose can be used instead of glucose for transcription induction (*see* **Note 1**).

BMM/Glu/Leu: 1X BMM with 2% glucose and 0.2 µg/mL leucine.

BMM/Glu/Leu/Trp: 1X BMM with 2% glucose, 0.2 µg/mL leucine, and 0.2 µg/mL tryptophan.

BMM/Gal: 1X BMM with 2% galactose, 1% raffinose, and 0.2 µg/mL leucine.

BMM/Gal/Leu: 1X BMM with 2% galactose, 1% raffinose, and 0.2 µg/mL leucine.

3. Complete medium for HEK293T cells: Dulbecco's modified Eagle's medium with high glucose (Irvine Scientific, Santa Ana, CA) supplemented with 10% bovine fetal serum (Irvine Scientific).

2.5. Cells

1. Yeast strains: *Saccharomyces cerevisiae* EGY48 (MATa, *trp*1, *ura*3, *his*, *leu*2:*plexApo*6-*leu*2), EGY191 (MATa, *trp*1, *ura*3, *his*, *leu*2:*plexApo*6-*leu*2).
2. Mammalian cell line: HEK293T (human epithelial kidney cancer cell line, ATCC).

2.6. Plasmids

1. pGilda, pJG4-5, pYESTrp, pSH18-34, and pRB1840 (Invirogen).
2. cDNA library plasmids (Invitrogen or OriGene Technology Inc. Rockville, MD).

3. Methods
3.1. Yeast Transformation

Transformation is performed using a LiOAc method that is more convenient in terms of speed and simplicity than the protoplast method using zymolyase *(13,14)*. The efficiency of this transformation method should be 1×10^3 to 1×10^4 transformants per microgram plasmid DNA. Here, we show a method developed by Brent and Golemis *(6,7)* that uses the pGilda plasmid encoding a bait protein fused to Lex A (DBD), and the pJG4-5(pYESTrp) plasmid that encodes a prey protein fused to B42 (AD), to transform either the EGY48 or EGY191 strain. However, these methods are generally applicable to similar yeast two hybrid systems based on expression of fusion proteins that utilize other components of transcription factors, different reporter genes, and different yeast strains (*see* **Notes 2** and **4**).

3.1.1. Small-Scale Transformation

1. Inoculate 2 mL of YPD media with yeast strains (EGY48 or EGY191) and grow to stationary phase (OD_{600} >1.0) at 30°C with shaking.
2. Transfer 2 mL of stationary culture into 100 mL YPD media and grow at 30°C with shaking for approx 4–6 h or until the OD_{600} = 0.3–0.5 (optimally 0.4).
3. Transfer the culture to 50-mL tubes and centrifuge at 1500*g* for 5 min. Discard the supernatant, wash the pellet once with dH_2O, and resuspend the pellet in 4 mL 0.1 *M* LiOAc/0.5X TE.
4. Let the suspension stand at room temperature at least for 10 min. Competent cells can be left at room temperature overnight before proceeding without significant loss of transformation efficiency.
5. In an Eppendorf tube, mix 1 µg plasmids (pGilda-bait, pJG4-5-prey, and pSH18-34 or pRB1840 reporter) and 10 µg ssDNA, previously boiled for 5 min and chilled on ice. Vortex well.

6. Add 100 μL competent cells to the Eppendorf tubes, vortex well, and incubate for 30 min at 30°C.
7. Add 0.7 mL 40% PEG3350/0.1 *M* LiOAc/1xTE to the tubes, vortex well, and incubate for 1 h at 30°C.
8. Add 88 μL DMSO and mix by gentle vortexing.
9. Heat shock at 42°C for 7 min. Quickly cool the tubes in room temperature water.
10. Microcentrifuge the tubes at top speed for 15 s and remove the supernatant by aspiration.
11. Resuspend the pellets in 1 mL sterile dH$_2$O and by gentle pipetting.
12. Microcentrifuge the tubes for 15 s at top speed and decant the supernatant. Resuspend the pellets in the remaining liquid and spread onto a BMM/Glu/Leu plate (100 mm).
13. Incubate at 30°C for 3–4 d until colonies appear.

3.1.2. Large-Scale Transformation (Library Transformation)

1. Inoculate the yeast strain that was transformed with pGilda (bait) and pSH18-34 (reporter) into 2 mL BMM/Glu/Leu/Trp media and grow to stationary phase (OD$_{600}$>1.0) at 30°C.
2. Transfer 2 mL of the culture into 500 mL BMM/Glu/Leu/Trp media and grow with shaking overnight.
3. When the OD$_{600}$ reaches 0.3–0.5 (0.4 is optimal), transfer the culture to 250-mL tubes and centrifuge at 1500*g* for 5 min. Discard the supernatant, wash the pellet once with dH$_2$O, and resuspend the pellet in 5 mL 0.1 *M* LiOAc/0.5X TE.
4. Let stand at room temperature at least for 10 min.
5. Mix 100 μg library plasmids (pJG4-5 or pYESTrp) and 5 mg of ss DNA. Vortex well.
6. Add 5 mL competent cells to the 50-mL tubes, vortex well, and incubate for 30 min at 30°C.
7. Add 35 mL of 40% PEG3350/0.1 *M* LiOAc/1X TE to the tubes, vortex well, and incubate for 1 h at 30°C.
8. Add 4.4 mL DMSO and mix by gentle vortexing.
9. Heat shock at 42°C for 7 min. Quickly cool the tubes in room temperature water.
10. Centrifuge the tubes for 15 s at top speed and remove the supernatant by aspiration.
11. Gently resuspend the pellets in 50 mL dH$_2$O.
12. Centrifuge the tubes at 1500*g* for 5 min and decant the supernatant. Resuspend the pellets with 5 mL sterile dH$_2$O and spread 1 mL each of suspension onto BMM/Gal plates (150 mm × 15).
13. Incubate at 30°C for 3–4 d until colonies appear.

3.2. Assay for LEU2 Reporter Activity

The *LEU2* reporter activity is measured by plating transformants onto plates that lack leucine. On such plates, protein–protein interactions will correlate with colony growth.

1. Pick 5–10 independent colonies from each BMM/Glc/Leu plate and resuspend in 10 mL sterile water. The suspension is plated onto BMM/Gal plate.
2. Incubate for 3–4 d at 30°C. Colonies on minus leucine plate will be visible within 2 d.

3.3. Assay for LacZ Reporter Activity (X-gal Assay)

The activity of *LacZ* reporter is measured by using X-gal as a substrate for the *LacZ* gene product (β-galactosidase) in a filter assay *(15,16)*. The β-galactosidase cleaves the X-gal and produces a product that turns the yeast colony blue.

1. Pick at least five independent colonies from each pair of constructs, place on BMM/Glu/Leu plate (100 mm) and incubate for 1–2 d at 30°C.
2. Lift colonies by gently placing a nylon membrane on the BMM/Glu/Leu plate, making sure the filter is completely wet, and then pull off. Place the filter membrane on BMM/Gal/Leu plate, colony side up, and incubate overnight at 30°C in order to induce protein expression.
3. Place the filter membrane in liquid nitrogen for 30 s to lyse the cells.
4. Remove the membrane from liquid nitrogen using forceps and let it thaw at room temperature.
5. Cut two pieces of Whatman 3MM filter paper slightly larger than the filter membrane and soak them in 5 mL of Z-buffer containing 0.5 mg/mL X-gal. Place the thawed membrane atop the soaked filter papers.
6. Incubate at 30°C and monitor for color changes at 1, 2, 3, and 6 h, respectively.

3.4. Mammalian 2-Hybrid Assay

The same strategy for studying protein interactions in yeast can be adapted for essentially any type of cells, including mammalian cells. The disadvantage of mammalian cells is transfection efficiency and the lack of convenient methods for analyzing millions of individual transfectants as colonies derived from independent clones. However, use of mammalian cells may enable detection of interactions between mammalian proteins that do not fold properly in yeast, posttranslational modification or additional interacting proteins or cellular factors for the interaction to occur. The methods for performing mammalian two-hybrid experiments were described using reagents from Stratagene, Inc., but similar systems of comparable utility have been reported and are available conveniently.

1. Seed HEK293T cells ($0.5–1 \times 10^5$) in 1 mL complete medium per well in a 24-well plate.
2. Incubate cells at 37°C in a 5% CO_2 incubator for 24 h.
3. Dilute plasmids (0.2 μg each) encoding bait (pBD vector, Stratagene, La Jolla, CA), prey (pAD vector, Stratagene) and reporter (pFR-Luc vector, Stratagene) in 50 μL Dulbecco's modified Eagle's medium without serum (*see* **Note 3**). Mix gently.

Table 2
Host Strains

Strain	Reporter gene	LexA operator sites
EGY 48	*LEU2*	6
EGY 195	*LEU2*	4
EGY 191	*LEU2*	2
L 40	*HIS3*	4

Table 3
Reporter Plasmids

Vector	Reporter gene	LexA operator sites
pSH18-34	*LacZ*	6
pRB1840	*LacZ*	1
pJK103	*LacZ*	2

4. Add 2 µL of Lipofectamine 2000 Reagent (Invitrogen) into the diluted DNA. Mix gently and incubate for 20 min at room temperature.
5. Add the 50 µL of the DNA–lipofectamine complex into each well.
6. Incubate the cells at 37°C in a 5% CO_2 incubator for 24 h.
7. Remove the medium from the transfected cells and wash the cells once with 0.5 mL 1X PBS.
8. Remove PBS from the wells and add 150 µL lysis buffer to each well.
9. Swirl the plate gently and incubate for 15 min at room temperature.
10. Transfer the cells and all liquid by pipet to an Eppendorf tube, vortex the tube for 10–15 s, then centrifuge at 12,000g for 1 min.
11. Transfer the supernatant to a new tube and measure the light emitted from the reaction with a luminometer using an integration time of 10 s. Luciferase activity is expressed in relative light units as detected by the lunimometer.

4. Notes

1. Difco Yeast Nitrogen Base (Difco, Sparks, MD) may be used instead of BMM. Dissolve 7 g Difco Yeast Nitrogen Base without amino acids and 20 g glucose in 1 L dH_2O. Autoclave at 120°C for 20 min and add the appropriate amino acid solution.
2. The plasmid markers are the same as those used in yeast, and can be interchanged with other systems. For example, the pVP16 (prey) plasmid can be used with both pBTM116 (bait) and pAS1 (bait).
3. The plasmids of the mammalian two-hybrid system are also interchangeable. For example, pM and pVP16 vectors (BD Biosciences) can be used with pRF-Luc (Stratagene).

4. The strength of the protein interaction predicted by the two-hybrid assay generally correlates with the number of LexA operator sites upstream of the reporter genes. For example, weaker interactions can be detected by having more LexA operator sites. However, increasing the number of operator sites may increase the assay background and consequently the number of false positive results. On the other hand, fewer LexA operators reduce the background and the number of false positives. Thus, selecting the appropriate strain and reporter plasmid allows optimization of the system to suit specific needs. Several available strains and vectors are listed in **Tables 2** and **3**.

References

1. Fields, S. and Song, O. (1989) A novel genetic system to detect protein–protein interactions. *Nature* **340**, 245–246.
2. Aronheim, A., Zandi, E., Hennemann, H., Elledge, S. J., and Karin, M. (1997) Isolation of an AP-1 repressor by a novel method for detecting protein–protein interactions. *Mol. Cell Biol.* **17**, 3094–3102.
3. Mohler, W. A. and Blau, H. M. (1996) Gene expression and cell fusion analyzed by lacZ complementation in mammalian cells. *Proc. Natl. Acad. Sci. USA* **93**, 12,423–12,437.
4. Blakely, B. T., Rossi, F. M., Tillotson, B., Palmer, M., Estelles, A., and Blau, H. M. (2000) Epidermal growth factor receptor dimerization monitored in live cells. *Nat. Biotechnol.* **18**, 218–222.
5. Galarneau, A., Primeau, M., Trudeau, L. E., and Michnick, S. W. (2002) β-lactamase protein fragment complementation assays as in vivo and in vitro sensors of protein protein interactions. *Nat. Biotechnol.* **20**, 619–622.
6. Gyuris, J., Golemis, E. A., Chertkov, H., and Brent, R. (1993) Cdi1, a human G1 and S phase protein phosphatase that associates with Cdk2. *Cell* **75**, 791–803.
7. Estojak, J., Brent, R., and Golemis, E. A. (1995) Correlation of two-hybrid affinity data with in vitro measurements. *Mol. Cell Biol.* **15**, 5820–5829.
8. Vojtek, A. B., Hollenberg, S. M., and Cooper, J. A. (1993) Mammalian *Ras* interacts directly with the serine/threonine kinase *Raf. Cell* **74**, 205–214.
9. Harper, J. W., Adami, G. R., Wei, N., Keyomarsi, K., and Elledge, S. J. (1993) The p21 Cdk-interacting protein Cip1 is a potent inhibitor of G1 cyclin-dependent kinases. *Cell* **75**, 805–816.
10. Sadowski, I., Bell, B., Broad, P., and Hollis, M. (1992) GAL4 fusion vectors for expression in yeast or mammalian cells. *Gene* **118**, 137–141.
11. Dang, C. V., Barrett, J., Villa-Garcia, M., Resar, L. M., Kato, G. J., and Fearon, E. R. (1991) Intracellular leucine zipper interactions suggest c-Myc hetero-oligomerization. *Mol. Cell Biol.* **11**, 954–962.
12. Fagan, R., Flint, K. J., and Jones, N. (1994) Phosphorylation of E2F-1 modulates its interaction with the retinoblastoma gene product and the adenoviral E4 19 kDa protein. *Cell* **78**, 799–811.
13. Ito, H., Fukuda, Y., Murata, K., and Kimura, A. (1983) Transformation of intact yeast cells treated with alkali cations. *J. Bacteriol.* **153**, 163–168.

14. Schiestl, R. H. and Gietz, R. D. (1989) High efficiency transformation of intact yeast cells using single stranded nucleic acids as a carrier. *Curr. Genet.* **16,** 339–346.
15. Brent, R. and Ptashne, M. (1985) A eukaryotic transcriptional activator bearing the DNA specificity of a prokaryotic repressor. *Cell* **43,** 729–736.
16. Schneider, S., Buchert, M., and Hovens, C. M. (1996) An in vitro assay of β-galactosidase from yeast. *Biotechniques* **20,** 960–962.

15

Ribonomic and Short Hairpin RNA Gene Silencing Methods to Explore Functional Gene Programs Associated With Tumor Growth Arrest

Timothy E. Baroni, Michele T. Lastro, Aparna C. Ranganathan, Scott A. Tenenbaum, Douglas S. Conklin, and Julio A. Aguirre-Ghiso

Summary

In this chapter, we present an approach using genomic and ribonomic profiling to investigate functional gene programs in a tumor growth model. To reach this goal, ribonomic profiling was combined with RNA interference in a tumor dormancy model. Strategies merging functional genomic technologies are outlined for the identification of novel posttranscriptionally regulated targets of p38 to show that they are functionally linked to the induction or interruption of cellular growth in cancer. In the first section of this chapter, we describe a method for the detection of mRNA subsets associated with RNA-binding proteins such as hnRNP A1 using (1) immunopurification of mRNA–protein complexes, from either whole cell lysates or subcellular fractions and (2) gene expression arrays to find those mRNAs bound to hnRNP A1. In the second section, short hairpin RNA technology was used to create a library of shRNAs that target p38 induced mRNAs expression libraries are utilized to "knockdown" the genes identified in the first section. Finally, this library of gene candidates is evaluated in vivo to address their functional role in the induction or maintenance of dormancy.

Key Words: Expression profiling; gene silencing; immunoprecipitation (IP); RNA-binding Protein (RBP); short hairpin RNA (shRNA); tumor dormancy.

1. Introduction

Cancer dormancy is a stage wherein the cancer cells enter a protracted slow or nondividing state, which can persist for years until the appearance of clinically detectable metastases *(1)*. Cellular dormancy in human cancers has been frequently suggested as a major mechanism by which cancer cells evade chemotherapy *(1,2)*. However, lacking well-defined model systems, the molecular

From: *Methods in Molecular Biology, vol. 383: Cancer Genomics and Proteomics: Methods and Protocols*
Edited by: P. B. Fisher © Humana Press Inc., Totowa, NJ

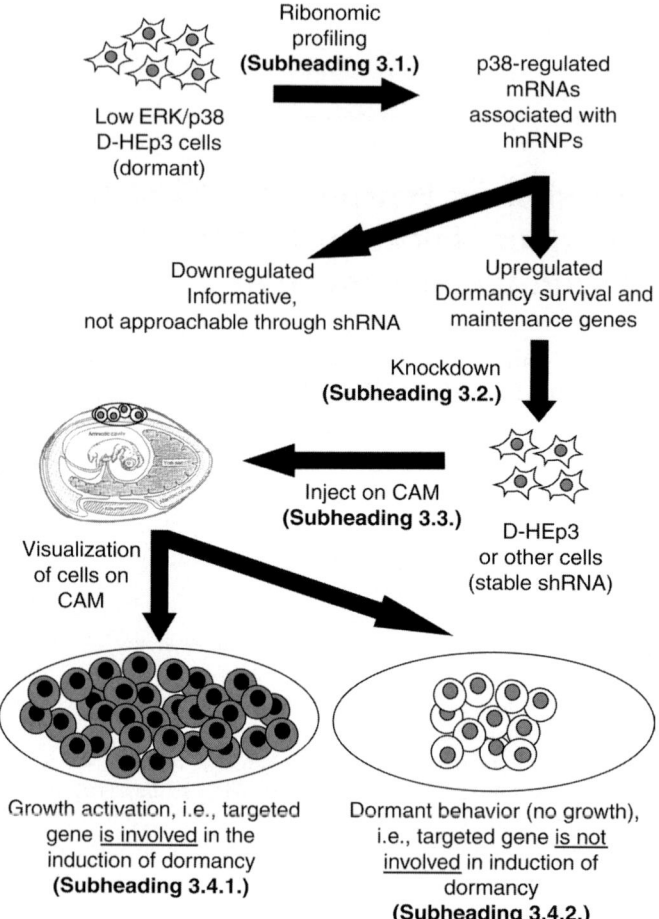

Fig. 1. Overview of the strategy. Immunoprecipitation of hnRNPs from dormant/ non-tumorigenic cells allows for the identification of p38-target mRNAs (**Subheading 3.1.**). The upregulated messages may be knocked down via shRNA expressing vectors (**Subheading 3.2.**). The cells expressing shRNAs are introduced onto the chorioallantoinc membrame of a chick embryo (**Subheading 3.3.**). Growth of the shRNA cells above that of the parental cell line indicates that the knocked down gene is involved in tumor dormancy or survival (**Subheading 3.4.**). The other possible outcome (not shown, *see* **Subheading 3.4.**) is that the gene is required for survival and/or growth in vivo, so no tumor nodule will be detected.

mechanisms underlying metastatic dormancy and their chemoresistance remain poorly understood. Here, we investigate the mechansims of tumor dormancy model using ribonomic profiling and short interfering RNA technologies for target identification and validation, respectively (**Fig. 1**).

A model of cancer dormancy was established in human squamous carcinoma cells (HEp3) where the ERK/p38 ratio determines their proliferative or dormant phenotype. This ratio is predictive of the proliferative behavior of various cancer cell lines *(3,4)*. A high ERK/p38 ratio favors growth, while a low ERK/p38 ratio leads to dormancy in different cell types *(3)* and a strong activation of p38 dictates the induction and maintenance of tumor dormancy *(3)*. Furthermore, we discovered that p38α regulates approx 300 genes that have been linked to the induction or interruption of dormancy. Recent evidence demonstrates that p38 regulates gene expression posttranscriptionally through p38-responsive regulatory elements in the 3′-UTRs of specific mRNAs *(5)*. Notably, it was found that p38 activation can regulate the nuclear to cytoplasmic shuttling of mRNA binding proteins such as hnRNP A1 (i.e., heterogeneous nuclear ribonucleoproteins A1), which is implicated in the posttranscriptional regulation of mRNAs *(6*, Rangnathan et al., unpublished results).

Messenger RNA-binding proteins (mRBPs) play an essential role in posttranscriptional gene regulation, potentially by influencing the integrity and/or translation of cognate messages. Such posttranscriptional gene regulation has been shown to be important for partitioning messages spacially and temporally and is the basis for the functional genomic study of mRNAs, called *ribonomics (7–9)*. For example, studies in yeast have shown that the Puf family of mRBPs differentially regulates expression of subsets of mRNAs to direct the function of their encoded proteins *(10)*. Although 87% of the transcripts associated with Puf3 are annotated to have mitochondrial function, the mRNA population associated with Puf4 shows less than 5% annotated as mitochondrial and 27% as nucleolar in nature. The concept that mRBPs can manage functional networks has been reviewed extensively *(9,11)*. Therefore, the immunoprecipitation of mRNA–protein complexes (mRNPs) and subsequent analysis of the associated mRNAs can provide integral information about the function of a given transcript in a particular network. Since p38 appears to regulate specific signals through hnRNP A1 posttranscriptionally, we will identify target mRNA and link their function to the induction or maintenance of dormancy.

To aid in the functional analysis of mRNAs regulated through the p38-hnRNP A1 pathway, we describe a method of using RNA interference (RNAi) to effectively "knockdown" gene expression *(12,13)* and thereby verify its relationship to the dormant phenotype. DNA-encoded short hairpin RNAs (shRNAs) take advantage of the cell's endogenous RNAi machinery and offer a convenient, effective, and systematic method to knockdown differentially regulated genes identified by ribonomic approaches and thereby test their functional contribution to a dormant phenotype. Short hairpin RNAs designed against p38-upregulated targets may be used individually or they can be pooled to form a p38-target library to test the dormant phenotype.

Here, we outline a systematic approach to address the role of p38-regulated hnRNP A1 complexes and their associated transcripts in the induction/maintenance of tumor cell dormancy. First, we use ribonomic profiling to identify the hnRNP A1-associated messages that are regulated by p38 signaling in dormant cells. Second, stable cell lines were generated where genes found to be upregulated by p38 are "knocked down" by shRNA expression. By testing the growth capacity of shRNA expressing cells in the chick embryo chorioallantoic membrane (CAM) system, we access the functional contribution of knocked-downgenes to tumor dormancy in vivo to assess the functional contribution of knocked down genes to tumor dormancy in vivo (*14*).

2. Materials

2.1. Ribonomic Profiling

1. We recommend the use of distilled, DNase-RNase-free water (Invitrogen, Carlsbad, CA; cat. no. 10977-015) to prepare buffers and solutions described next (*see* **Note 1**).
2. All tips and tubes must be RNase free (*see* **Note 2**).
3. Antibody to RNA binding protein of interest.
4. Tissue and/or cells of interest from which RNA will be harvested.
5. Proteinase K. Prepare a solution at a concentration of 20 mg/mL. Store aliquots of 50 µL at –20°C and avoid freeze–thaw.
6. 1 *M* Dithiothreitol. Store aliquots of 20 µL at –20°C and avoid freeze–thaw.
7. 1X Phosphate-buffered saline (PBS).
8. Glycogen, molecular biology grade (Roche, cat. no. 0901393). Store at –20°C.
9. Complete tablets, proteinase inhibitor (Roche, cat. no. 1697498). Store at –20°C.
10. Bovine serum albumin (BSA)-fraction V proteinase free (Roche, cat. no. 8100350). Store at 4°C.
11. RNase OUT™ (Invitrogen, cat. no. 10777-019, 40 U/µL). Store at –20°C.
12. Superase IN™ (Ambion cat. no. 2696, 20U/µL). Store at –20°C.
13. Protein A Sepharose beads: either from Sigma (P3391) or Amersham Biosciences (Fast Flow, cat. no. 17-0974-01). Store at 4°C.
14. Protein G Agarose beads (Sigma, cat. no. P4691). Store at 4°C.
15. Polysome lysis buffer (PLB): 10 m*M* 4-(2-hydroxyethyl)-1-piperazineethanesulfonic acid (HEPES), pH 7.0, 100 m*M* potassium chloride, 5 m*M* magnesium chloride, 25 m*M* ethylenediaminetetraacetic acid (EDTA), 0.5%. Additional components are for lysis buffers I and II to be added to PLB immediately before use to give these concentrations: 2 m*M* dithiothreitol, one tablet of complete proteinase inhibitor per 50 mL of buffer, 50 U/mL RNase OUT, 50 U/mL Superase IN. Note that some of these components are added to the NT2 buffer before immunoprecipitation as noted in **step 3** of **Subheading 3.1.4.1.**

16. NT2 buffer: 50 m*M* Tris-HCl, pH 7.4, 150 m*M* sodium chloride, 1 m*M* magnesium chloride, 0.05% Nonidet P-40.
17. Proteinase K digestion buffer: 100 m*M* Tris-HCl, pH 7.5, 10 m*M* EDTA, 50 m*M* sodium chloride, 1% sodium dodecyl sulfate.
18. Lysis buffer I: 40 m*M* Tris, pH 7.4, 100 m*M* sodium chloride, and 2.5 m*M* magnesium chloride.
19. Lysis buffer II: 10 m*M* Tris, pH 7.4, 100 m*M* sodium chloride, 2.5 m*M* magnesium chloride, and 0.5% Triton X-100.
20. A microtip sonicator. We use a Fisher Scientific Model F60 sonicator in these experiments.

2.2. shRNA Gene Targeting

1. Retroviral vector containing shRNA sequence against target mRNA (*see* **Note 22**).
2. Competent *Escherichia coli* (*see* **Note 23**).
3. Bacterial culture media.
4. Plasmid purification reagents/kits such as Qiagen midi columns (*see* **Note 24**).
5. Packaging cell lines capable of producing retroviral particles (several available from American-Type Culture Collection).
6. Media for mammalian cell lines (packaging cells and experimental cell line).
7. Transfection reagents (*see* **Note 25**).
8. Experimental cell line.
9. Hexadimethrine bromide (Polybrene, 8 mg/mL) (Sigma, cat. no. H9268).
10. Antibiotic for selection of retrovirus infected experimental cells.
11. Primers for RT-PCR analysis of mRNA of interest or antibody against corresponding protein.

2.3. Tumor Growth Assay (CAM System)

1. 1X PBS containing 2 m*M* EDTA.
2. PBS-plus: 1X PBS supplemented to contain 1.0 m*M* magnesium chloride, 0.5 m*M* calcium chloride, 100 U/mL penicillin, and 100 µg/mL streptomycin.
3. Nine- to ten-day-old embryonated eggs (Charles River Laboratories).
4. Collagenase solution containing BSA: 0.15 g collagenase (Sigma, cat. no. C-9891), 2.5 g BSA (Sigma, cat. no. A-2153) in 100 mL PBS-plus (*see* reagent 2 above).

3. Methods
3.1. Ribonomic Profiling

In this chapter we present two different protocols for the immunopurification of RNA-binding proteins. The procedure described in **Subheading 3.1.4.1.** is for whole cell lysate immunopurification of RNA-binding proteins, similar to that presented previously *(15)*. Another variation of the whole cell lysate immunoprecipitation of RNA-binding proteins is presented elsewhere *(16)*. The second method in **Subheading 3.1.4.2.** is tailored to specifically immunopurify hnRNP bound mRNP complexes from cellular fractions.

3.1.1. Preparation of Whole Cell RNP Lysates From Cultured Cells

1. Use 150 or 100 m*M* dishes to grow desired cell line.
2. Wash twice with ice-cold PBS, harvest using a scraper, and pellet the cells by centrifugation at 3000*g* for 5 min at 4°C.
3. Generate the RNP cell extract by estimating the pellet volume and adding approx 1.5 vol PLB for isolating cytoplasmic RNPs or whole cell lysis buffer II for isolating both nuclear and cytoplasmic RNPs.
4. Pipet several times until the extract looks uniform, and spin in a microcentrifuge at 14,000*g* for 10 min at 4°C.
5. Remove the supernatant and save.
6. Freeze the cell extracts in aliquots of 200–400 µL and store at –80°C (*see* **Note 3**). The extracts are ready to proceed to mRNP isolation (*see* **Subheading 3.1.4.1.**).

3.1.2. Preparation of Cellularly Fractionationed mRNP Lysates for hnRNP Immunoprecipitation

This protocol is a modified version of a previously published protocol for immunopurification of mRNP complexes from cytoplasmic and nuclear fractions *(17)*.

1. Perform all fractionation steps on ice.
2. Grow the cells to confluence in 10-cm dish.
3. Wash the cells three times in ice-cold PBS and collect with a cell scraper in 0.75–1 mL of lysis buffer I (*see* **Notes 4** and **5**).
4. Transfer to a clean microcentrifuge tube and vortex vigorously for 5–10 s.
5. Incubate on ice for 10 min.
6. Centrifuge the sample at 2000*g* for 8 min at 4°C. Remove the soluble cytosolic fraction (i.e., the supernatant) from the nuclear fraction (i.e., the pellet). This cytosolic fraction can then be carried onto the mRNP immunoprecipitation **(Subheading 3.1.4.2.).** Proceed to **step 7** with the pellet.
7. Resuspend the pellet from **step 6** in lysis buffer II.
8. Incubate on ice for 5 min.
9. Centrifuge the sample at 2000*g* for 8 min at 4°C (*see* **Note 5**). Remove the Triton-extracted material by pipet. Proceed to **step 10** with the pellet.
10. Resuspend the pellet from **step 9** in lysis buffer II and sonicate on ice for 10 s using a microtip sonicator.
11. Centrifuge the sonicated material at 4000*g* for 15 min at 4°C.
12. The supernatant (representing the nuclear RNP fraction) can then be used (if desired) in the mRNP immunoprecipitation **(Subheading 3.1.4.2.).**

3.1.3. Antibody Coating of Bead Matrix

1. Swell or resuspend the desired amount of protein G agarose beads (for monoclonal antibodies) or protein A Sepharose beads (for rabbit serum or rabbit polyclonal antibodies) in 5–10 volumes of NT2 containing 5% BSA or lysis buffer I. IMPORTANT! (*see* **Notes 5–8**).
2. Add the immunoprecipitating antibody or serum and incubate for one hr at room temperature on a rotating device or from 1 to 12 h at 4°C (*see* **Notes 9–11**).
3. Beads coated with antibodies can be stored for several months at 4°C when buffers are supplemented to contain 0.02% sodium azide.

3.1.4. Immunoprecipitation of mRNPs

3.1.4.1. WHOLE CELL MRNP ISOLATION

1. Centrifuge the RNP lysate in a microcentrifuge at 14,000g for 10 min at 4°C, then transfer the supernatant to a new tube on ice.
2. Calculate the amount of antibody-coated beads necessary to perform the appropriate number of immunoprecipitations you are planning and wash the beads several times at room temperature with NT2 buffer (*see* **Note 11**).
3. Resuspend the antibody-coated beads in NT2 buffer supplemented with 50 U/mL RNase OUT, 50 U/mL Superase IN, 1 mM dithiothreitol, and 20–30 mM EDTA.
4. The volume of resuspended beads in NT2 buffer should correspond to approx 10 times the volume of the RNP lysate being used (*see* **Note 12**).
5. Mix the resuspended antibody-coated beads several times by inversion, add the RNP lysate and tumble the immunoprecipitation reactions end-over-end at room temperature for at least 2–4 h at 4°C (but preferably overnight). A sample of the supernatant can be collected at the beginning of the incubation to serve as a total RNA control, which can assess RNase activity (*see* **Notes 13–15**).
6. After the incubation, spin the beads down and wash with approx 10–20 bed volumes of ice-cold NT2 buffer, vigorously mixing between each rinse. Repeat four to six times (*see* **Note 16**).
7. Resuspend the washed beads in 600 µL proteinase K digestion buffer plus 25 µL proteinase K stock solution and incubate for 30 min at 50°C, mixing occasionally.
8. Add 600 µL phenol:chloroform:isoamyl alcohol (v/v 25:24:1) to the bead suspension, vortex for 1 min, and centrifuge at 14,000g for 10 min at 4°C.
9. Remove the upper phase and repeat the extraction with one volume of chloroform.
10. Precipitate the RNA by adding one volume of isopropanol, 60 µL 4 M ammonium acetate, 3 µL 1 M magnesium chloride, and 8 µL glycogen.
11. Store samples at −80°C until ready for gene expression analysis.
12. To recover RNA, centrifuge samples at 14,000g for 30 min at 4°C and wash with 100 µL 80% ethanol (*see* **Note 17**).

3.1.4.2. Fractionated Cellular mRNP Isolation

This protocol is specifically designed to the immunopurification of RNP complexes bound to hnRNP proteins. The addition of protease and RNase inhibitors is to be avoided during the lysis step (*see* **Note 18**).

1. Typically, it is recommended to use 10 µg of antibody per immunopurification from 10-cm plate.
2. Wash the antibody-protein A complex five to six times with lysis buffer I.
3. Resuspend the antibody-protein A bead complex in lysis buffer I (*see* **Note 6**).
4. Mix the antibody-protein A complex gently by inverting the tube several times.
5. Add the RNP extracts from the cytoplasm or nucleus to the antibody-protein A-agarose complex. Incubate the complex on an orbital shaker at 4°C for no more than 10–20 min (*see* **Note 19**).
6. Following incubation, spin the RNP-antibody bound beads by centrifuging at 1000 rpm for 10–20 s.
7. Wash the beads five to six times with lysis buffer I. Mix the beads after each wash by inverting the tube several times.
8. Resuspend the immunopurified complexes in 200 µL Tris-EDTA buffer containing 1% SDS and incubate at 65°C for 5 min.
9. Purify RNA through phenol/chloroform extraction and precipitate with isopropanol (*see* **Subheading 3.1.4.1.**, **steps 7–12**).

3.1.5. DNase Treatment of Isolated RNA (Optional)

1. Following RNA extraction, incubate the RNA with 4 U of RQ1 DNase (Promega, Madison, WI) for 30 min at 37°C and recover the RNA by phenol extraction and ethanol precipitation.

3.1.6. Expression Analysis of mRNA From mRNPs

1. Targets of mRNA binding proteins can be identified and quantified by several methods **(Fig. 2)**. Techniques in which mRNAs can be identified directly, without amplification *(15)* are preferred.
2. Multiprobe-based RNase protection assays (PharMingen, San Diego, CA) are an ideal alternative for the optimization and high-throughput analysis of mRNP immunoprecipitations *(15)*.
3. Many RNP-associated mRNAs were identified using cDNA/genomic arrays. It was found that Affymetrix tiling array platform and the BD Pharmingen Atlas Nylon cDNA Expression Array platforms were *(15)* excellent for conducting ribonomic analysis. Irrespective of the array platform used, apropriate controls and normalization of signals should be considered on a case by case basis (*see* **Notes 20** and **21**, and **refs.** *18–20*).
4. If gene expression analysis is performed using glass arrays that use Cy3 and Cy5 labeling or on Affymetrix arrays, typically it is recommended to increase the amount of extract by three to five times that required for Atlas Nylon arrays.

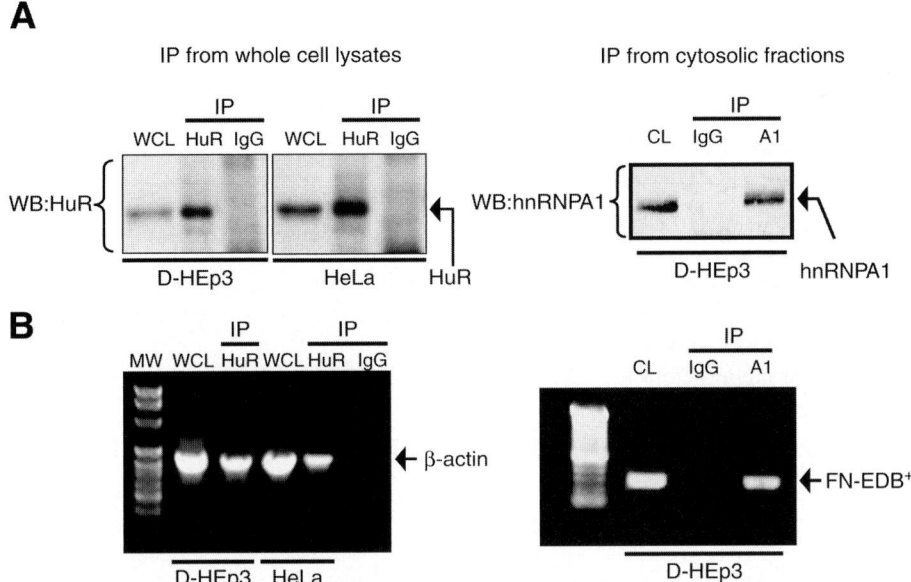

Fig. 2. **(A)** Immunopurification of mRNA binding proteins HuR and hNRNP A1. Whole cell lysates **(Subheading 3.1.1.)** and cytoplasmic lysates **(Subheading 3.1.2.)** were immunopurified following the approprate procedure **(Subheadings 3.1.4.1.** and **3.1.4.2.**, respectively) using anti-HuR and anti-IgG and anti-hnRNP A1antibodies accordingly. A fraction of the immunopurified protein was analysed by Western blotting using anti-HuR antibody (left panel), or anti-hnRNP A1 antibody (right panel). Both HuR and hnRNP A1 are specifically immunopurified compared with their respective IgG controls. WCL, whole cell lysates; CL, cytoplasmic lysates; WB, Western blot. **(B)** RT-PCR analysis of mRNA associated with HuR and hnRNP A1 mRNP complex. RNA extracted from HuR and hnRNP A1 mRNP complexes **(Subheadings 3.1.4.1.** and **3.1.4.2.**, respectively) was subjected to RT-PCR analysis for the targets β-actin (left panel) or fibronectin-EDB (FN-EDB+, right panel). Fibronectin-EDB is the alternative spliced form of fibronectin mRNA. As hnRNP A1 is involved in alternative splicing, we tested for the association of EDB mRNA with hnRNP A1.

3.2. shRNA Gene Silencing for Functional Analysis of Cancer Cell Growth

The results from ribonomic profiling will provide a list of putative genes involved in the induction of cancer cell dormancy or survival. Plasmid encoded shRNAs may be tested individually or, alternatively, the system can be easily expanded by pooling the plasmids to form a library of shRNAs targeting genes upregulated by p38. Modifiers of tumor dormancy are then screened in a high throughput manner and targets are identified via a unique barcode sequence associated with each shRNA *(21)*.

Hairpin sequences for a gene are designed and inserted into a plasmid capable of retroviral packaging (**Fig. 3**, *see also* **Note 26**). For space and simplicity, the single target approach is presented here and the library approach has been described previously in **ref. *21***.

3.2.1. Growth and Preparation of shRNA-Encoding Plasmid DNA

1. Once an oligonucleotide has been designed that will form a short hairpin structure complimentary to a target message it can be ligated into a plasmid capable of retroviral packaging *(21)*.
2. Transform competent *E. coli.* with shRNA containing plasmid following standard protocols.
3. Pick a single colony and seed 250 mL liquid culture for plasmid preparation (*see* **Note 24**).

3.2.2. Production of Stable Cell Lines Expressing shRNAs

1. After the plasmid containing the appropriate shRNA has been purified, we may proceed to the production of stable cell lines.
2. *Day 1.* Plate exponentially growing packaging cells (several lines are available from American Type Culture Collection) so that they will be approx 30–50% confluent the following day (*see* **Note 27**).
3. *Day 2.* One to two hours before transfection replace the old media with fresh serum containing media.
4. For transfection use about 15 µg plasmid DNA per 10-cm plate or 1–4 µg per well in a six-well dish, *see* **Note 28**). Several methods are available for the introduction of plasmid DNA into cells (*see* **Note 25**).
5. Following transfection incubate the cells at 37°C for 72 h (*see* **Note 29**).
6. *Day 4.* Plate experimental cell line such that they can continue to grow for 3 d without reaching confluency.
7. *Day 5.* Collect the viral supernatant from packaging cells (*see* **Note 30**) and filter through a 0.45-µm syringe filter. Store at 4°C if supernatant will be used immediately or –80°C for later use (*see* **Note 31**).
8. Replace experimental cell line media with 0.5 mL fresh media.
9. Add supernatant/polybrene mixture (8 µg polybrene per milliliter viral supernatant) to each well (*see* **Notes 32–34**).
10. Change media 12–24 h after infection.
11. Begin antibiotic selection 3 d postinfection (*see* **Note 35**).
12. Perform the following steps in order to assess the degree of shRNA-mediated reduction of expression.
13. Prepare RNA from the stable cell line clone, and then perform RT-PCR.
14. Alternatively, immunoblots may be performed to check for the presence and/or level of the corresponding protein.

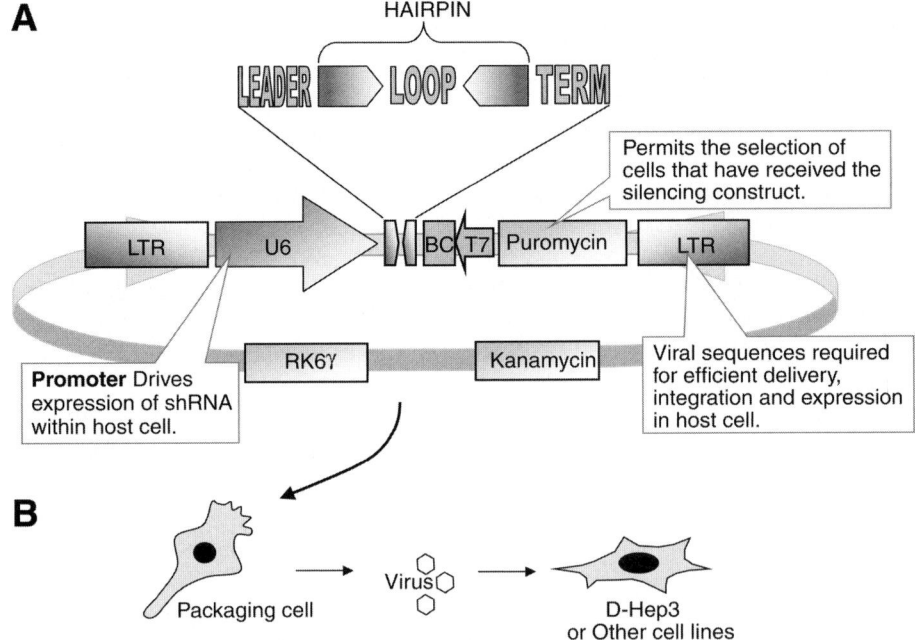

Fig. 3. Schematic representation of the retrovirus producing pSHAG-MAGIC vector. **(A)** Hairpin constructs are designed and ligated into pSHAG-MAGIC. The region flanked by two long terminal repeats contains the shRNA and the puromycin selection cassette. The vector contains the RK6γ origin which requires expression of *PIR1* for replication. **(B)** Once the hairpin has been inserted into pSHAG-MAGIC, the vector is used to transfect packaging cells capable of producing retrovirus. Viral particles are then used to infect the target cell line. The barcode sequence can be used to follow hairpin construct in complex population via microarray hybridization as described in **ref. 21**.

3.3. Tumor Growth Assay (CAM System)

In this section we present a method to successfully knockdown the expression of those messages that were found to be upregulated and associated with hnRNP A1. After having obtained dormant/non-tumorigenic (D-HEp3) cells stably expressing the shRNA (*see* **Note 36**), they will be inoculated onto the CAM of embryonated eggs (*see* **Note 37**). The functional assay to test the role of these messages in the model of cancer dormancy is tumor formation after innoculation. Although expression of some of these genes may be involved in the modulation of the dormant phenotype, others may be essential for the survival of the dormant tumor cells in vivo. There may be other genes whose increase or decrease in expression may not have any effect on the tumor cell dormancy, but they cannot be tested.

1. Using 9- or 10-d-old embryonated eggs are recommended.
2. The CAMs were prepared as previously described in **ref. *14***. Puncture the egg shell on the long side of the egg (*see* **Note 39**).
3. Apply suction though an opening over the eggshell of the natural air sac, so as to create an artificial air chamber. This results in the separation of the CAM from the eggshell (*see* **Note 39**).
4. Open a window over the displaced CAM and seal it with a piece of sterile tape.
5. Once the CAM is ready, detach stable shRNA expressing dormant/non-tumorigenic D-HEp3 cells from culture using PBS containing 2 m*M* EDTA (*see* **Notes 36** and **40**).
6. Pellet the cells by centrifuging at 3000*g* for 10 min at 4°C. Discard supernatant.
7. Wash the cells twice with 1X PBS.
8. Use 5×10^5 cells per inoculation per CAM (*see* **Note 41**).
9. Resuspend cells in **step 6** in 50 µL of PBS-plus.
10. Inoculate the cells gently onto the CAM using a sterile pipet (*see* **Note 39**).
11. Seal the opening of the eggshell over the CAM with transparent tape.
12. Incubate the eggs (tape window up) in a stationery incubator at 37°C for 3–7 d.
13. Excise the tumor nodule and weigh it.
14. At the end point of the assay (*see* **Note 42**), mince the tumor into small pieces in a clean Petri dish.
15. Incubate in type IA collagenase for 30 min at 37°C (*see* **Note 43**).
16. Count the number of tumor cells with a hemocytometer (*see* **Note 44**).
17. Analyze cell viability using Trypan blue exclusion test.

3.4. Functional Analysis of shRNA Mediated Knockdown of Messages

shRNA mediated downregulation of genes that were found to be upregulated in the dormant cells may have three phenotypic outcomes upon inoculation on the CAM (**Fig. 4**).

1. *Increased tumor growth:* if the expression of a gene required for the induction/ maintenance of dormancy is downregulated using shRNA, this would result in interruption of dormancy and enhanced tumor growth. Therefore, a tumor growth rate higher than the control sample suggests the involvement of that gene's expression in the dormancy process (*see* **Note 45**).
2. *Dormant nodule:* some genes identified by our technique could be dispensable either to the dormancy program or in vivo viability. Hence downregulation of these would show no effect on tumor growth (*see* **Note 45**).
3. *No detectable tumor:* some of the genes that were found to be upregulated by ribonomic profiling could be essential for the survival of the tumor cells in vivo. Therefore, downregulation of these gene products in the dormant cells would lead to small or no tumor growth (*see* **Note 46**).

4. Notes

1. Generally, solutions which are certified DNase-free and RNase-free from the manufacturer will make for easier solution preparation and allow for faster

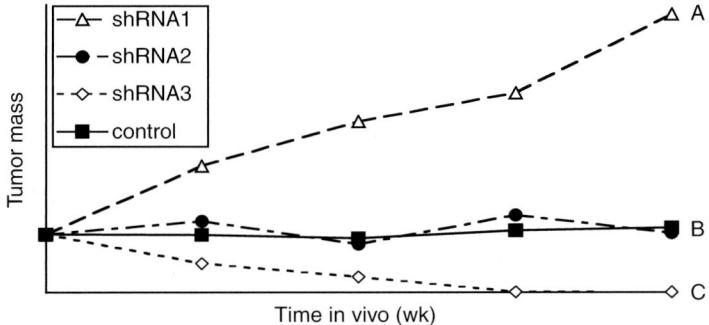

Fig. 4. A graph portraying three possible phenotypic outcomes of shRNA mediated downregulation of gene expression. (**A**) Growing nodule (open triangles, shRNA1): Genes are required for induction/maintenance of dormancy. (**B**) Dormant nodule (closed circles, shRNA2): genes are not required for induction/maintenance of dormancy. (**C**) No detectable growth (open diamonds, shRNA3): genes are required for survival in vivo.

    troubleshooting if they are handled properly. Ambion's buffer kit (cat. no. 9010) contains concentrated solutions of Tris (pH 7.0 and pH 8.0), EDTA, sodium chloride, magnesium chloride, potassium choride, ammonium acetate, and DEPC-treated water.

2. All instruments, glassware and plasticware that touch cells or cell lysates should be certified DNase-free and RNase-free or should be prewashed with RNaseZap (Ambion, cat. no. 9780; 9782) or RNaseAway (Molecular BioProducts cat. no. 7001) followed by DEPC water and allowed to air-dry.

3. Extracts typically range in concentration from 10 to 50 μg/mL of total protein, depending on the cytoplasmic volume of the cell type being used.

4. The number of cells required for each immunopurification can vary from cell type to cell type. Typically we recommend one confluent 10-cm dish per immunopurification. In the case of HEp3 cells, this corresponds to approx 8×10^6 cells.

5. Again, the amount of lysis buffer used typically depends on the amount of protein. Typically we recommend resuspending in a volume of lysis buffer that will have a protein concentration of 1–2 mg/mL. Lysis buffer I is used to isolate cytoplasmic fraction while lysis buffer II is used to isolate nuclear material upon sonication. However in the process we recommend an intermediate step of lysis with lysis buffer II, which partially solubilizes the nuclear membrane. This selectively releases soluble nuclear components as well as Triton X-100 soluble organellar material, leaving behind a purely nuclear pellet.

6. Important! The whole cell variant (**Subheading 3.1.1.**) uses NT2 buffer from **Subheading 2.1.** to resuspend and wash the antibody coated beads. The cellular fractionation variant (**Subheading 3.1.2.**) uses lysis buffer I from **Subheading 2.1.** for bead resuspension and washing.

7. The antibody that is used for immunopurification of hnRNP A1 is a mouse monoclonal antibody (4B10 *[22]*). However, it functions best with protein A and not protein G agarose.

8. As a general rule, we recommend Protein-A Sepharose 4 Fast Flow beads (Amersham Biosciences) or Protein-A Sepharose CL-4B (Sigma) if you plan to use rabbit serum and protein-6 Agarose beads (Sigma) if you plan to use murine monoclonal antibodies.

9. Check the binding capacity of the beads and the antibody concentration in order to introduce the antibody in excess and therefore minimize background-binding problems. Typically 2–20 µL serum per immunoprecipitation reaction are used, depending on the concentration and specificity of the antibody.

10. Antibody-coated beads can be prepared in bulk and stored at 4°C with 0.02% sodium azide.

11. Depending on antibody titer and mRNP concentration, use 50–100 µL packed antibody-coated beads and 100–400 µL RNP lysate (~2–5 mg total protein) for each immunoprecipitation reaction.

12. Performing the immunoprecipitation reactions in larger volumes can decrease background problems.

13. We have noted that the temperature and duration of incubation can influence the efficacy and/or quality of the immunoprecipitation reaction. Longer incubation times may result in better RNP recovery. However, carrying the reaction for too long may result in RNA repartitioning and/or degradation *(23)*.

14. Provided there is no RNA degradation or problems related to postlysis protein mRNA exchange *(23)*, immunoprecipitations should be performed for a minimum of 2–3 h at room temperature or overnight at 4°C. A low background is occasionally observed, which is presumably the result of nonspecific binding to the beads.

15. A concern when isolating mRNP complexes is the possibility of exchange of proteins and mRNAs. In principle, crosslinking agents, such as formaldehyde, could prevent this *(24)*. However, we have found mRNA exchange to occur at a minimal level using these methods and crosslinking therefore to be unnecessary. In some cases, formaldehyde actually can interfere with subsequent mRNA detection methods *(25)*.

16. Several additional washes with NT2 buffer supplemented with 1–3 *M* urea can increase specificity and reduce background *(8)*. However, it is important to first determine whether urea disrupts binding of the antibody to the target protein.

17. RNA pellets from isopropanol precipitations can detach from the centrifuge tube very easily. Extra care should be taken when resuspending the RNA pellet.

18. The addition of RNase inhibitors may interfere with the binding of RBPs to their cognate targets (communication between J.A. Aguirre-Ghiso and S. Piñol-Roma).

19. Immunopurification of hnRNP bound mRNP complexes are typically performed for a maximum of 30 min as longer immunopurification times can result in the disruption of the mRNP complex *(23)*.

20. Depending on the quality of the antibody being used for ribonomic profiling, results and background can vary. Although nonspecific binding can occur, background also may arise from specific RNA–antibody interactions *(26)*.

21. Informative comparisons between total mRNA profiles and mRNP-associated RNAs are frequently limited by the dramatic differences in signal intensity. There

can be a large difference in the number of mRNA species detected in the total RNA as compared with mRNP complexes, which makes most normalization approaches misleading. For this reason, we have typically not compared mRNP profiles with total RNA and suggest that totals be compared with other totals and mRNP immunoprecipitations compared with other mRNP immunoprecipitations. We have found that poly-A binding protein is an useful control RBP.

22. We recommend constructing at least three hairpins per gene. Hairpins may be purchased commercially from vendors such as Open Biosystems; Ambion, Inc.; or Santa Cruz Biotechnology, Inc. Alternatively, shRNA sequences may be generated using the software on their websites.

23. The strain of competent *E. coli* and corresponding cell culture media will vary depending on which plasmid is chosen.

24. Purity of plasmid DNA is crucial for optimal results. We suggest preparing the DNA using Qiagen columns or cesium chloride gradients.

25. Lipid based transfection reagents, like FuGENE (Roche, cat. no. 1 814 443) or Lipofectamine (Invitrogen, cat. no. 18324-111) were used. Calcium chloride transfection methods also work well.

26. Plasmid design and construction for shRNA expression is beyond the scope of this manuscript and has been previously reviewed in **ref. *21***.

27. In our expericence, transfections generally work better when packaging cells are freshly passaged. General protocols and notes for retrovirus production can be found at the following website http://www.stanford.edu/group/nolan/protocols/pro_helper_dep.html.

28. We also recommend performing in parallel some form of positive control transfection using an easily identifiable marker such as green fluorescent protein as a means of assessing the degree of transfection. If the plasmid is capable of being packaged as a retrovirus, it can also aid in assessing the degree of infection.

29. Viral production is greatest 48–72 h after addition of transfection mixture.

30. Retroviruses are potentially dangerous and infectious to humans. Follow proper procedures and strictly observe biosafety guidelines when using retroviruses.

31. Repeated freeze–thaw cycles significantly reduce viral titer. We recommend freezing in 1-mL aliquots if multiple uses are required.

32. Reducing the total volume of media (for the first 24 h) or centrifuging the cell culture dishes may help improve efficiency of infection.

33. Remember to always include an uninfected control. This is especially important when you are performing antibiotic selections.

34. Infection of experimental cell lines with retrovirus essentially involves mixing a cationic polymer such as polybrene with viral supernatant and adding it to cells.

35. Allow 24 h for infection and 48 h to allow for the expression of the shRNA before antibiotic addition. The concentration of antibiotic needed to kill cells varies widely. We recommend testing a range of antibiotic concentrations.

36. The central hypothesis of these experiments is that cancer dormancy program is modulated when expression of genes in the program are disrupted. Therefore, testing of the dormancy program must be performed with cell lines shown to

have a dormant/non-tumorigenic or tumorigenic behavior on CAM such as D-HEp3, MCF-7, MDA-MB-231, PC3, and so on *(27)*.

37. The CAM is a very efficient, timesaving and cost-effective system for the purpose of intital screening purposes. Further, owing to the absence of a mature immune system and the presence of a rich vascular bed, the CAM provides a favorable mileu for the growth of several human tumor cell lines *(14)*.

38. Usually the eggshell is punctured over the area of the CAM rich in embryonic blood vessels.

39. Be extremely careful during the preparation of CAM. Use of excessive pressure during this process can result in the breakage of the eggs. In addition, care should be taken while inoculating the cells on the CAM. Piercing and damaging of the CAM with the pipet tip can be avoided by approaching the CAM at a low angle to inoculate the cells.

40. Detachment of cells using 2 m*M* EDTA in PBS instead of trypsin is important. We do not recommend the use of trypsin as it may disrupt cell surface molecules required for cell–cell interaction.

41. We recommend using this number of cells, but you may use anywhere from 2×10^5 to 10×10^5 cells, depending on the growth rate of the cell line used.

42. The tumor cells can be grown on a given CAM for a maximum period of 7 d from the time of inoculation. However, after 1 wk of incubation the tumor nodule formed at the site of inoculation can be excised, minced, inspected microscopically for the presence of tumor cells and reinoculated on a fresh CAM. Thus, the same tumor population can be serially passaged week after week from one CAM to another.

43. The treatment with collagenase causes the dissociation of the CAM tissue and the tumor nodule to produce a single cell suspension.

44. Tumor cells are recognized by their larger diameter compared to chicken cells.

45. Normally in our system a dormant tumor nodule weighs approx 25–50 mg. Therefore tumors whose size is >50 mg and continues to grow upon serial passaging, it is indicative of a gene required for the induction/maintenance of the dormancy process.

46. Although some of the genes found to be upregulated by ribonomic profiling could be essential for the survival of the tumor cells in vivo, others may be necessary for its viability in vitro. Downregulation of these genes using shRNA would result in loss of viability in culture. This can be overcome by using shRNA controlled by an inducible promoter system.

References

1. Pantel, K. and Otte, M. (2001) Occult micormetastasis: enrichment, identification and characterization of single disseminated tumor cells. *Semin. Cancer Biol.* **11**, 327–337.

2. Naumov, G. N., Townson, J. L., MacDonald, I. C., et al. (2003) Ineffectiveness of doxorubicin treatment on solitary dormant mammary carcinoma cells or late-developing metastases. *Breast Cancer Treat.* **82**, 199–206.

3. Aguirre-Ghiso, J. A., Liu, D., Mignatti, A., Kovalski, K., and Ossowski, L. (2001) Urokinase receptor and fibronectin regulate the ERK(MAPK) to p38(MAPK) activity ratios that determine carcinoma cell proliferation or dormancy in vivo. *Mol. Biol. Cell* **12**, 863–879.

4. Aguirre-Ghiso, J. A., Estrada, Y., Liu, D., and Ossowski, L. (2003) ERK(MAPK) activity as a determinant of tumor growth and dormancy; regulation by p38(SAPK) *Cancer Res.* **63**, 1684–1695.

5. Clark, A. R., Dean, J. L., and Saklatvala, J. (2003) Post-transcriptional regulation of gene expression by mitogen-activated protein kinase p38. *FEBS Lett.* **546**, 37–44.

6. van der Houven van Oordt, W., Diaz-Meco, M. T., Lozano, J., Krainer, A. R., Moscat, J., and Caceres, J. F. (2000) The MKK(3/6)-p38-signaling cascade alters the subcellular distribution of hnRNP A1 and modulates alternative splicing regulation. *J. Cell Biol.* **149**, 307–316.

7. Tenenbaum, S. A., Carson, C. C., Lager, P. J., and Keene, J. D. (2000) Identifying mRNA subsets in messenger ribonucleoprotein complexes by using cDNA arrays. *Proc. Natl. Acad. Sci. USA* **97**, 14,085–14,090.

8. Tenenbaum, S. A., Lager, P. J., Carson, C. C., and Keene, J. D. (2002) Ribonomics: identifying mRNA subsets in mRNP complexes using antibodies to RNA-binding proteins and genomic arrays. *Methods* **26**, 191–198.

9. Hieronymus, H. and Silver, P. A. (2004) A systems view of mRNP biology. *Genes Dev.* **18**, 2845–2860.

10. Gerber, A. P., Herschlag, D., and Brown, P. O. (2004) Extensive association of functionally and cytotpically related mRNAs with Puf family RNA-binding proteins in yeast. *PLOS Biol.* **2**, 342–354.

11. Keene, J. D. and Tenenbaum, S. A. (2002) Eukaryotic mRNPs may represent posttranscriptional operons. *Mol. Cell* **9**, 1161–1167.

12. Fire, A., Xu, S., Montgomery, M. K., Kostas, S. A., Driver, S. E., and Mello, C. C. (1998) Potent and specific genetic interference by double-stranded RNA in *Caenorhabditis elegans*. *Nature* **391**, 806–811.

13. Elbashir, S. M., Harborth, J., Lendeckel, W., Yalcin, A., Weber, K., and Tuschl, T. (2001) Duplexes of 21-nucleotide RNAs mediate RNA interference in cultured mammalian cells. *Nature* **411**, 494–498.

14. Scher, C., Haudenschild, C., and Klagsbrun, M. (1976) The chick chorioallantoic membrane as a model system for the study of tissue invasion by viral transformed cells. *Cell* **8**, 373–382.

15. Penalva, L. O., Tenenbaum, S. A., and Keene, J. D. (2004) Gene expression analysis of messenger RNP complexes. *Methods Mol. Biol.* **257**, 125–134.

16. Tenenbaum, S. A., Lager, P. J., Carson, C. C., and Keene, J. D. (2002) Ribonomics: identifying mRNA subsets in mRNP complexes using antibodies to RNA-binding proteins and genomic arrays. *Methods* **26**, 191–198.

17. Mili, S., Shu, H. J., Zhao, Y., and Pinol-Roma, S. (2001) Distinct RNP complexes of shuttling hnRNP proteins with pre-mRNA and mRNA: candidate intermediates in formation and export of mRNA. *Mol. Cell Biol.* **21**, 7307–7319.

18. Brown, V., Jin, P., Ceman, S., et al. (2001) Microarray identification of FMRP-associated brain mRNAs and altered mRNA translational profiles in fragile X syndrome. *Cell* **107,** 477–487.

19. Hieronymus, H. and Silver, P. A. (2003) Genome-wide analysis of RNA-protein interactions illustrates specificity of the mRNA export machinery. *Nat. Genet.* **13,** 13.

20. Roy, P. J., Stuart, J. M., Lund, J., and Kim, S. K. (2002) Chromosomal clustering of muscle-expressed genes in *Caenorhabditis elegans*. *Nature* **418,** 975–979.

21. Hannon, G. J., Sun, P., Carnero, A., et al. (1999) MaRX: an approach to genetics in mammalian cells. *Science* **283,** 1129–1130.

22. Piñol-Roma, S., Choi, Y. D., Matunis, M. J., and Dreyfuss, G. (1988) Immunopurification of heterogeneous nuclear ribonucleoprotein particles reveals an assortment of RNA-binding proteins. *Genes Dev.* **2,** 215–227.

23 Mili, S. and Steitz, J. A. (2004) Evidence for reassociation of RNA-binding proteins after cell lysis: implications for the interpretation of immunoprecipitation analyses. *RNA* **10,** 1692–1694.

24. Niranjanakumari, S., Lasda, E., Brazas, R., and Garcia-Blanco, M. A. (2002) Reversible cross-linking combined with immunoprecipitation to study RNA–protein interactions in vivo. *Methods* **26,** 182–190.

25. Penalva, L. O., Burdick, M. D., Lin, S. M., Sutterluety, H., and Keene, J. D. (2004) RNA-binding proteins to assess gene expression states of co-cultivated cells in response to tumor cells. *Mol. Cancer* **3,** 24–35.

26. Lipes, B. D. and Keene, J. D. (2002) Autoimmune epitopes in messenger RNA. *RNA* **8,** 762–771.

27. Naumov, G. N., Bender, E., Zurakowski, D., et al. (2006) A model of human tumor dormancy: an angiogenic switch from the nonangiogenic phenotype. *J. Natl. Cancer Inst.* **98,** 316–325.

16

Surface-Epitope Masking (SEM)

An Immunological Subtraction Approach for Developing Monoclonal Antibodies Targeting Surface-Expressed Molecules

Neil I. Goldstein and Paul B. Fisher

Summary

An immunological subtraction approach, surface-epitope masking (SEM), is described that permits the efficient and selective production of monoclonal antibodies (MAbs) reacting with both known and unknown molecules expressed on the cell surface. The tenet underlying SEM involves blocking (masking) of shared antigens between two target populations, a "driver" and a "tester," and using appropriately modified surface-masked "tester" cells to generate MAbs reacting with surface antigens unique to the "tester population" that differentiate the two antigen sources. SEM has been employed to develop MAbs that react with the multidrug resistance surface-expressed P-glycoprotein (MDR-1) and the human interferon-γ receptor and two potentially novel tumor-associated antigens (TAAs) expressed on the surface of prostate carcinoma and breast carcinoma cells. In principle, the SEM approach provides an uncomplicated and effective means of developing MAbs, which can also be used to identify genes, associated with important cellular processes involved in normal physiology, such as growth, aging, differentiation, and development. In addition, this strategy is amenable to produce MAbs and identify genes associated with specific disease states, including cancer, neurodegeneration, autoimmunity, and infection with pathogenic agents.

Key Words: Immune subtraction; monoclonal antibodies; surface-epitope masking; tester; driver; p-glycoprotein; SEM; tumor-associated antigens.

1. Introduction

Development of the hybridoma/monoclonal antibody (MAb) technology by Kohler and Milstein in 1975 (*1*) ushered in a new era in immunology and revived the concept initially proposed by Ehrlich and colleagues in 1904 (*2*) that antibodies might serve as "magic bullets" for the treatment of cancer. In the 1980s significant research was focused on developing murine MAbs that recognize a variety of tumor-associated antigens (TAAs) (as revewied in **refs. 3–6**).

From: *Methods in Molecular Biology, vol. 383: Cancer Genomics and Proteomics: Methods and Protocols*
Edited by: P. B. Fisher © Humana Press Inc., Totowa, NJ

Early studies highlighted a number of problems relative to using murine MAbs targeting TAAs for potential therapeutic applications, including production of human anti-murine immunoglobulin antibodes (HAMA), inability to deliver therapeutically effective doses of MAbs to tumors, lack of adequate activation of effector functions in patients, delayed blood compartment clearance of MAbs, failure to target appropriate target organs, poor antibody affinity and avidity, heterogeneous expression of TAAs on tumor cells, and insufficient penetration of antibodies in tumors (as revewied in **refs. 7–9**).

Recent techniques employing genetic engineering resulting in chimerized, humanized, or completely human antibodies have changed the landscape, decreasing some of the obstacles to effectively using MAbs for cancer therapy (as revewied in **refs. 8,10**, and **11**). In this context, alterations in existing murine MAbs (chimeric MAbs) or development of fully or partially humanized MAbs have now resulted in specific MAbs entering the clinic. Some of them are showing early promise of providing therapeutic benefit to patients with diverse cancers. Examples of MAbs now in clinical trials, include: rituximab (chimeric) and epratuzumab (humanized) targeting CD20 or CD22, respectively, for treatment of non-Hodgkin lymphoma; tratuzumab (humanized) targeting HER2/NEU (ERBB2) for treatment of breast cancer; gemtuzumab ozogamicin targeting CD33 for treatment of acute myeloid leukemia; alemtuzunab (humanized) targeting CD52 for treatment of B-cell chronic lymphocytic leukemia; edrecolomab (murine) targeting epithelial cellular-adhesion molecule for adjuvant therapy of colorectal cancer; cetuximab (chimeric), erbitux (chimeric), h-R3 (humanized); and ABX-EGF (human, transgenic mouse) targeting the epidermal growth-factor receptor for the therapy of colorectal, head, and neck and non-small-cell lung carcinomas; huA33 (humanized) targeting A33 for the therapy of colorectal cancer; G250 (chimeric) targeting G250/MN for the therapy of renal cancer; and KW-2871 (chimeric) targeting GD3 for the therapy of melanoma (as revewied in **refs. 11** and **12**). MAbs currently being evaluated in patients represent the tip of the iceberg and a plethora of new MAbs targeting cancer awaits entry into the clinical arena.

Production of murine MAbs to surface-expressed molecules is frequently a tedious and inefficient procedure requiring repeated antigen injection in mice and subsequent testing of a large number of hybridomas for production of appropriate antibodies (as revewied in **refs. 5** and **7**). An approach to simplify and maximize the ability to produce MAbs reacting with defined and unknown molecules expressed on the cell surface, a strategy called surface-epitope masking (SEM) *(13,14)* was developed. A general schematic of the SEM approach is shown in **Fig. 1**. In principle, this strategy involves immunological subtraction, in which high titer polyclonal antibodies are prepared against the repertoire of antigens immobilized on the surface of one cell population, called the "driver," and these are used to coat (and presumably block) antigenic sites shared by another cell population, called the "tester." Injection of antibody-complexed cells into mice results in the production of hybridomas producing MAbs interacting with antigens expressed differentially on the "tester" vs the "driver" cell populations *(13)*. As with genomic subtraction, this immunological subtraction approach has very broad applications and in theory can be used to produce MAbs distinguishing antigens between closely related as well as disparate cellular genotypes. In these contexts, SEM is applicable for a wide array of applications in which the desired end point is to develop MAbs that distinguish different cellular phenotypes, which can then be used to define the genes regulating these changes. Among its diverse applications, SEM has potential for the targeted development of MAbs: distinguishing normal from cancer cells (including early stage cancer and advanced cancer and metastasis) *(14,15)*; identifying specific stages in cellular differentiation; distinguishing different stem cell lineages; discriminating surface markers unique to cancer stem cells; recognizing cellular alterations associated with components of the tumor/cell microenvironment; recognizing antigens involved in immunological recognition; identifying stages of cell growth and senescence; distinguishing antigens characteristic of atypical MDR; interacting with antigens unique to specific autoimmune diseases; distinguishing different types and stages of neurodegeneration; and characterizing cellular changes associated with infectious diseases.

Two applications of the SEM approach to produce MAbs reacting with both known and unknown surface-expressed molecules, providing proof-of-principles for this selective immunological subtraction strategy *(13)* were described. In its initial application, a specific cloned rat embryo fibroblast target cell *(16)*, CREF-Trans 6 *(17,18)* provided the "driver" cell to produce high titer polyclonal antibodies to be employed for masking epitopes on a "tester" cell population *(13,14)* **(Fig. 1)**. To provide the "tester," CREF-Trans 6 cells were genetically modified to become resistant to several structurally unrelated drugs (MDR) by transfection with the human MDR-1 gene *(13)*, which encodes a 170-kDa surface-localized protein functioning as an efflux pump. Following masking with polyclonal antibodies developed against CREF-Trans 6 and application of SEM a series of MAbs were developed that reacted with MDR-1-expressing CREF-Trans 6 and human breast carcinoma (MCF-7) cells engineered to express MDR-1, while not reacting with unmodified CREF-Trans 6 or MCF-7 cells **(Fig. 2)**.

A second application of SEM involved a rapid expression cloning system (RExCS) developed to identify dominant-acting oncogenes *(18)*. The RExCS process involves (1) cotransfecting either sheared high molecular weight human tumor DNA or a high quality cDNA library in combination with a selectable antibiotic resistance gene into a new, efficient DNA acceptor cell line, CREF-Trans 6; (2) selecting antibiotic resistant cells; and (3) injecting these cells into

Surface-epitope masking

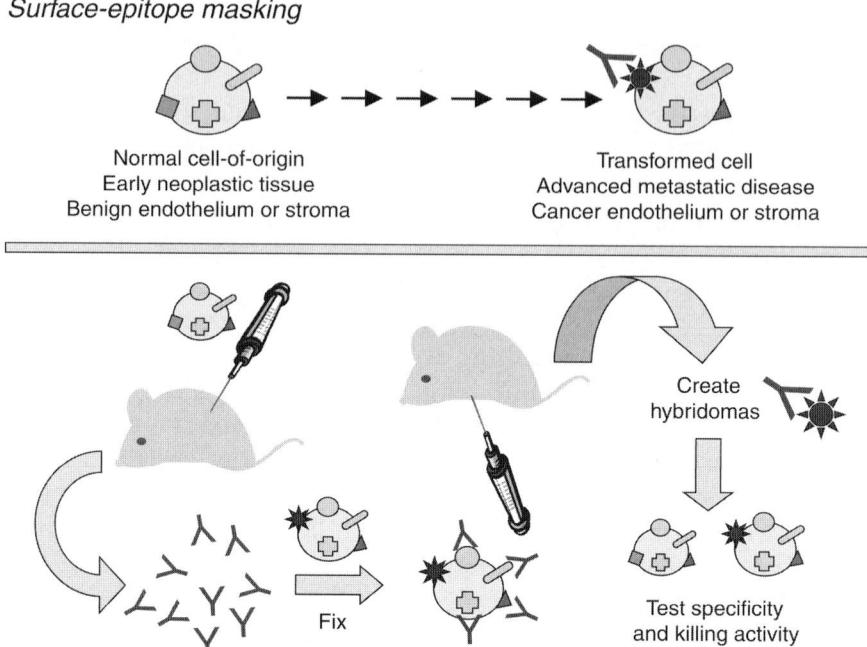

Fig. 1. Schematic of the SEM approach. Polyclonal antibodies are generated against the spectrum of antigenic epitopes expressed on one cell type (the "driver") and these are used to mask a second cell type (the "tester") that contains additional new antigens expressed on the cell surface. Antigen–antibody complexes are fixed on the surface of genetically altered tester cells and these cells are used to immunize mice to create hybridomas producing antibodies reacting with new surface-expressed molecules.

nude mice. Tumors developing in these animals are placed in culture for use as a "tester" population for the SEM approach and for identifying human genes that may mediate or be affected by tumor transformation *(18–20)*. The RExCS approach has been used to transfer transforming human genetic elements from prostate, breast, and small-cell lung carcinoma, glioblastoma multiforme and from a primary metastasis from a patient with colorectal cancer. A novel gene, designated prostate tumor inducing gene (PTI)-1, was identified using differential RNA display from CREF-Trans 6:4NMT (nude mouse tumor-derived clone), a RExCS-derived cell line generated by transfecting high molecular weight DNA from the human prostate cell line LNCaP into CREF-Trans 6 followed by selection for tumor formation in nude mice *(18–20)*. Tumor-derived cells were isolated, the presence of human DNA sequences was confirmed by probing with *Alu* repetitive human sequences and these cells were used as the "tester" as part of the SEM approach. Using this strategy a series of MAbs that react with the

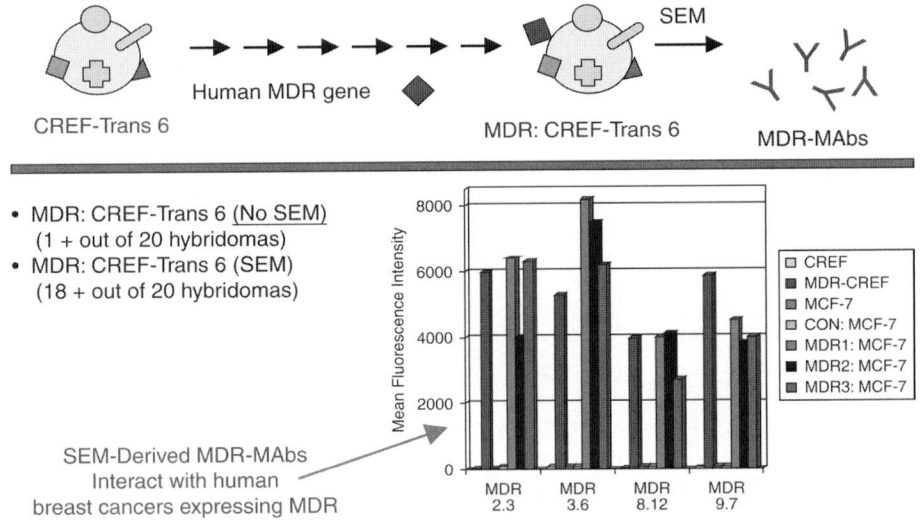

Fig. 2. Proof-of-principle for the SEM approach: MDR gene. CREF-Trans 6 cells were genetically modified to express human MDR gene (MDR-CREF-Trans 6). The SEM approach was applied resulting in the production of MDR-MAbs (MDR 2.3, MDR 3.6, MDR 8.12, and MDR 9.7) reacting with the MDR gene product in CREF-Trans 6 cells and in genetically altered human MDR MCF-7 breast cancer cells, data provided for three clones (MDR1: MCF-7, MDR2: MCF-7 and MDR3: MCF-7). The frequency of MDR-positive hybridoma formation greatly exceeded what was obtained without SEM.

surface of human prostate cancer cells and tissue samples from patients with prostatic intraductal neoplasia (PIN) and prostate carcinomas have been developed (Pro series of MAbs) **(Fig. 3)** *(19)*. These test situations confirm that SEM can be used to target MAb production for molecules expressed on the cell surface and in this context will have very broad applications for developing immunological reagents and identifying genes involved in diverse biological contexts.

2. Materials

2.1. Media and Buffers

1. Cells (established cultures or tumor- or metastasis-derived) are routinely grown in media specific for each cell line *(13,16–19)*. For example, cells such as DU-145 (prostate cancer), are grown in RPMI 1640 medium supplemented with 10% fetal bovine serum, 2 mM L-gluatmine, and antibiotics. Medium and supplements can be purchased from various vendors such as Sigma (St. Louis, MO).
2. Tissue culture grade buffers and reagents such as Trypsin-ethylene diamine tetra acetic acid (EDTA) and Dulbecco's phosphate-buffered saline (DPBS) can be purchased from many vendors including Sigma.

*Surface-epitope masking (known or unknown targets,
native context): CREF-Trans 6*

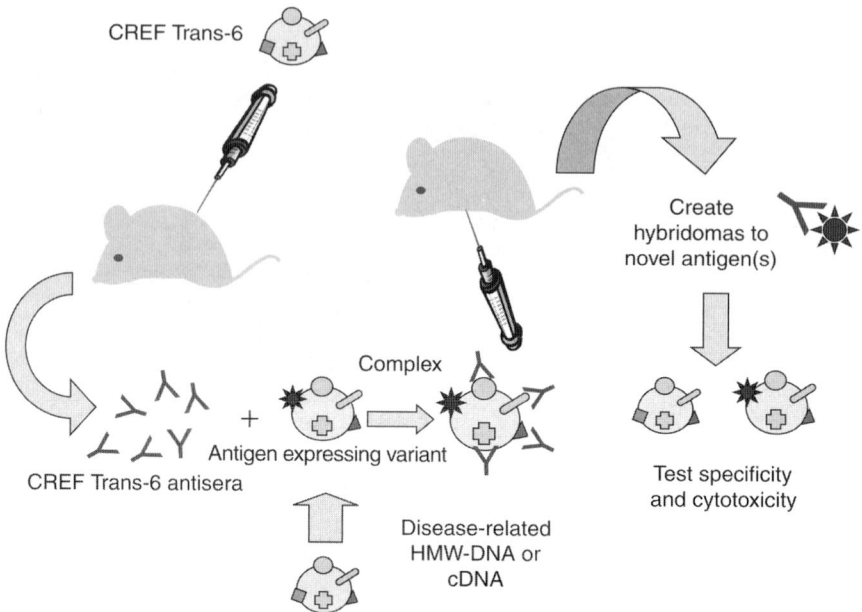

Fig. 3. Schematic of the SEM approach to develop antibodies recognizing surface-expressed molecules using the CREF-Trans 6 cell line as the "driver" and genetically modified CREF-Trans 6 cell lines as the "tester." CREF-Trans 6 cells are used to develop high titer CREF-Trans 6 polyclonal antisera in mice, which is then complexed with CREF-Trans 6 cells containing disease-related high molecular weight (HMW)-DNA or cDNAs. The immobilized cells are then used to immunize mice resulting in an immune response and the subsequent development of hybridomas (by standard procedures) producing MAbs reacting with the disease-related target cells that were the source for the HMW-DNA or cDNAs. These antibodies can then be tested for specificity and potential cytotoxic properties.

3. Neutral-buffered formalin used for the masking protocol is prepared by diluting 37% formaldehyde reagent (Sigma) to the appropriate concentration in DPBS.
4. Hybridoma reagents such as polyethelene gycol (PEG) and hypoxanthine, aminopterin, thymidine (HAT) can be purchased from various vendors such as Sigma.
5. Immunological reagents such as secondary antibody for enzyme-linked immunosorbent assay (ELISA) can be purchased from various vendors such as Sigma.
6. 2′-azino-bis (3-ethylbenzthiazoline-6-sulfonic acid) (ABTS) solution for cell-based ELISAs can be purchased from BioFX (Owings Mills, MD).

2.2. Cell Lines

1. Cells lines are dependent upon the type of SEM experiment being performed. Cell lines can be purchased from the American-Type Culture Collection (ATCC; www.attcc.org).
2. CREF-Trans 6 is a specific clone of Fischer cloned rat embryo fibroblast (CREF) *(16)* cells that display elevated transformation with high molecular weight DNA, cDNAs, or plasmid DNAs. This cell line is described in **refs. *17*** and ***18***.
3. Normal cells, such as normal mammary epithelial cells or normal human prostate epithelial cells can be purchased from many suppliers, including Cambrex.
4. Fusion partners for hybridomas (e.g., NS1) can be purchased from the ATCC.

3. Methods

3.1. Surface-Epitope Masking

1. Identify a "driver" cell type for the experiment. The "driver" should express the antigens that are to be "masked." The driver can be one cell type (such as CREF-Trans 6) that will be used to mask a genetic derivative of the first cell type (such as CREF-Trans 6 MDR, or CREF-Trans 6:NMT). Additionally, one could use as a driver a normal cell vs a cancer cell of the same lineage, an early-stage tumor vs a late stage tumor, a cell type from the normal vasculature vs the same cell type from a tumor vasculature, a cancer cell vs a cancer stem cell, a normal cell vs a diseased cell (nervous system, immune system, cardiovascular system), and so on **(Fig. 1)**. (*see* **Notes 1** and **2**).
2. Immunize immunocompetent mice with the "driver" line, in which antigens have been immobilized by fixation. This usually involves six to eight injections of the cell line into the animals over a 2- to 3-mo time frame **(steps 3–9)**.
3. Remove confluent cells with EDTA and wash twice in Dulbecco's PBS.
4. Resuspend the cells at a concentration of 1×10^6 in 0.1 mL of DPBS.
5. Add neutral-buffered formalin to a final concentration of 0.1% (v/v) and incubate for 20 min at room temperature.
6. Inject a minimum of 10 mice (BALB/c or other mouse strain) with the resuspended cells by the intraperitoneal route.
7. Animals are immunized biweekly for 2 mo (four injections). At this time, the animals are bled and the sera tittered by ELISA (*see* **Subheading 2.**) (*see* **Note 3**).
8. If the titer is not high enough, continue the injections for an additional 1–2 mo.
9. The titer of the driver antisera should be a minimum of 1:10,000 for efficient use in SEM.
10. Prepare the "tester" cells. In the case of CREF-Trans 6, this would involve adding either a known gene that expresses on the cell surface, for example, MDR-1, epidermal growth factor receptor, interferon-γ receptor, and so on. For performing RExCS details can be found in **refs. *17*** and ***18***. Briefly, this approach involves isolating HMW tumor DNA (or preparing cDNA libraries), shearing HMW tumor DNA and mixing with a selectable plasmid DNA (such as one expressing the neomycin or hygromycin resistance gene), transfect CREF-Trans 6 cells and

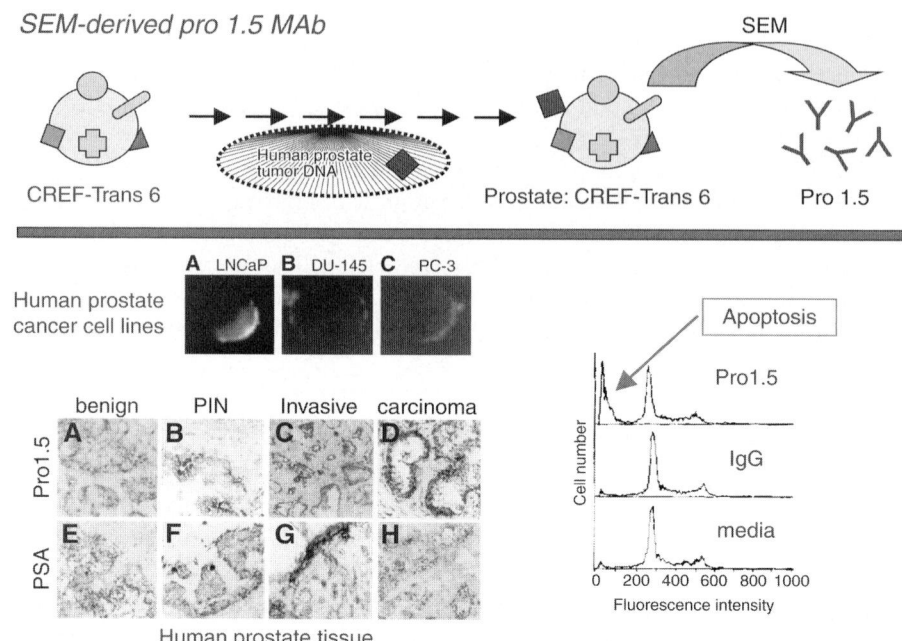

Fig. 4. SEM-Derived Pro 1.5 MAb. Genetically modified CREF-Trans 6 cells, resulting from tumors induced in mice following transfection with HMW-DNA from the LNCaP human prostate cancer cell line (prostate: CREF-Trans 6), were used as the "tester" to develop MAbs using the SEM approach (upper panel). A predicted outcome was the development of MAbs that would react with antigens expressed on the surface of prostate cancer cells, both cell lines and patient-derived prostate carcinomas. A SEM-derived MAb, Pro 1.5, reacts with LNCaP, the initial source of the antigen, as well as DU-145 and PC-3 cells (data from **ref. *19*.** Additionally, the Pro 1.5 MAb reacts immunohistochemically with tissue sections from patients with PIN and invasive carcinoma, with marginal staining of normal prostate epithelial cells or samples from patients with benign prostatic hypertrophy. In contrast, antibodies to human prostate-specific antigen react with patient-tissue samples from normal, benign prostatic hypertrophy, PIN, and invasive carcinoma of the prostate (data from **ref. *19*.** Addition of the Pro 1.5 MAb to DU-145 prostate carcinoma cells decreases the DNA content of cells (increase in hypodiploidy (A_0 DNA content), as monitored by fluorescence-activated cell sorter, indicating apoptosis–induction in these cancer cells. These effects are not apparent when using matched IgG control antibody or just media.

select for antibiotic resistance, inject pooled antibiotic resistant cells into athymic nude mice, when tumors develop, isolate and establish in culture. These cells become "tester" cells (**Fig. 4**).

11. Mask the "tester" cells with the "driver" polyclonal antisera (**steps 12–16**) (*see* **Note 4**).

12. Remove the cells with EDTA and wash two times with DPBS.
13. Resuspend the cells at a concentration of 1×10^6 cells in 0.1 mL of DPBS.
14. Add neutral-buffered formalin to a final concentration of 0.1% (v/v) and incubate for 20 min at room temperature.
15. Add 10 mL of DPBS and centrifuge the cells at 1000g for 5 min.
16. Wash twice with DPBS and resuspend the fixed cells at a concentration of 1×10^6 cells/0.1 mL of DPBS.
17. Use the masked "driver" cells for immunization of immunocompetent mice (**steps 10–20**).
18. Inject immunocompetent mice with 1×10^6 masked "driver" cells in 0.1 mL of DPBS. The cells can be mixed with adjuvant (e.g., Ribi) to ensure a higher titer of splenocytes for fusion.
19. Animals are usually immunized bi-weekly for 2 mo and bled after the fourth injection to determine titer by cell-based ELISA.
20. Animals are sacrified for hybridoma formation when their titers are >1:10,000 (*see* **Note 7**).

3.2. Hybridoma Formation

1. Spleens are ascepticaly removed from sacrificed animals.
2. Splenocytes are recovered by mincing the spleens with a bent sterile needle into medium.
3. Splenocytes are collected by centrifugation and washed two times with cold serum-free medium.
4. Cell fusion is performed by mixing 1×10^7 splenocytes with 1×10^6 myeloma cells. The cells are collected by centrifugation and resuspended in the residual medium. One milliliter of hybridoma-grade PEG is added drop-wise over 2 min with shaking. The cells are then pelleted and resuspended in 60 mL of growth medium containing HAT.
5. One hundred microliters of the cell suspension are plated per well in six 96-well plates. A splenocyte feeder layer is added to each well prior to plating the fused cells (*see* **Note 5**).
6. Plates are incubated at 37°C in HAT containing medium for 1–2 wk. Plates are refed twice weekly by semidepleting and replacing each well with fresh medium.
7. Visible hybridoma colonies are usually observed within 2 wk and are expanded for screening purposes.
8. Positive hybridomas are cloned at <1 cell per well (96-well plate) over a feeder-layer of splenocytes.

3.3. Cell-Based ELISA Assay

1. Cells are grown to confluency in 96-well plates and fixed with 0.1% neutral-buffered formalin for 20 min at room temperature (*see* **Note 6**).
2. Plates are washed two times with DPBS and wells are blocked with 300 μL of 2% nonfat milk in DPBS for 60 min at room temperature.
3. One hundred microliters of each hybridoma supernatant are added to each well and the plates incubated for 2 h at room temperature.

4. Plates are washed three times with DPBS and probed with goat antimouse IgG conjugated to horseradish peroxidase.
5. Plates are incubated for 60 min, washed three times with DPBS and binding determined by incubating with ABTS solution and reading after 30–60 min at 405 nM.

3.4. Internalization Assay (to Define Internalization of Specific SEM-Derived MAbs)

1. Cells are grown to near confluency in the appropriate medium.
2. Cells are removed with 2 mM EDTA, washed twice with cold serum-free medium containing 1% BSA (SF-BSA) and resuspended in several plastic test tubes at a final concentration of 0.5–1 × 10^6 cells/tube.
3. One hundred microliters of the test and control monoclonal antibody in SF-BSA are added to each tube at a final concentration of 10 µg/mL.
4. Cells were incubated on ice for 60 min, washed with SF-BSA and stained with goat antimouse IgG (Tago, Burlingame CA) for 30 min in the cold.
5. Cells are then washed, resuspended in 100 µL of SF-BSA, and incubated at either 4°C for an additional 60 min (cell surface staining) or at 37°C in a water bath for 5, 15, 30, and 60 min (internalization).
6. At the end of each time period, the cells are fixed by the addition of 200 µL of cold 1% paraformaldehyde in PBS for 15 min on ice.
7. After washing, the stained cells are viewed with a fluorescence microscope and photographed.

3.5. FACS Analysis

1. Cells are removed with 2 mM EDTA, washed and resuspended in test tubes at a concentration of 0.5–1 × 10^6 cells/ tube in a final volume of 100 µL of cold SF-BSA.
2. Test or control MAbs are added and the cells are incubated for 60 min on ice.
3. After washing, goat antimouse IgG-FITC in SF-BSA is added for an additional 30 min on ice.
4. Cells are washed, fixed in cold 1% paraformaldehyde for 15 min, and analyzed by flow cytometry.
5. Results can be presented as the mean fluorescence intensity (MFI), which is an indirect measure of antigen density and defined as the mean channel fluorescence × the percentage of positive cells).

3.6. Apoptosis Assays (for SEM-Derived MAbs That Induce Programmed Cell Death)

3.6.1. Propidium Iodide and FACS Analysis

1. Test cells in the log phase are harvested with trypsin-EDTA and replated at a concentration of 2 × 10^5 cells/100 mM^2 plate in 10 mL of RPMI supplemented with 2% fetal bovine serum (FBS) (test medium).
2. Test and control monoclonal antibody are added to cells at a final concentration of 100 nM.

3. Cells are incubated for 72 h at 37°C at which time 10 mL of the appropriate medium is added to each plate and the cells incubated for an additional 24 h. For longer assays including those for 7 d, plates are fed on day 6.

4. On days 4 or 7, adherent cells are removed with trypsin-EDTA, washed three times and counted.

5. The cells are fixed and permeabilized with cold 70% ethanol for 30 min on ice.

6. The cells are washed with PBS containg 0.1% Triton X-100, resuspended at $1-2 \times 10^6$ cells per tube in 1 mL of PBS-T containing 20 µg of DNase-free RNase (Boehringer Mannheim) for 30 min in a 37°C water bath.

7. Treated cells are stained with propridium iodide (50 µg/mL) for 1 h at 4°C and analyzed for DNA content within 24 h using flow cytometry.

8. Cell cycle distribution is determined using manual gating. The hypodiploid ($<2n$) or A_0 region has been shown to be associated with cells undergoing apoptosis.

3.6.2. TUNEL Methods

1. Apoptosis can be directly measured using the APO-DIRECT kit (Pharmigen) and the *in situ* Cell Death Detection kit (BM), or other commercially available kits. Both indicated kits utilize the TUNEL (TdT-mediated dUTP nick end labeling) method for determining apoptosis.

2. Cells are treated as in **Subheading 3.1.** and processed as per the kit instructions.

3. Results for the APO-DIRECT kit are determined using flow cytometry using the positive and negative controls supplied by the manufacturer to validate each assay.

4. Results for the *in situ* Death Detection kit are determined by immunofluorescence. Three different fields (100 cells/field) are counted and the percentage of apoptotic cells defined as: number of apoptotic cells/total cells × 100. Statistical significance is determined by ANOVA using the statistical program Sigma Stat (Jandel). Cells are also stained with DAPI to visualize the extent of nuclear damage. Photomicrographs are taken using an Olympus microscope equipped with epifluorescence.

3.7. Immunoblot Analysis

1. Cells are grown for 4 d in medium, test, or an isotype matched control IgG.

2. Cells are solubilized in RIPC buffer (20 mM Tris pH 7.5, 0.5 M NaCl, 0.5% NP-40, and 100 µg/mL of PMSF) and separated by gel electrophoresis on 4–20% SDS gradient gels.

3. The proteins are immunoblotted onto nitrocellulose and probed with a monoclonal antibody to the protein being tested as previously described in **ref. *13***. Blots are stained with the appropriate secondary antibody and developed using enhanced chemiluminescence (ECL).

3.8. Immunohistochemistry (to Define SEM MAb Activity in Human Tissue Sections)

1. Tissue sections are prepared from fresh tissues frozen in liquid nitrogen or tissues fixed in formalin.

2. Sections are fixed in acetone, blocked with 3% hydrogen peroxide for 7 min at room temperature and incubated for 20 min in PBS containing 3% FBS.

3. Sections are incubated with monoclonal antibody or control antibody for 45 min at room temperature.
4. Samples are washed five times with PBS and incubated with biotinylated secondary antibody for 45 min at room temperature.
5. Samples are washed five times with PBS and incubated with horseradish peroxidase-conjugated streptavidin in PBS.
6. Reactivity is detected with diaminobenzidine for 15–30 min at room temperature.
7. Samples are counterstained with hematoxylin for 5 min at room temperature.

3.9. Animal Studies (to Define Potential Antitumor Biological Activity of SEM-Derived MAbs)

1. Male athymic nude mice (NIH Swiss, nu/nu) are purchased from appropriate approved vendors (such as Taconic Farms) and injected subcutaneously with 1×10^6 tumor cells mixed 1:1 with matrigel.
2. When tumors in each animal reach about 100–200 mm^3, mice are randomly divided into two groups to be treated with either test SEM-derived MAb or an irrelevant isotype matched control IgG.
3. Treatments consist of twice weekly intraperitoneal injections of the antibodies over a 5-wk period (10X) at a dose of 200 µg/injection.
4. Tumors are measured two to three times per week during the course of the study and tumor volumes are determined using the formula pi/6 × larger diameter × (smaller diameter^2) *(15,20)*.
5. Animals are followed for an additional 4 wk following the termination of the injections. Statistical significance between the groups is determined by ANOVA with a p-value of <0.005 considered significant.
6. Data can be presented as tumor volume +/– the standard deviation over time or weight +/– S.D. over time. The tumor volume and weight are determined as described in **refs. *15*** and ***20***.

4. Notes

1. The SEM technology can be used in any scenario where a test parameter ("tester" cell type) can be evaluated against a control parameter ("driver" cell type). For example, SEM has been used to identify antigens expressed on human tumor cells by expressing the tumor DNA in a driver cell line (CREF-Trans 6), masking the control antigens with an antisera to CREF-Trans 6 ("driver" antisera) and using the masked cells (tester) to immunize mice. The approach has worked in several proofs-of-principles, including prostate cancer, breast cancer, MDR, and human interferon-γ receptor (*13,14,19,* and unpublished data). However, there are several factors to consider when using the SEM technique.
2. The "tester" cell is critical to the process. It must have a similar repertoire of antigens as the "driver" cell so that the masking antisera will block all antigens except those specific to the driver cell line.
3. It is important to develop high-titer antisera against the "driver" cell line. This ensures that the entirety of normal antigens are masked.

4. It is important to properly mask the "tester" cell line under the correct conditions. This involves using the correct fixative (e.g., 0.1% neutral-buffered saline) and incubation time (e.g., 20 min at room temperature). This fixation is critical and other fixatives and incubation time will likely prove effective also in masking and employing the SEM approach.

5. It is important to use a feeder layer when plating hybridomas to increase the number of MAb secreting hybridoma colonies isolated.

6. It is important to fix the "driver" cells in the same manner used for the immunizations for the cell-based ELISA.

7. There have not been any definitive studies regarding the mechanistic basis concerning why SEM has worked with such great efficiency. The "weak" fixation allows the antigens to remain in the correct conformation so as to allow the generation of MAbs that can bind to native structures expressed on the cell surface was hypothesized. Also, the use of syngeneic antisera to mask the normal antigens probably increases the immunogenicity of the driver–antisera complex and allows efficient uptake and processing of this complex by macrophages and other accessory cells.

Acknowledgments

The valuable input of Dr. Ronald A. DePinho, Harvard Medical School, Boston, MA to our studies is greatly appreciated. The research underlying the present study was supported in part by National Institutes of Health Grants CA035675, CA074468, CA097318, CA098712, P01 CA104177, GM068448, and P01 NS31492; the Samuel Waxman Cancer Research Foundation; and the Chernow Endowment. P.B.F. is the Michael and Stella Chernow Urological Cancer Research Scientist in the Departments of Pathology and Urology, College of Physicians and Surgeons of Columbia University, and a SWCRF Investigator.

References

1. Kohler, G. and Milstein, C. (1975) Continuous cultures of fused cells secreting antibody of predefined specifity. *Nature* **256,** 495–497.
2. Ehrlich, P., Herta, C. A., and Shigas, K. (1904) Ueber einige verwendungen der naphtochinosuflsaure. *Z. Phisiol. Chem.* **61,** 379–392.
3. Schlom, J., Greiner, J. W., Horan Hand, P., et al. (1984) Monoclonal antibodies to breast cancer associated antigens as potential reagents in the management of breast cancer. *Cancer* **54,** 2777–2794.
4. Schlom, J., Greiner, J. W., Horan Hand, P., et al. (1985) Human breast cancer markers as defined by monoclonal antibodies, in *Monoclonal Antibodies in Cancer, Cancer Markers III,* (Sell, S. ed.), Humana Press, Clifton, NJ, pp. 247–278.
5. Schlom, J., Colcher, D., Horan Hand, P., et al. (1985) Monoclonal antibodies reactive with breast tumor associated antigens. *Adv. Cancer Res.* **43,** 143–173.

6. Greiner, J. W., Schlom, J., Pestka, S., et al. (1985) Modulation of tumor associated antigen expression and shedding by recombinant human leukocyte and fibroblast interferons. *Pharmacol. Therapeut.* **31,** 209–236.

7. Leon, J. A., Goldstein, N. I., and Fisher, P. B. (1994) New approaches for the development and application of monoclonal antibodies for the diagnosis and therapy of human cancer. *Pharmacol. Therapeut.* **61,** 237–278.

8. Sanz, L., Blanco, B., and Alvarez-Vallina, L. (2004) Antibodies and gene therapy: teaching old "magic bullets" new tricks. *Trends Immunol.* **25,** 85–91.

9. Christiansen, J. and Rajasekaran, A. K. (2004) Biological impediments to monoclonal antibody-based cancer immunotherapy. *Mol. Cancer Ther.* **3,** 1493–1501.

10. Roque, A. C. A., Lowe, C. R., and Taipa, M. A. (2004) Antibodies and genetically engineered related molecules: production and purification. *Biotechnol. Prog.* **20,** 639–654.

11. Harris, M. (2004) Monoclonal antibodies as therapeutic agents for cancer. *Lancet Oncology* **5,** 292–302.

12. Rayzman, V. and Scott, A. (2002) Monoclonal antibodies for cancer therapy. *Cancer Forum* **26,** 104–108.

13. Shen, R., Su, Z. -Z., Olsson, C. A., Goldstein, N. I., and Fisher, P. B. (1994) Surface-epitope masking: a strategy for the development of monoclonal antibodies specific for molecules expressed on the cell surface. *J. Natl. Cancer Inst.* **86,** 91–98.

14. Fisher, P. B. (1995) A new technology for preparing monoclonal antibodies to molecules expressed on the cell surface. *Pharmacol. Tech.* **19,** 42–48.

15. Gopalkrishnan, R. V., Kang D. -C., and Fisher, P. B. (2001) Molecular markers and determinants of human prostate cancer metastasis. *J. Cell. Physiol.* **189,** 245–256.

16. Fisher, P. B., Babiss, L. E., Weinstein, I. B., and Ginsberg, H. S. (1982) Analysis of type 5 adenovirus transformation with a cloned rat embryo cell line (CREF). *Proc. Natl. Acad. Sci. USA* **79,** 3527–3531.

17. Su, Z. -Z., Olsson, C. A., Zimmer, S. G., and Fisher, P. B. (1992) Transfer of a dominant-acting tumor-inducing oncogene from human prostatic carcinoma cells to cloned rat embryo fibroblast cells by DNA-transfection. *Anticancer Res.* **12,** 297–304.

18. Shen, R., Su, Z. -Z., Olsson, C. A., and Fisher, P. B. (1995) Identification of the human prostatic carcinoma oncogene PTI-1 by rapid expression cloning and differential RNA display. *Proc. Natl. Acad. Sci. USA* **92,** 6778–6782.

19. Su, Z. -Z., Lin, J., Shen, R., Fisher, P. E., Goldstein, N. I., and Fisher, P. B. (1996) Surface-epitope masking and expression cloning identifies the human prostate carcinoma tumor antigen gene PCTA-1 a member of the galectin gene family. *Proc. Natl. Acad. Sci. USA* **93,** 7252–7257.

20. Su, Z. -Z., Goldstein, N. I., and Fisher, P. B. (1998) Antisense inhibition of the PTI-1 oncogene reverses cancer phenotypes. *Proc. Natl. Acad. Sci. USA* **95,** 1764–1769.

17

Approaches for Monitoring Signal Transduction Changes in Normal and Cancer Cells

Paul Dent, Philip B. Hylemon, Steven Grant, and Paul B. Fisher

Summary

This chapter will describe methods to assess the activities of protein kinases. Initial studies in the 1950s and 1960s in the field of glucose metabolism examined the activities of several highly specific protein and carbohydrate kinases in cell lysates or isolated cell fractions. As more protein kinases were discovered in the 1980s and 1990s, coupled with the development of immunoprecipitating antibodies, in vitro kinase assays of isolated kinase proteins using γ-^{32}P ATP became a standard procedure. In the 1990s, antibodies were developed that recognize specific sites of regulatory phosphorylation on a variety of protein kinases (phospho-specific antibodies), which have been used to assess kinase activity indirectly through immunoblotting. In this chapter, Methodologies to perform immune complex protein kinase assays and immunoblotting using phospho-specific antibodies against regulatory sites of phosphorylation in protein kinases will be described. The strengths and weaknesses of each approach in determining protein kinase activity will also be discussed.

Key Words: Kinase; phosphorylation; phosphor-specific; immunoprecipitation; ERK1/2; MAP kinase; JNK1/2; p38; AKT; Raf-1; B-Raf; cdk.

1. Introduction

Protein and lipid phosphorylations play a vital role in the regulation of cell metabolism, cell growth, and cell viability. However, only 35 yr ago, protein phosphorylation was thought to be restricted to relatively few proteins, specifically those involved in the regulation of glycogen synthesis/breakdown and in the regulation of the tricarboxcylic acid cycle. The number of protein kinases and the number of proteins whose function(s) are regulated by reversible phosphorylation is now, 35 yr later, listed in the thousands.

Early investigations examining the regulation protein phosphorylation were fortunate, in as much as the protein kinase being studied, phosphorylase kinase, was

From: *Methods in Molecular Biology, vol. 383: Cancer Genomics and Proteomics: Methods and Protocols*
Edited by: P. B. Fisher © Humana Press Inc., Totowa, NJ

highly specific for its substrate, glycogen phosphorylase, and was colocalized with glycogen phosphorylase on glycogen particles. Subsequently, discovery of the cAMP dependent protein kinase demonstrated that protein kinases can phosphorylate multiple substrates, for example, phosphorylase kinase and glycogen synthase. This, in turn, lead to the discovery of specific amino acid sequences, which define substrate specificity for each protein kinase, demonstrating that many proteins contain multiple putative sites of (regulatory) phosphorylation. Thus, peptide substrates corresponding to the (minimal) primary sequence phosphorylated by a protein kinase were identified and developed to assess kinase activity directly in cell lysates, using γ-^{32}P-labeled ATP and measuring the incorporation of ^{32}P into the "specific" peptide substrate. However, as greater numbers of protein kinases were discovered, it became clear that in cell lysates the specificity of one protein kinase for a specific sequence of amino acids in a kinase assay over another protein kinase in the same assay was often marginal.

To circumvent the problem of substrate specificity, several approaches were undertaken. First, antibodies were developed that could immunoprecipitate protein kinases in their active state, thereby permitting isolation/purification of the enzyme from other kinase proteins. Second, incorporation of peptide or protein substrates within a polymerized sodium dodecyl sulfate-polyacrylamide gel electrophoresis (SDS-PAGE), followed by protein renaturation in the gel after electrophoresis and then incubation with γ-^{32}P-labeled ATP, resulted in the appearance of phosphorylated bands within the gel. Based on prior knowledge of the molecular mass of a specific protein kinase, the change in band radioactivity could then be correlated to the activity of the protein kinase of a given molecular mass. Based on ease of use, the technique involving immunoprecipitation of the active kinase became the standard procedure for measuring protein kinase activity by the late 1980s/early 1990s.

The generation of antibodies that recognize specific sites of phosphorylation in a protein was initially developed to examine tyrosine phosphorylation of growth factor receptors and tyrosine phosphorylated proteins in the 1980s. Subsequently, with the discovery of cascades of protein kinases, each kinase catalyzing the phosphorylation and activation/inactivation of the protein kinase below it in a defined signal transduction pathway, leads to the development of "phospho-specific" antibodies that recognize distinct amino acid sequences wherein the site of phosphorylation (containing phosphate) was specifically recognized by the antibody. As the 1990s progressed and into the new Millennium, more phospho-specific antibodies became commercially available, generated against an expanding range of phospho-proteins, the use of immunoprecipitation and in vitro assays using γ-^{32}P-labeled ATP diminished. This was, in part, again owing to the ease of use for phospho-specific antibodies and also resulting from the negative connotations of radioactivity use by investigators. In response to the disinclination of biological scientists to use γ-^{32}P-labeled ATP, some biotechnology companies have developed in vitro assays using nonradioactive forms of ATP coupled to dyes, which generate fluorescent phosphorylated substrates whose fluorescence can be measured in a spectrophotometer.

This chapter will examine the relative merits of determining protein kinase activity by either immunoprecipitation and in vitro kinase assay using γ-^{32}P-labeled ATP or through cell lysis, SDS-PAGE, transfer to nitrocellulose and determination through immunoblotting of the phosphorylation status in activating sites of phosphorylation in a protein kinase using phospho-specific antibodies.

2. Materials

1. RPMI 1640 medium (Invitrogen Life Technologies, Inc., Carlsbad, CA).
2. Williams E medium (Invitrogen Life Technologies).
3. Fetal calf serum (Invitrogen Life Technologies).
4. Penicillin–streptomycin was purchased from Invitrogen Life Technologies.
5. Insulin (Novo Nordisk Pharmaceuticals, Princeton, NJ).
6. Dexamethasone (Novo Nordisk Pharmaceuticals).
7. Thyroxine (Novo Nordisk Pharmaceuticals).
8. Triazma base (Sigma Chemicals, St. Louis, MO).
9. β-Glycerophosphate (Sigma Chemicals).
10. Sodium pyrophosphate (Sigma Chemicals).
11. Sodium orthovanadate (Sigma Chemicals).
12. Triton X-100 (Sigma Chemicals).
13. Deoxycholic acid (DCA) (Sigma Chemicals).
14. Ethylenediaminetetra acetic acid (Sigma Chemicals).
15. Ethylene glycol bis (2-aminoethyl ester)-N, N, N′N′-tetraacetic acid (EGTA) (Sigma Chemicals).
16. Benzamidine (Sigma Chemicals).
17. Phenylmethylsulphonyl fluoride (Sigma Chemicals).
18. Glycerol (Sigma Chemicals).
19. Adenosine triphosphate (ATP) (Sigma Chemicals).
20. Myelin basic protein (MBP) (Sigma Chemicals).
21. Microcystin-LR (Sigma Chemicals).
22. GST-c-Jun (aa 1–169) (synthesized in the Dent laboratory by standard protocols using IPTG-inducible plasmids expressed in *Escherichia coli*).
23. GST-mitogen-activated extracellular regulated kinase 1 (MEK1) and (His)$_6$-MEK1 (synthesized in the Dent laboratory by standard protocols using IPTG-inducible plasmids expressed in *E. coli*).
24. Peptide comprising of the first 11 amino acids of glycogen synthase kinase 3, with two additional NH$_2$-terminal receptor residues (synthesized by the Virginia Commonwealth University peptide synthesis core laboratory RRGRPRTSSFAEG).
25. Anti-T308 protein kinase B (also called AKT) (Cell Signaling Technologies, Worcester, MA).

26. Anti-S473 AKT (Cell Signaling Technologies).
27. Total AKT antibodies (Cell Signaling Technologies).
28. Antiphospho-/total extracellular regulated kinase (ERK1)/2, phospho-/total-p38α/β, phopho-/total-JNK1/2 (Santa Cruz Biotechnology Santa Cruz, CA).
29. β-Actin (Santa Cruz Biotechnology).
30. Secondary antibodies (antirabbit horse-radish peroxidase [HRP], antimouse-HRP, and antigoat-HRP) (Santa Cruz Biotechnology).
31. Protein A/G agarose (Santa Cruz Biotechnology).
32. Secondary antibodies (antimouse, antigoat, antirabbit) labeled with infrared fluorescent tags (Molecular Probes, Eugene, OR, and Rockland Immunochemicals, Gilbertsville, PA).
33. Western immunoblotting performed using either the Perkin-Elmer Life Sciences Enhanced Chemi-luminescence (ECL) System (Boston, MA) or the Odyssey Infrared Imaging System (LI-COR Biosciences, Lincoln, NE).
34. γ-^{32}P ATP (NEN Life Science Products, NEN Life Science Products, Boston, MA) used at a specific activity of 5000 cpm/pmol, radioactivity incorporated into substrate.
35. Immunoblot transfer cassette containing in order:
    a. One Scotch-Brite pad.
    b. Two pieces of Whatman 3M filter paper.
    c. Acrylamide gel.
    d. 0.2-μm Nitrocellulose paper.
    e. Two pieces of Whatman 3M filter paper.
    f. One Scotch-Brite pad.
36. Nondenaturing lysis buffer:
    a. 25 m*M* β-glycerophosphate (pH 7.4, final).
    b. 25 m*M* NaF.
    c. 10% (v/v) Glycerol.
    d. 1% (v/v) Triton X-100.
    e. 1 m*M* Phenylmethylsulphonyl fluoride (*freshly* made in 100% EtOH).
    f. 1 μ*M* Leupeptin.
    g. 1 μ*M* Antipain.
    h. 1 μ*M* Aprotinin.
    i. 5 m*M* Sodium pyrophosphate.
    j. 5 m*M* Sodium orthovanadate.
37. Whole-cell lysis buffer:
    a. 0.5 *M* Tris-HCl.
    b. pH 6.8, 2% (v/v) SDS.
    c. 10% (v/v) Glycerol.
    d. 1% (v/v) β-mercaptoethanol.
    e. 0.02% (v/v) Bromophenol blue.
38. Transfer buffer:
    a. 5.8 g Triazma-base.

　　b. 2.9 g Glycine.

　　c. Milli-Q water to 800 mL.

　　d. 1.8 mL 20% (w/v) SDS.

　　e. 200 mL Methanol.

39. 1X Tris-buffered saline (TBST) buffer:

　　a. 1 dm^3 10X TBS: 24.2 g Tris-base, 292.2 g NaCl, 0.1% (v/v) Tween, pH 7.5.

40. Blocking buffer:

　　a. 5% (w/v) nonfat milk in 1X TBST.

41. PBS containing 0.1% (v/v) Tween-20.

42. Kinase reaction buffer:

　　a. 25 mM HEPES, pH 7.4.

　　b. 15 mM MgCl$_2$.

　　c. 0.1 mM Na$_3$VO$_4$.

　　d. 0.1% (vol/vol) 2-mercaptoethanol.

43. Kinase reaction buffer A:

　　a. 0.2 mM [γ-^{32}P] ATP (5000 cpm/pmol).

　　b. 1 mM Microcystin-LR.

　　c. 0.5 μg/μL Myelin basic protein.

44. Kinase reaction buffer B:

　　a. 2 μL (10 μg) GST-c-Jun (aa 1–169).

45. Buffer B:

　　a. 0.2 mM [γ-^{32}P]ATP (5000 cpm/pmol).

　　b. 1 μM Microcystin-LR.

46. Kinase reaction buffer C:

　　a. 0.2 mM [γ-^{32}P]ATP (5000 cpm/pmol).

　　b. 1 μM Microcystin-LR.

　　c. 200 μM Peptide substrate.

47. Kinase reaction buffer D:

　　a. 2 μL (10 μg) GST-MEK1 (or alternatively [His]$_6$-MEK1).

　　b. 2 mM MnCl$_2$.

48. Kinase reaction buffer E:

　　a. 0.2 mM [γ-^{32}P]ATP (5000 cpm/pmol).

　　b. 1 μM Microcystin-LR.

3. Methods

The following sections describe procedures to determine the activity of a protein kinase by either immune-complex kinase assay in vitro or by immunoblotting using phospho-specific antibodies. The **ref. *1–20*** comprise a list of studies that have examined ERK1/2, JNK1/2, Raf-1, B-Raf, AKT, p70 S6 kinase, GSK3, p38 mitogen activated protein kinase (MAPK), Cdk2 and Cdk4 activities by either immune-complex kinase assays (after immunoprecipitation) in vitro and/or by use of phospho-specific antibodies through immunoblotting (*see* **Note 1**).

3.1. Primary Culture of Rodent Hepatocytes

1. Hepatocytes were isolated from adult male Sprague Dawley rats by the two-step collagenase perfusion technique.
2. The freshly isolated hepatocytes were plated on rat-tail collagen (Vitrogen)-coated plate at a density of 2×10^5 cells/well, and cultured in Williams E medium supplemented with 0.1 μM dexamethasone, 1 μM thyroxine, and 100 μg/mL of penicillin/streptomycin, at 37°C in a humidified atmosphere containing 5% CO_2.
3. The initial medium change was performed 3 h after cell seeding to minimize the contamination of dead or mechanically damaged cells.
4. Unless otherwise indicated, cells were treated with 100 μM DCA approx 24 h after isolation.

3.2. Culture of the HCT116 Colon Carcinoma Cell Line

1. Asynchronous HCT116 carcinoma cells were cultured in RPMI 1640 media, supplemented with 10% (v/v) fetal calf serum at 37°C in 95% (v/v) air/5% (v/v) CO_2.
2. Cells were plated at a density 3×10^3 cells/cm^2 plate area and all cells were plated from log phase cultures.
3. For radiation-induced activations of protein kinases, cells were cultured for 4 d in this media, and for 24 h before irradiation were cultured in serum free DMEM medium.
4. Cells were irradiated: $t = 0$ at the end of the radiation exposure.

3.3. Cell Treatments, SDS-PAGE, and Western Immunoblot Analysis

1. Hepatocytes were exposed to DCA (100 μM). Carcinoma cells were irradiated using a Co^{60} source (1–6 Gy, as indicated).
2. For SDS-PAGE and immunoblotting, at various time points after indicated treatment, hepatocytes and tumor cells were lysed in 1 mL of either a nondenaturing lysis buffer, and prepared for immunoprecipitation as described below or in whole-cell lysis buffer, and the samples were boiled for 30 min.
3. After immunoprecipitation and immune complex kinase assays, samples were also boiled in an equal volume of 2X stock whole cell lysis buffer. The protein concentration of the samples was determined by the Method of Bradford and boiled samples were loaded onto 10–14% SDS-PAGE.
4. Based on the molecular mass of the protein, electrophoresis gels were run for 4 h or overnight.
5. Proteins were electrophoretically transferred onto 0.22 μm nitrocellulose by the Method of Towbin, and immunoblotted with various primary antibodies against different proteins.
6. All hepatocyte immunoblots were visualized by ECL. Tumor cell immunoblots were visualized using both the ECL method and the Odyssey Infrared Imaging System (LI-COR Biosciences, Lincoln, NE).
7. Briefly, the transfer of proteins from the gel to the nitrocellulose consisted of the gel being placed in an immunoblot transfer cassette. It is essential in this process that no air bubbles be present between the acrylamide gel and the nitrocellulose.

8. The transfer cassette was then placed into a transfer box containing transfer buffer, with the nitrocellulose side oriented to the anode for protein transfer. Proteins were transferred overnight at a constant 15V.

9. The nitrocellulose was removed from the transfer cassette and rinsed in 1X TBST buffer for 5 min, then placed in blocking buffer for 4 h at 4°C with gentle mixing.

10. The blocking buffer was removed and the desired primary antibody, diluted in 5% (w/v) nonfat milk in 1X TBST, for 4 h at room temperature or overnight at 4°C (depending on the manufacturer's notes on antibody specifications). The nitrocellulose was then washed with blocking buffer three times for 15 min each.

11. A secondary antibody specific to the animal source of the primary antibody, conjugated to either HRP or an infrared fluorescent dye was diluted 1:2000 in blocking buffer, and the nitrocellulose incubated with this solution for 1 h.

12. After incubation with the secondary antibody, the nitrocellulose was washed three times (15 min per wash) in TBST. For enhanced chemiluminescent immunoblotting, the nitrocellulose was removed from the TBST and taken, with film, ECL reagent and plastic sheet to a dark room.

13. A 1:1 solution of enhanced luminol reagent and oxidizing reagent was prepared and placed on the nitrocellulose for 2 min, with agitation. The nitrocellulose was then taken out of the solution, covered in clear polyvinyl plastic, and placed in an X-ray film cassette.

14. Kodak X-Ray film was placed in the cassette and left for exposure times ranging from 10 s to 20 min, depending on the intensity of the signal obtained.

15. For studies using fluorescent dye conjugated antibodies, the nitrocellulose was removed from the TBST and placed into the Odyssey Infrared Imaging System. The image is obtained by computer-directed laser scanning the nitrocellulose, at two infrared wavelengths.

16. Visualization of immunoblots using ECL usually involves the use of several pieces of Kodak X-OMAT X-ray film, in which the intensity of the chemi-luminescent bands has to be judged by the investigator so that a film is generated containing bands of variable intermediate intensity. In this *art form*, an exposure of x min generates a signal on the film with intensity y, and an exposure of $2x$ min generates a signal on the film with intensity $2y$, and so on. The inherent problem with quantifying band intensity by the use of X-ray film to define signal potency (a.k.a. kinase activity) is that approximately only a 5- to 10-fold difference in signal strength using this method is on this *linear range* of the film and that, in addition, a portion of signal corruption will occur during scanning into an electronic format. Thus measuring kinase activity using phospho-specific antibodies and X-ray film has many potential data interpretative difficulties. In addition to these issues, the affinity of a phospho-specific antibody for a given phospho-specific site in a protein (antigen) may also influence whether basal phosphorylation or a change in phosphorylation is detected by this method.

17. The use of an Odyssey Infrared Imaging System, in part, overcomes many of the problems associated with the limited linear range of X-ray film and data digitization.

The Odyssey system has a linear range more than three orders of magnitude, and the digital image of the immunoblot is already scanned for direct quantitation. In addition, because the Odyssey system scans at two wavelengths it is possible to perform two immunoblots (one appearing as red bands, the other as green bands) on the same piece of nitrocellulose at the same time. The major drawback of the Odyssey system is the high price for an individual laboratory to purchase the system.

3.4. Immunoprecipitation From Cell Lysates

1. Antibodies may be purchased either in solution or coupled to agarose beads. In general, although more expensive, the purchase of an agarose bead-conjugated antibody makes for an easier and more reproducible assay system (conjugated antibodies generated against Raf-1, ERK1/2, JNK1/2, p38, and AKT are commercially available from several vendors).
2. Assuming that an antibody in solution has been purchased, 50 µL of protein A/G-agarose slurry (25-µL bead volume) was placed into a Dolphin-nosed Eppendorf tube (Fisher Scientific, Pittsburgh, PA), washed twice with 1 mL of PBS containing 0.1% (v/v) Tween-20 and was resuspended in 100 µL of the same buffer.
3. Antibodies (2 µg, 20 µL) or serum from an animal bleed containing the antibody of interest (20 µL) was added to each agarose containing tube and incubated (for 3 h, at 4°C) with gentle rocking.
4. After conjugation, beads were separated from unbound proteins by centrifugation (10,000g, 10 min) and the supernatant removed using a vacuum aspirator. The final approx 20 µL of supernatant should be removed using a pipet. Using a Dolphin nosed tube permits the removal of virtually all supernatant from the beads without disturbing or losing to aspiration, any of the beads.
5. Cell lysates were clarified by centrifugation (10,000g, 10 min, 4°C) and their protein concentration measured by the Method of Bradford. Clarified cell homogenates (0.5 mL, 1 mg of total protein) were mixed with protein A/G agarose-conjugated antibody in duplicate using gentle agitation (for 2.5 h, at 4°C).
6. After antigen–antibody complexing, protein A/G agarose was recovered by centrifugation (10,000g, 10 min) and the supernatant was either discarded or boiled with an equal volume of 2X stock whole cell lysis buffer. Boiled supernatant can be used for total loading protein control assays when processed on SDS-PAGE followed by immunoblotting for invariant proteins such as β-actin, total ERK2 and total AKT1.
7. Immunoprecipitates were washed (10 min each wash) sequentially with 0.5 mL of nondenaturing lysis buffer (twice), PBS containing 0.1% (v/v) Tween-20, and kinase reaction buffer. HEPES buffer can be substituted by β-glycerophosphate buffer, which can help suppress contaminating ser/thr phosphatase activities in immunoprecipitates, although at concentrations > 25 µ*M*, β-glycerophosphate also has the potential to inhibit the activity of many protein kinases.

3.5. In Vitro Immune-Complex Kinase Assays: ERK1/2 and p38 MAPK

1. Immunoprecipitates to determine ERK1/2 or p38 MAPK activity were incubated (final volume 50 µL) with kinase reaction buffer A at 37°C, which initiated reactions (*see* **Notes 2** and **3**).

2. Beads/tubes are mixed 10 min after reaction initiation to maintain the agarose beads in suspension. In some cell types, for example, primary cells, the length of reaction may need to be extended: tubes should be mixed every 10 min.
3. After 20 min at 37°C, 40 μL of the reaction mixtures were spotted onto a 2-cm circle of P81 paper (Whatman, Maidstone, UK) and immediately placed into 180 mM phosphoric acid.
4. Papers were washed four times (10 min each) with phosphoric acid and once with acetone, and ^{32}P-labeled incorporation into myelin basic protein was quantified by liquid scintillation spectroscopy.
5. Preimmune controls should be performed at least once by an investigator new to the field to ensure that the phosphorylation of myelin basic protein was dependent on specific immunoprecipitation of ERK1/2 or p38.

3.6. In Vitro Immune-Complex Kinase Assays: JNK1/2

1. Immunoprecipitates were incubated kinase reaction buffer B (final volume 50 μL, 37°C) and reactions were initiated with 48 μL of buffer B. After 30 min, reactions were terminated by addition of an equal volume of 2X whole-cell lysis buffer (*see* **Note 4**).
2. The bead mixture is boiled for 10 min in a dry bath before SDS-PAGE (10% gel: the system of best choice is the Bio-Rad Protean II system, 20 cm length for maximum resolution of substrate protein, IgG heavy chain at approx 50 kDa and IgG light chain at approx 25 kDa).
3. Unbound [γ-^{32}P]ATP and Pi are present in the acrylamide gel around the gel dye-front: this material must be carefully removed by cutting the gel before the staining/fixation procedures (1.0% [w/v] Coomassie brilliant blue G250, 40% [v/v] methanol, 10% [v/v] acetic acid, 50% [v/v] water).
4. Stained/destained gels are washed overnight in water, with several changes, to lower background nonspecific radioactivity levels. JNK1/2 activity is quantified by determining ^{32}P-labeled incorporation into excised, Coomassie blue-stained GST-c-Jun (aa 1–169) bands by liquid scintillation spectroscopy.
5. As previously noted, preimmune control assays should be performed to ensure that GST-c-Jun (aa 1–169) phosphorylation was dependent on specific immunoprecipitation of JNK1/2 in the kinase assay.

3.7. In Vitro Immune-Complex Kinase Assays: AKT

1. AKT was immunoprecipitated and assayed vs a peptide substrate derived from GSK3 (RRGRPRTSSFAEG; termed by some investigators "Crosstide") (*see* **Note 5**).
2. Immunoprecipitates were incubated (final volume 50 μL, 37°C) with 50 μL of kinase reaction buffer C, which initiated reactions.
3. After 20 min, 40 μL of the reaction mixtures were spotted onto a 2-cm circle of P81 paper (Whatman) and immediately placed into 180 mM phosphoric acid. Papers were washed four times (10 min each) with phosphoric acid and once with acetone, and ^{32}P-labeled incorporation into peptide substrate compared to identical samples using a nonspecific antibody.

3.8. In Vitro Immune-Complex Kinase Assays: Raf-1/B-Raf

1. Immunoprecipitates were incubated with kinase reaction buffer D (final volume 50 µL, 37°C). Reactions were initiated with 48 µL of kinase reaction buffer E (*see* **Note 6**).
2. After 30 min, reactions were terminated by addition of an equal volume of 2X whole-cell lysis buffer. The bead mixture was boiled for 10 min in a dry bath before SDS-PAGE (10% gel).
3. Unbound [γ-^{32}P]ATP and P*i* are present in the acrylamide gel around the gel dye-front: this material must be carefully removed before the staining/fixation (1.0% [w/v] Coomassie brilliant blue G250, 40% [v/v] methanol, 10% [v/v] acetic acid, 50% [v/v] water).

   Quantification of ^{32}P-labeled incorporation into excised, Coomassie blue-stained MEK1 protein bands is determined by liquid scintillation spectroscopy. As noted earlier, preimmune control assays should be performed to ensure that MEK1 phosphorylation was dependent on specific immunoprecipitation of *Raf-1* or *B-Raf* in the assay.
4. The specificity of MEK1 phosphorylation by *Raf* family proteins can also be tested in three other ways:
   a. Perform coupled assays in which, 10 min after initiation of the first reaction 2 µL (10 µg) of kinase inactive ERK2 (K52R) is added to the reaction mixture. Measuring the phosphorylation of ERK2 by activated MEK1 is highly specific technique to assess MEK1 activity as it indicates that phosphorylation of MEK1 occurred on the activating sites S217/S221. Assays using further coupling of ERK2 to MBP have also been used in the literature to amplify the "activity" signal further, but require multiple controls to ensure that MBP phosphorylation is completely dependent on the presence of both MEK1 and ERK2 in the Raf-1 activity kinase assay.
   b. Perform assays in the presence of the MEK1/2 inhibitor PD98059 (50 µ*M*), which inhibits the phosphorylation of MEK1 by Raf-1 with an IC50 of 5 µ*M*.
   c. Heat MEK1 to 50°C for 1 h to cause partial MEK1 protein denaturation. Raf-1 and B-Raf, unlike many other protein kinases, appear to recognize secondary and tertiary protein structure within MEK1, and *Raf* kinases poorly phosphorylate isolated MEK1 peptides containing S217/S221.

4. Notes

1. The data in the following figures provides an indication to the reader of "what they can expect" when performing analyses to determine protein kinase activity. ERK1/2, JNK1/2, and p38 MAPK activity is regulated by dual phosphorylation at threonine and tyrosine residues catalyzed by a MAP kinase kinase (also called a MEK protein: this should *not* be confused with MEKK proteins). Phosphorylation occurs at a sequence TxY, where x is, for example, E in ERK1/2, P in JNK1/2, G in p38 MAPK. As a result, measurement of phosphorylation at these sites by phospho-immunoblotting can be judged to be a good approximation of kinase catalytic activity.

2. **Figure 1** shows activation of ERK1/2 in primary hepatocytes by the bile acid deoxycholic acid (DCA) as determined by phospho-immunoblotting against the phosphorylated tyrosine and threonine residues in ERK1/2 (upper panel) and by immune-complex kinase assay measuring catalytic activity in vitro (lower panel). As mentioned in **Subheading 3.**, which discussing about the linearity of the observed response using these techniques of analysis, it is noticeable that activation of ERK1/2 follows a modest time dependent path (0–30 min) before reaching a plateau of high activity, compared with parallel analyses using the phospho-specific ERK1/2 antibody which argues that the bile acid caused an "all or nothing response" for ERK1/2 activation. In part, this may be because the immunoblot was measuring ERK1/2 "activity" in the nonlinear range of the X-ray film.

3. The activity of p38 MAPK in primary hepatocytes is also a good example of where modest activation of an enzyme, as measured by catalytic activity, is not reflected in the data generated using phospho-immunoblotting (**Fig. 2**). As judged by phospho-immunoblotting, neither basal nor bile acid-stimulated phosphorylation of p38 MAPK was observed in primary hepatocytes (**Fig. 2**, upper panel). However, in immune complex kinase assays, a small but reproducible activation of the p38 pathway was noted (**Fig. 2**, lower panel). Thus direct measurements of kinase activity also permits investigators to determine with greater sensitivity the low basal and stimulated kinase activities that are frequently found in nontransformed cells.

4. Assays to measure JNK1/2 activity can also be complicated owing to the expression of several c-Jun NH_2-terminal kinase (JNK) isoforms. In general, the molecular mass of JNK1 is usually stated in the literature as approx 46 kDa and that of JNK2 as approx 54 kDa. However, a truncated form of JNK2 (~47kDa) and a larger version of JNK1 (~53 kDa) are also known to be expressed in cells, and in neuronal tissues an additional JNK3 isoform is present. Collectively, this can make a precise determination of JNK1/2/3 activity by phospho-immunoblotting difficult if the activities of each isoform of JNK are being differentially modulated by a given stimulus (it has been noted in several cell systems). Thus, in hepatocytes JNK1/2 activation by DCA can be measured by immune complex kinase assay in vitro (**Fig. 3**, upper panel) and by phospho-immunoblotting (**Fig. 3**, lower panel). Note that in cells exposed to DCA for long periods, or exposed to DCA in the presence of the MEK1/2 inhibitor PD98059, additional JNK1/2 bands appear, indicative that other JNK1/2 isoforms have become activated. Identical experiments to those examining JNK1/2 activity can be performed to examine Raf-1 and B-Raf activity, in which Raf-1 or B-Raf is immunoprecipitated and the activity of these kinases measured against GST-MEK1 (e.g., **ref. 20**).

5. In contrast to MAPK family enzymes such as ERK1/2 and JNK1/2 which are activated by dual phosphorylation at threonine and tyrosine, the AKT protein kinase is regulated by multisite phosphorylation, at multiple distant/independent sites. In general, AKT activity is thought to be primarily regulated by phosphorylation at T308 catalyzed by PDK-1 and at S473, potentially via T308 stimulated-autophosphorylation, although other sites of regulatory phosphorylation, including

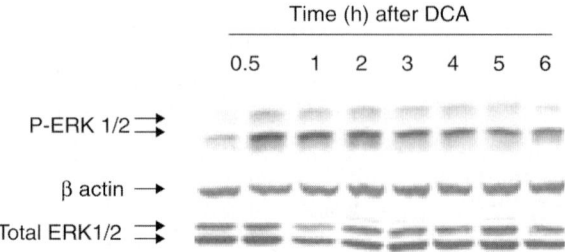

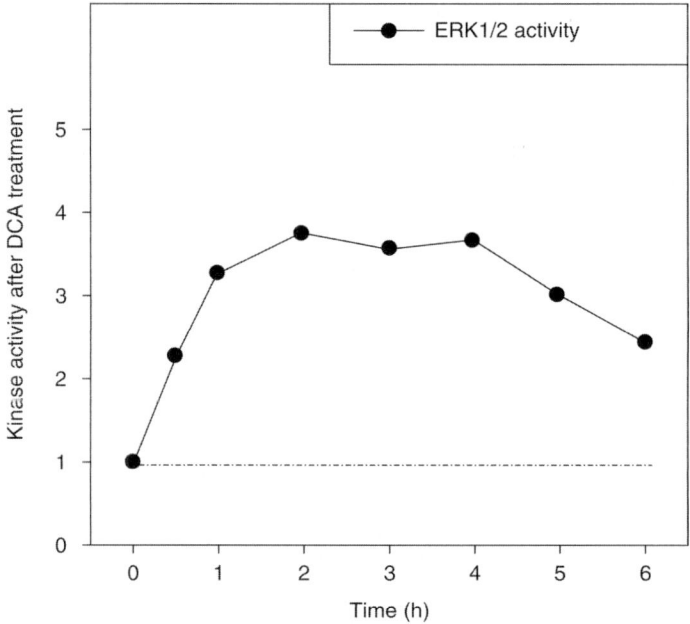

Fig. 1. Activation of ERK1/2 by the bile acid deoxycholic acid (DCA) in primary rat hepatocytes. Primary rat hepatocytes were treated with DCA (100 μ*M*) and cells isolated/lysed at various time-points after exposure as noted in Methods. Lysates were processed for either: upper panel; SDS-PAGE and immunoblotting to determine the phosphorylation status of ERK1/2 using ECL (**Subheading 3.4.**). Lower panel; lysates were immunoprecipitated using an anti-ERK1/2 antibody and in vitro immune-complex kinase assays were performed. Data are expressed as the –fold increase in MBP phosphorylation (**Subheadings 3.5.** and **3.6.**).

tyrosine, have been noted. Many groups have monitored AKT activity in a variety of cells using phospho-immunoblotting for increases in T308 and in S473 phosphorylation (**Fig. 4**, upper panel). Alternatively, using immune-complex kinase assay

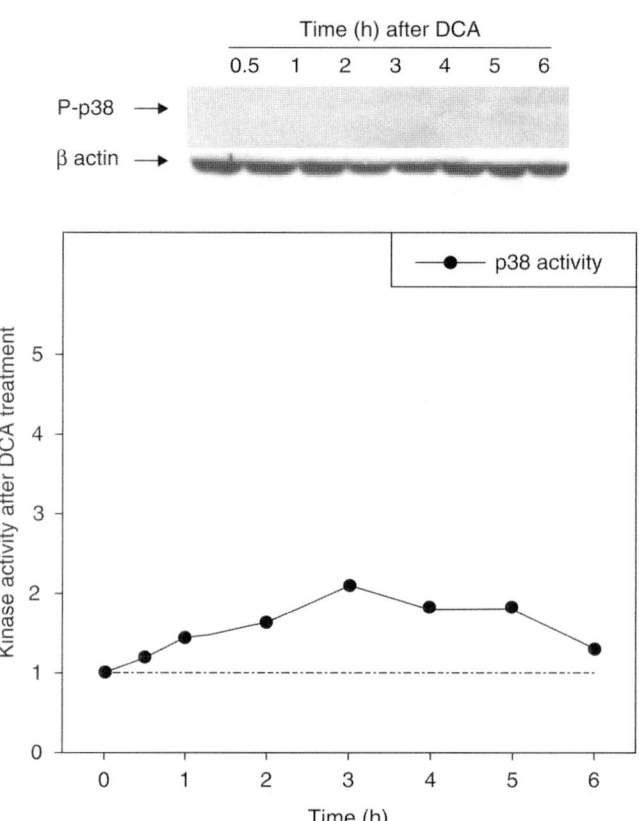

Fig. 2. Activation of p38 MAPK by the bile acid deoxycholic acid (DCA) in primary rat hepatocytes. Primary rat hepatocytes were treated with DCA (100 μ*M*) and cells isolated/lysed at various time-points after exposure as noted in Methods. Lysates were processed for either: upper panel; SDS-PAGE and immunoblotting to determine the phosphorylation status of p38 MAPK using ECL (**Subheading 3.4.**). Lower panel; Lysates were immunoprecipitated using an anti-p38 MAPK antibody and in vitro immune-complex kinase assays were performed. Data are expressed as the –fold increase in MBP phosphorylation (**Subheadings 3.5.** and **3.6.**).

with a peptide substrate derived from GSK3, AKT catalytic activity can be measured (**Fig. 4**, lower panel). As multiple sites can potentially impact on AKT catalytic activity, immune-complex kinase assays may represent the best method to assess AKT function. Of note, GSK3, a primary downstream target of AKT activity can also be used as a corroborative read-out for AKT activity by measuring the phosphorylation of S9 (GSK3α) or S21 (GSK3β) (*see* **refs.** *2* and *14*). In tumor cells (e.g., HCT116) the basal activities of ERK1/2 and AKT are often much higher than those observed in primary cell types (**Fig. 5**, upper panel; compare with data in **Fig. 1–4**). Thus care is always needed in initial assays to assess "how much signal" is obtained comparing data from, for example, a weakly tumorigenic cell line to a

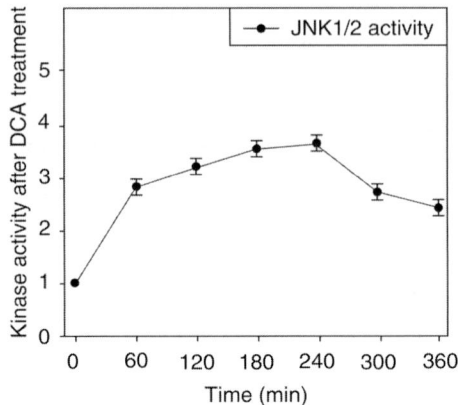

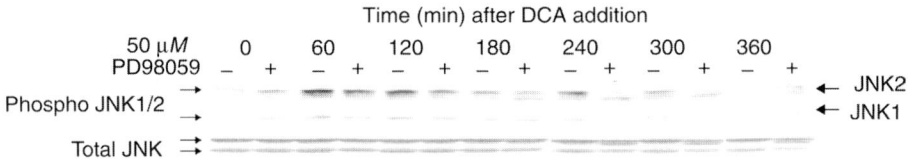

Fig. 3. Activation of JNK1/2 by the bile acid deoxycholic acid (DCA) in primary rat hepatocytes. upper panel; primary rat hepatocytes were treated with DCA (100 μ*M*) and cells lysed at various time-points after exposure as noted in Methods **(Subheadings 3.5. and 3.7.).** Lysates were immunoprecipitated using an anti-JNK1/2 antibody and in vitro immune-complex kinase assays were performed using GST-c-Jun (aa 1–169). Data are expressed as the –fold increase in GST-c-Jun phosphorylation **(Subheading 3.7.).** Lower panel; primary rat hepatocytes were treated with PD98059 (50 μ*M*) and/or DCA (100 μ*M*) and cells lysed at various time-points after exposure as noted in Methods. Cell lysates were subjected to SDS-PAGE and immunoblotting to determine the phosphorylation status of JNK1/2 isoforms **(Subheading 3.4.).**

highly tumorigenic cell line to primary isolates of nontransformed cells (**Fig. 5,** upper panel). Of note, in this panel, HCT116 cell phospho-blots are presented for parental HCT116 cells, HCT116 cells lacking expression of their single allele of K-RAS D13 owing to in vitro homologous recombination, and these cells expressing H-RAS V12. Thus, loss of K-RAS D13 abolishes radiation-induced activation of ERK1/2 and AKT, and expression of H-RAS V12 promotes radiation-induced activation of AKT (S473 phosphorylation) to a greater extent than activation of ERK1/2.
6. It should be noted, in cells where one can expect higher levels of kinase activity that the activity of protein kinases, assessed by immunoblotting, can be determined using methods independent of the ECL system. Two alternate methods include the use of conjugated secondary antibodies that convert colorless reagents

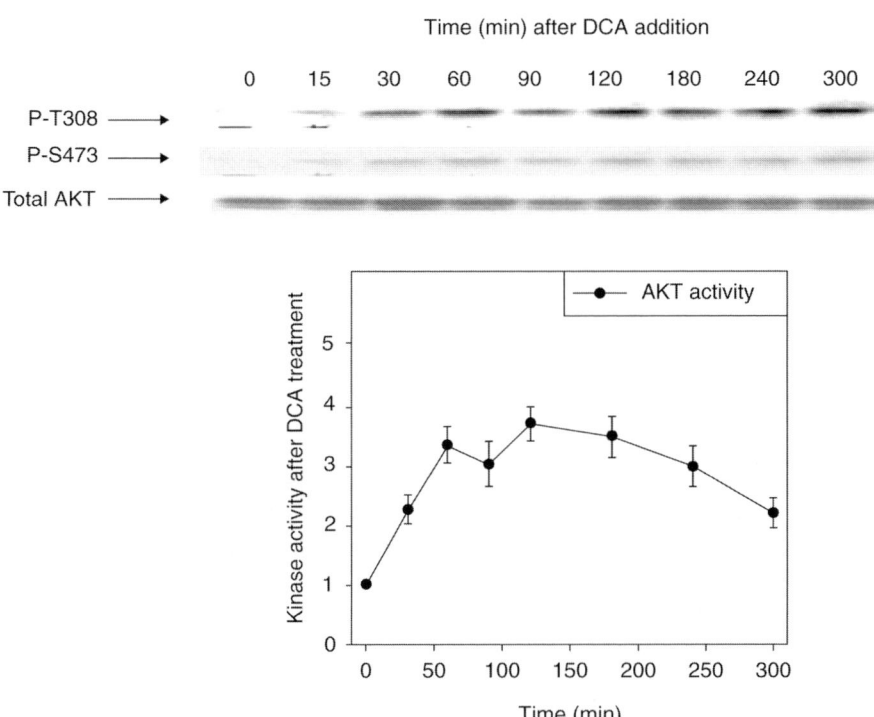

Fig. 4. Activation of AKT by the bile acid deoxycholic acid (DCA) in primary rat hepatocytes. Upper panel; primary rat hepatocytes were treated with DCA (100 μ*M*) and cells lysed at various time-points after exposure as noted in Methods. Cell lysates were subjected to SDS-PAGE and immunoblotting to determine the phosphorylation status of AKT at T308 or of AKT at S473 (**Subheading 3.4.**); lower panel; primary rat hepatocytes were treated with DCA (100 μ*M*) and cells lysed at various time-points after exposure as noted in Methods (**Subheadings 3.5.** and **3.8.**). Lysates were immunoprecipitated using an anti-AKT1/2 antibody and in vitro immune-complex kinase assays were performed using a substrate peptide derived from GSK3. Data are expressed as the –fold increase in ^{32}P radioactivity incorporated into the peptide (**Subheading 3.7.**).

into brown/purple products, thereby permitting visualization of protein bands (an "older" technique, *see* **ref. 7**), or alternatively, using fluorescent conjugated secondary antibodies in a state of the art Odyssey Infrared Imaging System (**Fig. 5**, lower panel), which records immunoblots digitally.

Acknowledgments

This work was funded from PHS grants (R01-DK52825; P01-DK38030; R01-CA35675; P01-CA72955; R01-CA88906; R01-CA097318; R01-CA098712; R01-CA107326; P01-CA177104; P01-CA104177; R01-CA108520), Department

Dent et al.

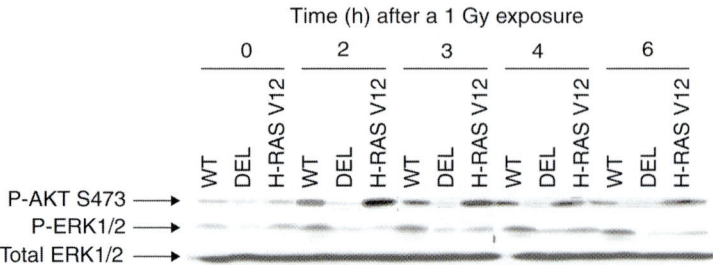

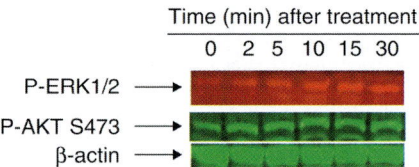

Fig. 5. Activation of ERK1/2 and AKT by ionizing radiation in parental HCT116 colon carcinoma cells (WT), HCT116 cells lacking expression of K-RAS D13 (DEL) and HCT116 cells lacking K-RAS D13 transfected with H-RAS V12 (H-RAS V12). Upper panel; HCT116 cell lines were exposed to radiation (1 Gy) and lysed at various time-points after exposure as noted in **Subheading 3.** The cell lines used are parental HCT116 cells (WT) that express an activated K-RAS D13 protein; HCT116 cells that have been manipulated by homologous recombination so that the allele of K-RAS D13 has been deleted (DEL); HCT116 DEL cells that have been stably transfected to express activated H-RAS V12 (H-RAS V12). Cell lysates were subjected to SDS-PAGE and immunoblot-ting to determine the phosphorylation status of ERK1/2 and AKT at S473 using the ECL technique **(Subheading 3.4.).** Lower panel; HCT116 cells (H-RAS V12) were exposed to heregulin (5 ng/mL) and cells lysed at various time-points after exposure as noted in Methods **(Subheading 3.4.).** Cell lysates were subjected to SDS-PAGE and immunoblot-ting to determine the phosphorylation status of ERK1/2 and of AKT at S473 using the Odyssey infrared imaging system (LI-COR Biosciences, Lincoln, NE). Phosphorylated ERK1/2 is displayed as red bands on the nitrocellulose whereas AKT S473 appears as a green band. A total protein loading control (β-actin) is also shown as a green band. Thus on the same piece of nitrocellulose, blotted at the same time, two or more signaling pro-teins of interest can be imaged.

of Defense Awards (BC980148, BC020338), the Lustgarten Foundation for Pancreatic Cancer Research, the Samuel Waxman Cancer Research Foundation and the Chernow Endowment; P.D. is the holder of the Universal Inc. Professorship in Signal Transduction Research; P.B.F. is the Michael and Stella Chernow Urological Cancer Research Scientist and a SWCRF Investigator.

References

1. Auer, K. L., Park, J. S., Seth, P., et al. (1998) Prolonged activation of the mitogen-activated protein kinase pathway promotes DNA synthesis in primary hepatocytes from p21Cip-1/WAF1-null mice, but not in hepatocytes from p16INK4a-null mice. *Biochem. J.* **336,** 551–560.
2. Auer, K. L., Contessa, J., Brenz-Verca, S., et al. (1998) The Ras/Rac1/Cdc42/SEK/JNK/c-Jun cascade is a key pathway by which agonists stimulate DNA synthesis in primary cultures of rat hepatocytes. *Mol. Biol. Cell* **9,** 561–573.
3. Carter, S., Auer, K. L., Reardon, D. B., et al. (1998) Inhibition of the mitogen activated protein (MAP) kinase cascade potentiates cell killing by low dose ionizing radiation in A431 human squamous carcinoma cells. *Oncogene* **16,** 2787–2796.
4. Contessa, J. N., Hampton, J., Lammering, G., et al. (2002) Ionizing radiation activates Erb-B receptor dependent Akt and p70 S6 kinase signaling in carcinoma cells. *Oncogene* **21,** 4032–4041.
5. Cross, D. A., Alessi, D. R., Cohen, P., Andjelkovich, M., and Hemmings, B. A. (1995) Inhibition of glycogen synthase kinase-3 by insulin mediated by protein kinase B. *Nature* **378,** 785–789.
6. Cross, D. A., Watt, P. W., Shaw, M., et al. (1997) Insulin activates protein kinase B, inhibits glycogen synthase kinase-3 and activates glycogen synthase by rapamycin-insensitive pathways in skeletal muscle and adipose tissue. *FEBS Lett.* **406,** 211–215.
7. Dent, P., Haser, W., Haystead, T. A., Vincent, L. A., Roberts, T. M., and Sturgill, T. W. (1992) Activation of mitogen-activated protein kinase kinase by v-Raf in NIH 3T3 cells and *in vitro*. *Science* **257,** 1404–1407.
8. Dent, P., Chow, Y. H., Wu, J., Morrison, D. K., Jove, R., and Sturgill, T. W. (1994) Expression, purification and characterization of recombinant mitogen-activated protein kinase kinases. *Biochem. J.* **303,** 105–112.
9. Dent, P., Jelinek, T., Morrison, D. K., Weber, M. J., and Sturgill, T. W. (1995) Reversal of Raf-1 activation by purified and membrane-associated protein phosphatases. *Science* **268,** 1902–1906.
10. Dent, P., Reardon, D. B., Morrison, D. K., and Sturgill, T. W. (1995) Regulation of Raf-1 and Raf-1 mutants by Ras-dependent and Ras-independent mechanisms *in vitro*. *Mol. Cell. Biol.* **15,** 4125–4135.
11. Dent, P., Reardon, D. B., Park, J. S., et al. (1999) Radiation-induced release of transforming growth factor alpha activates the epidermal growth factor receptor and mitogen-activated protein kinase pathway in carcinoma cells, leading to increased proliferation and protection from radiation-induced cell death. *Mol. Biol. Cell* **10,** 2493–2506.
12. Gupta, S., Natarajan, R., Payne, S. G., et al. (2004) Deoxycholic acid activates the c-Jun N-terminal kinase pathway via FAS receptor activation in primary hepatocytes. Role of acidic sphingomyelinase-mediated ceramide generation in FAS receptor activation. *J. Biol. Chem.* **279,** 5821–5828.
13. Gupta, S., Stravitz, R. T., Dent, P., and Hylemon, P. B. (2001) Down-regulation of cholesterol 7alpha-hydroxylase (CYP7A1) gene expression by bile acids in

primary rat hepatocytes is mediated by the c-Jun N-terminal kinase pathway. *J. Biol. Chem.* **276,** 15,816–15,822.

14. Han, S. I., Studer, E., Gupta, S., et al. (2004) Bile acids enhance the activity of the insulin receptor and glycogen synthase in primary rodent hepatocytes. *Hepatology* **39,** 456–463.

15. McCubrey, J. A., Steelman, L. S., Hoyle, P. E., et al. (1998) Differential abilities of activated Raf oncoproteins to abrogate cytokine dependency, prevent apoptosis and induce autocrine growth factor synthesis in human hematopoietic cells. *Leukemia* **12,** 1903–1929.

16. Park, J. S., Carter, S., Reardon, D. B., Schmidt-Ullrich, R., Dent, P., and Fisher, P. B. (1999) Roles for basal and stimulated p21(Cip-1/WAF1/MDA6) expression and mitogen-activated protein kinase signaling in radiation-induced cell cycle checkpoint control in carcinoma cells. *Mol. Biol. Cell* **10,** 4231–4246.

17. Qiao, L., Studer, E., Leach, K., et al. (2001) Deoxycholic acid (DCA) causes ligand-independent activation of epidermal growth factor receptor (EGFR) and FAS receptor in primary hepatocytes: inhibition of EGFR/mitogen-activated protein kinase signaling module enhances DCA-induced apoptosis *Mol. Biol. Cell* **12,** 2629–2645.

18. Qiao, L., Leach, K., McKinstry, R., et al. (2001) Hepatitis B virus X protein increases expression of p21(Cip-1/WAF1/MDA6) and p27(Kip-1) in primary mouse hepatocytes, leading to reduced cell cycle progression. *Hepatology* **34,** 906–917.

19. Qiao, L., Han, S. I., Fang, Y., et al. (2003) Bile acid regulation of C/EBP beta, CREB, and c-Jun function, via the extracellular signal-regulated kinase and c-Jun NH2-terminal kinase pathways, modulates the apoptotic response of hepatocytes. *Mol. Cell Biol.* **23,** 3052–3066.

20. Tombes, R. M., Auer, K. L., Mikkelsen, R., et al. (1998) The mitogen-activated protein (MAP) kinase cascade can either stimulate or inhibit DNA synthesis in primary cultures of rat hepatocytes depending upon whether its activation is acute/phasic or chronic. *Biochem. J.* **330,** 1451–1460.

18

PKR in Innate Immunity, Cancer, and Viral Oncolysis

Siddharth Balachandran and Glen N. Barber

Summary

The mammalian innate immune system provides a first line of defense against microbial pathogens and also serves to activate an antigen specific acquired immune program. Key components of innate immunity are the interferons (IFNs), a family of related cytokines with potent antimicrobial and immuno-modulatory activities. The IFNs exert their effects through the induction of numerous genes, one of which is the double-stranded RNA-dependent protein kinase (PKR), a pivotal antiviral protein found in most human cells. Following activation by double stranded (ds) RNAs produced during viral replication, PKR phosphorylates the α-subunit of eukaryotic translation initiation factor (eIF) 2, causing a severe inhibititon of cellular and viral protein synthesis. Phosphorylation of eIF2α and consequent inhibition of protein synthesis is a major cell growth checkpoint utilized by at least three other kinases, in addition to PKR, following exposure to such cellular stresses as amino acid deprivation and the presence of misfolded proteins in the endoplasmic reticulum. Indeed, it has been demonstrated that disruption of the eIF2α checkpoint can lead to the transformation of immortalized rodent and human cells, plausibly by increasing the protein synthesis rates of proto-oncogenes. Further, it has been shown that disregulation of the eIF2α checkpoint and consequent permissiveness to virus infection may be a common occurrence in tumorigenic mammalian cell lines. These findings have been exploited to develop potent oncolytic RNA viruses that can selectively replicate in and destroy a variety of neoplasias in vitro and in vivo. In this chapter, we describe some of the techniques commonly used in our laboratory to examine PKR activity and eIF2 regulation. Protocols for the generation and use of recombinant vesicular stomatitis virus variants are also described.

Key Words: eIF2α; eIF2B; oncolysis; PKR; translational control; VSV.

1. Introduction
1.1. Background

The innate (or natural) immune system provides a broad, relatively nonspecific primary line of defense against invading microbial pathogens, but lacks the properties of antigenic specificity and immunological memory that characterize

From: *Methods in Molecular Biology, vol. 383: Cancer Genomics and Proteomics: Methods and Protocols*
Edited by: P. B. Fisher © Humana Press Inc., Totowa, NJ

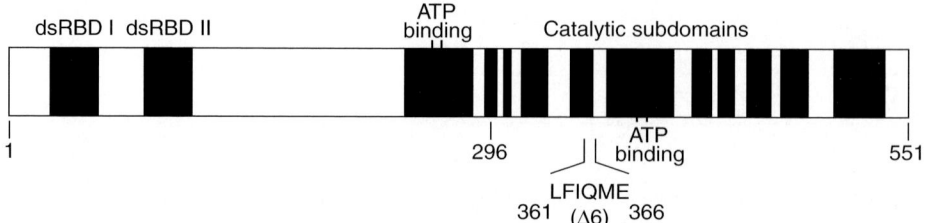

Fig. 1. Structure of PKR. PKR protein contains two conserved dsRNA binding domains in its amino-terminal half, and eleven serine–threonine catalytic sundomains toward its carboxyl terminus. Also depicted are two putative ATP binding regions, as well as the locations of the key lysine 296 and the six amino acid (LFIQME) deletion that results in the dominant-negative PKR Δ6 mutant.

acquired immunity *(1)*. The innate immune system includes physical barriers such as skin and mucous membranes, myeloid lineage cells (e.g., neutrophils and macrophages), and various cytokines secreted by these and other nonlymphoid cells in response to the invading pathogen *(2)*. Indispensable components of innate immunity are the IFNs, a family of cytokines referred to as type I (α/β) and type II (γ) *(3)*. Type I IFNs are produced by most cell types following virus infection, although type II IFN is synthesized by a subset of immune system cells (such as T cells and NK cells), and is not strongly virus-inducible. Both types of IFNs, however, have potent antiviral, immuno-modulatory and antitumor properties (reviewed in **ref. 3**). The IFNs manifest their pleiotropic activities by triggering Janus kinase/Signal transducer and activator of transcription (STAT)-1 and -2-dependent signaling cascades that culminate in the induction of several (>100) genes, many of which are of unknown function *(4)*.

1.2. Structure and Cellular Regulatory Roles of PKR

An important and well-studied component of the IFN-induced antiviral repertoire is the serine/threonine kinase PKR *(5)*. The human *pkr* gene, located on chromosome 2, encodes a 551-amino acid protein translated from a 2.5-kB mRNA *(5,6)*. PKR has two conserved dsRNA-binding motif-containing domains (dsRBDs) in its amino-terminal half, and a typical catalytic serine/threonine kinase domain, consisting of eleven subdomains, toward its carboxyl terminus. Within this region is also the ATP-binding pocket, with the ATP-accepting lysine at position 296 *(5)*. Mutation of this lysine to arginine, or deletion of six critical amino acids (positions 361–366) in the kinase domain render PKR inactive, and represent two well-studied mutants of PKR (termed PKR K296R and PKR Δ6, respectively) as described in **refs. 7–10** (**Fig. 1**).

Following virus infection, dsRNAs produced during viral replication activate PKR, inducing its autophosphorylation. Although the activation process of PKR remains to be fully clarified, autophosphorylation of this kinase has been found to occur through interaction with low concentrations of dsRNA, but not single-stranded (ss) RNA, ssDNA, or DNA:RNA hybrids, and neither is there any known dsRNA sequence dependence for its activation *(8,11–14)*.

Activated PKR can subsequently phosphorylate numerous cellular target proteins, but its best-characterized substrate remains eIF2α reviewed in **ref. 16**. In addition to its well-defined role as an eIF2α kinase, PKR has been reported to function in a variety of signal transduction pathways, including those involving interleukin-3, platelet-derived growth factor, IFN-β, nuclear factor (NF)-κB, p53, interferon regulatory factor-1, STAT1, and mitogen-activated protein kinases *(16,18–24)*. PKR has also been demonstrated to interact with several cellular proteins, including, among others, STAT1, p53, the ribosomal large subunit protein L18, melanoma differentiation-associated gene 7/interleukin-24, the influenza virus activated chaperone protein p58, the nuclear factors associated with dsRNA (NFARs)-1 and -2, and PACT, a protein capable of activating PKR in the absence of dsRNA *(16,18,23,25–28)* as reviewed in **refs. 15,16,29–31**. Taken together, these studies demonstrate a key role for PKR in the response to pathogenic invasion, as well as in multiple cellular homeostatic processes (reviewed in **ref. 16**).

PKR is also known to be targeted for inhibition by numerous viruses, presumably because of the deleterious effects that activation of this kinase would have on virus replication. Viral inhibition of PKR has been shown to take many forms. Viruses encode proteins that physically bind to and inactivate PKR (for e.g., hepatitis C virus NS5A), that mimic the PKR substrate eIF2α and act as pseudosubstrate decoys (vaccinia virus K3L), that sequester dsRNA and prevent it from activating PKR (vaccinia virus E3L and influenza virus NS1) and that operate downstream of PKR activation by inducing the dephosphorylation of eIF2α (herpervirus γ 34.5). Additionally, viruses also encode RNA inhibitors of PKR (such as the adenoviral VA$_I$ RNA) and might target it for degradation (a method employed by poliovirus) (reviewed in **refs. 32** and **33**). Indeed, the importance of PKR in controlling viral infection is underscored by the fact that mice lacking PKR are very susceptible to lethal intranasal infection by a number of viruses. For example, wild-type mice were able to survive up to 1×10^5 plaque forming units (PFU) vesicular stomatitis virus (VSV) intranasally, whereas PKR−/− mice succumbed to doses of VSV as low as 50 PFU/mouse *(34–36)*. Thus, PKR appears to be an essential component of the innate immune system.

In addition to experiments performed on PKR−/− embryonic fibroblasts and mice, efforts at delineating the role of PKR in various cellular processes have involved the use of several catalytically inactive variants of this kinase. Intriguingly, the ectopic overexpression of such variants of PKR causes the malignant transformation of immortalized rodent fibroblasts, and it appears that these mutants fall into two broad categories: *bona fide* dominant-negative inhibitors of PKR, which directly inhibit dimerization and/or autophosphorylation of this molecule (for e.g., PKR Δ6), and those that inhibit PKR activity mainly by sequestering dsRNA and making it unavailable for the activation of the kinase (such as PKR K296R) *(7,9,10,15,37)*. In vivo, NIH 3T3 cells transformed by dominant-negative inhibitors of PKR exhibit enhanced growth rates, diminished eIF2α phosphorylation, and aggressive tumorigenesis in immunocompromised mice, compared to NIH 3T3 cells transformed by dsRNA-sequestering mutants of PKR *(7)*. These results demonstrate the importance of PKR in maintenance of cellular homeostasis.

In contrast to the phenotype exhibited by mutants of PKR, overexpression of wild-type PKR in mammalian and insect cells, as well as in yeast, results in a profound inhibition of cellular growth *(10,38–40)*. In fact, attempts at generating mammalian cell lines constitutively expressing this kinase proved unsuccessful, and detailed studies of the impact of PKR on cell growth could only be performed under conditions in which PKR was expressed under control of an inducible promoter. We and others have subsequently shown that inducible overexpression of PKR is accompanied by the upregulation of proapoptotic genes, and triggered FADD- and caspase 8-dependent apoptosis in mammalian cells *(41–44)*. Paradoxically, PKR can also trigger the activation of NF-κB and consequent production of antiapoptotic genes in response to dsRNA *(44–46)*. It thus appears that PKR can activate two independent cellular programs: one triggered early on activation that results in the NF-κB dependent induction of survival genes and IFN-β, and a second pathway triggered by sustained PKR activation that depends on eIF2α phosphorylation and culminates in apoptosis *(45)*.

1.3. The eIF2α Checkpoint

PKR appears to primarily exert its antiviral activity through phosphorylation of the translation initiation factor eIF2α *(35,48)*. In its GTP-bound form, eIF2 (a trimeric protein complex, consisting of α, β, and γ subunits) is required for the transfer of initiator methionyl-tRNA (Met-tRNA$_i$) to the small (40S) ribosomal subunit. After deposition of Met-tRNA$_i$ onto the 40S ribosomal subunit, eIF2 is released in a GDP-bound form and must be recycled to eIF2-GTP in order to participate in subsequent rounds of protein synthesis initiation *(49)*. Replacement of GDP with GTP on eIF2 is carried out by the pentameric guanine nucleotide exchange factor, eIF2B (consisting of subunits termed α-ε). Phosphorylation of eIF2α by PKR converts eIF2 from a substrate to a competitive inhibitor of eIF2B function. Reduction of eIF2B activity causes a global decrease in translation rates, potently inhibiting both cellular metabolism and viral replication (reviewed in **ref.** *49*) (**Fig. 2**).

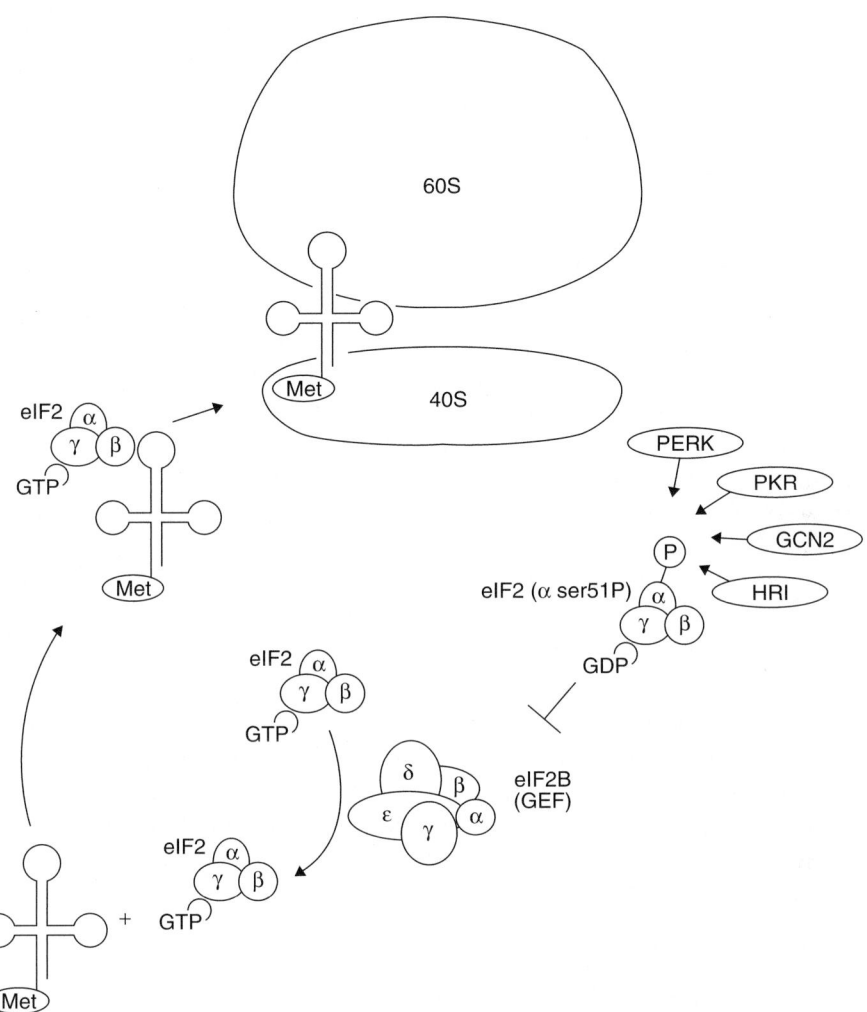

Fig. 2. eIF2 In initiation of translation and its inhibition by eIF2α phosphorylation. The ternary complex of eIF2-Met.tRNAi-GTP is required to position the initiatior methionyl tRNA on the 40S subunit of the ribosome. After this reaction, eIF2-GDP is recycled to its GTP-bound form by eIF2B. Phosphorylation of eIF2α on serine 51 by PKR and other eIF2α kinases severely inhibits eIF2B-mediated nucleotide exchange on eIF2.

In addition to PKR, at least three other kinases phosphorylate eIF2α in response to a diverse range of stresses (reviewed in **ref. 49**). The heme-regulated inhibitor (HRI) kinase phosphorylates eIF2α and impedes translation in reticulocytes under conditions of heme deficiency in these cells *(50)*. Similarly, the yeast kinase GCN2 (as well as its mammalian homolog) responds to reduced amino acid availability by attenuating translation rates via phosphory-lation of eIF2α *(51,52)*. Finally, the endoplasmic reticulum (ER)-resident kinase PERK responds to the accumulation of misfolded proteins in the lumen of the ER by phosphorylating eIF2α and reducing protein synthesis rates as the Unfolded Protein Response (UPR) (*see* **Note 1**) is activated and ER stress reduced *(53,54)*. Indeed, PERK-deficient animals manifest severe diabetes as a result of ER-stress-triggered apoptosis in the β-cells of the islets of Langerhans *(53)*. Similarly, mice in which both wild-type eIF2α alleles were replaced with a version of eIF2α in which the phosphoacceptor serine 51 was replaced with alanine (eIF2α A/A) are also severely diabetic, and die shortly after birth *(55)*. Thus, regulation of translation through eIF2α phosphorylation is important in glucose homeostasis and β-cell survival in vivo *(56)*.

Although these findings clearly establish the importance of the eIF2α check-point in decreasing protein synthesis rates and promoting cell survival following various specific stresses, other studies have shown an important role for eIF2α regulation under normal conditions as well. For example, the authors have shown that a nonphosphorylatable variant of eIF2α, in which serine 51 has been replaced with an alanine (eIF2α S51A), can potently transform rodent and human fibrob-lasts *(48,57)*. Recent results from our laboratory have since revealed that the majority of mammalian tumors appear to be significantly impaired in their ability to modulate protein synthesis rates in response to eIF2α phosphorylation. Several tumor cell lines studied, whereas displaying apparently normal eIF2α phosphorylation, exhibited greatly increased eIF2B-mediated guanine nucleotide exchange activity downstream of eIF2. Indeed, this aberrantly high eIF2B activ-ity in transformed cells appears to be sufficient to at least partially neutralize the inhibitory effects of eIF2α phosphorylation *(58)*.

1.4. VSV Oncolysis

Studies evaluating the susceptibility of PKR−/− mice to infection by various viruses led us to the discovery that VSV replicates poorly in primary cells with normal PKR-mediated translational control and IFN signaling *(35,36,59)*. In contrast, VSV can replicate well in the vast majority of commonly used tissue culture cell lines and tumor-derived cells, even in the presence of exogenously added IFN **(Fig. 3)**. Our group, as well as others, has since demonstrated that VSV possesses potent oncolytic activity against a wide variety of mammalian tumors in vitro and in vivo *(60–65)*. VSV is a particularly attractive candidate for viral therapy of tumors because it is a nonintegrating virus that does not undergo spontaneous genetic reassortment, possesses no known transforming properties, and is very malleable to genetic manipulation *(59,66,67)*. In fact, we have developed several recombinant variants of VSV that express high levels of immuno-modulatory and antitumor proteins *(62,63,68)*. These viruses display greatly increased specificity and oncolytic activity, compared to wild-type VSV.

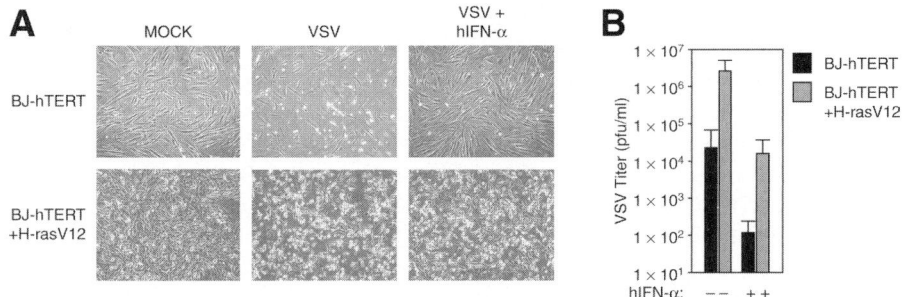

Fig. 3. VSV replicates efficiently in transformed human cells, even in the presence of IFN-α. (**A**) BJ human fibroblasts immortalized with the SV40 large T antigen and hTERT human telomerase catalytic subunit (BJ-hTERT), or transformed by the addition of an activated *ras* oncogene (BJ-hTERT+H-rasV12), were either mock-infected or infected with VSV (m.o.i. = 5) in the absence or presence of 200 U/mL hIFN-α for 36 h and subsequently photographed at ×200 magnification. (**B**) Virus progeny yield was determined from cells treated as in **A**. *See* **Subheading 3.9.**, for details on virus infection and titration procedures. Figure reproduced from **ref. 58**.

Since transformed cells possess defective IFN antiviral activity and are largely refractory to the translational inhibitory consequences of PKR activation, we have now developed recombinant versions of VSV that exploit these defects to specifically replicate in tumor cells without harming normal cells. For example, VSV expressing IFN-β has been generated and demonstrated to remain potently oncolytic in vitro and in vivo, while exhibiting significantly reduced cytopathicity in normal cells *(63)* . Thus VSV displays significant potential as an oncolytic agent.

In this chapter, the techniques commonly employed in the laboratory to study PKR and the eIF2 checkpoint, as well as the protocols to generate and use recombinant VSV were described.

2. Materials

1. Polyanionic activator heparin (Sigma, St. Louis, MO).
2. Lysis buffer A:
   a. 20 m*M* Tris-HCl (pH 8.0).
   b. 500 m*M* NaCl.
   c. 100 U/mL Aprotinin.
   d. 0.2 m*M* Phenylmethylsulfonyl fluoride (PMSF).
3. Buffer I:
   a. 20 m*M* Tris-HCl (pH 7.5).
   b. 50 m*M* KCl.
   c. 400 m*M* NaCl.
   d. 1 m*M* Ethylenediamine tetraacetic acid (EDTA).

   e. 100 U/mL Aprotinin.

   f. 1 m*M* Dithiothreitol (DTT).

   g. 0.2 m*M* Phenylmethylsulfonyl fluoride (PMSF).

   h. 1% (v/v) Triton X-100.

   i. 20% (v/v) Glycerol.

4. Buffer II:

   a. 10 m*M* Tris-HCl (pH 7.5).

   b. 100 m*M* KCl.

   c. 0.1 m*M* EDTA.

   d. 100 U/mL Aprotinin.

   e. 20% (v/v) Glycerol.

5. Neutralizing buffer:

   a. 20 m*M* 4-2-hydroxyethyl-1-piperazineethanesulfonic acid (HEPES, pH 7.4).

   b. 50 m*M* KCl.

   c. 0.1 m*M* EDTA.

   d. 1 m*M* DTT.

   e. 0.2 m*M* PMSF.

   f. 100 U/mL Aprotinin.

   g. 10% (v/v) Glycerol.

6. Lysis buffer B:

   a. 10 m*M* Tris-HCl (pH 7.5).

   b. 50 m*M* KCl.

   c. 1 m*M* DTT.

   d. 2 m*M* EDTA.

   e. 1 m*M* MgCl$_2$.

   f. 0.2 m*M* PMSF.

   g. 100 U/mL Aprotinin.

   h. 1% Triton X-100.

7. Sodium dodecyl sulfate-polyacrylamide gel electrophoresis (SDS-PAGE) loading buffer:

   a. 5% (w/v), 20% β-mercaptoethanol.

   b. 150 m*M* Tris-HCl (pH 6.8).

   c. 20% (v/v) Glycerol.

   d. 0.1% Before bromophenol blue.

8. Blocking buffer:

   a. PBS containing 5% (w/v) nonfat dry milk powder.

   b. 0.1% (v/v) Tween-20.

9. Buffer III:

   a. 20 m*M* Tris-HCl (pH 7.5).

   b. 1 m*M* DTT.

   c. 0.1 m*M* EDTA.

   d. 25 m*M* KCl.

   e. 1 m*M* PMSF.

   f. 5% (v/v) Glycerol.

10. Buffer IV:
    a. 10 m*M* Tris-HCl (pH. 7.5).
    b. 0.1 m*M* EDTA.
    c. 500 m*M* KCl.
    d. 20 m*M* MgCl$_2$.
    e. 20 m*M* MnCl$_2$.
    f. 12.5 μ*M* ATP.
    g. 100 U/mL Aprotinin.
    h. Bovine serum albumin (BSA, 3 mg/mL).
    i. 40% (v/v) Glycerol.
11. dsRNA: 0.01–0.5 μg/mL poly (I):poly (C) (Amersham, Piscataway, NJ).
12. poly (I):poly (C).
    a. 30 μg in 50 μL Lipofectamine 2000 (Invitrogen) in 5 mL phosphate-free DMEM containing 2% (v/v) dialyzed fetal bovine serum (FBS).
13. Buffer V:
    a. 20 m*M* Tris-HCl (pH 7.5).
    b. 50 m*M* KCl.
    c. 400 m*M* NaCl.
    d. 1% Nonidet P (NP)-40.
    e. 1 m*M* EDTA.
    f. 100 U/mL Aprotinin.
    g. 1 m*M* DTT.
    h. 25 m*M* NaF.
    i. Phosphatase inhibitor cocktail II (Sigma).
14. poly (I):poly (C) agarose beads:
    a. About 100 μL slurry, (Amersham) washed well in buffer V.
15. Homogenization buffer:
    a. 45 m*M* HEPES (pH 7.4).
    b. 0.375 m*M* Magnesium acetate.
    c. 0.075 m*M* EDTA.
    d. 95 m*M* Potassium acetate.
    e. 2.5 mg/mL Digitonin.
    f. 10% (v/v) Glycerol.
16. eIF2 about 10 μg/sample (gift of S. Kimball, Pennsylvania State University, Hershey, PA).
17. [^{3}H]GDP (2.5 m*M*, 10.9 Ci/mmol).
18. Assay buffer:
    a. 62.5 m*M* 4-morpholinepropanesulfonic acid (MOPS, pH 7.4).
    b. 125 m*M* KCl.
    c. 1.25 m*M* DTT.
    d. 0.2 mg/mL BSA.
19. Wash buffer:
    a. 50 m*M* MOPS.
    b. 2 m*M* Magnesium acetate.

    c. 100 m*M* KCl.

    d. 1 m*M* DTT.

20. Cellulose nitrate filters (Whatman, Maidstone, UK).
21. Vacuum drum (Millipore, Billerica, MA).
22. Agarose (LMTA, Gibco BRL/Invitrogen).
23. T7 polymerase (vTF7-3) (gift of B. Moss, NIH, Bethesda, MD).
24. 2.5 µg pVSV-XN2 (encoding the transgene of choice), along with 0.7 µg pB1-N, 1.25 µg pB1-P and 0.5 µg pB1-L, plasmids encoding the N, P, and L proteins, respectively (gift of J. Rose, Yale University, New Haven, CT).
25. Lipofectamine Plus (Invitrogen).
26. 0.2 µ*M* Cellulose acetate filters (Corning, Corning, NY).
27. DMEM/10% (v/v) FBS containing 1% low melting temperature agarose (LMTA, Gibco BRL, now Invitrogen).
28. Crystal violet-based fixative:
    a. 0.1% (w/v) Crystal violet.
    b. 30% Methanol.
29. Tryptone/yeast extract (TY) medium.
30. Ampicillin.
31. IPTG.
32. DNase 1.
33. CNBr-activated beads.
34. KSCN.
35. Centriprep columns.
36. Protein G agarose.
37. Enhanced chemiluminescence (ECL) solutions.
38. Heparin.
39. DMEM.
40. FBS.
41. Tunicamycin.
42. Bradford assay protein quantitation kit.
43. GDP.
44. Low melting temperature agarose (LMTA).
45. 0.25% Trypsin.

3. Methods

3.1. Expression of Recombinant PKR in E. coli

1. Expression of wild-type mammalian PKR has been found to be toxic in a variety of eukaryotic expression systems *(15)*. For example, in *S. cerevisiae*, human PKR can phosphorylate the yeast eIF2α homolog SUI-2 to inhibit protein synthesis *(69)*.
2. For production of PKR protein in prokaryotic expression systems, plasmids (such as the pET series of vectors, Novagen, Madison, WI) encoding wild-type or

mutant PKR under control of the T7 promoter are transformed into *E. coli* strain BL21(DE3) pLysS (Novagen) and grown overnight at 37°C in tryptone/yeast extract (TY) medium containing ampicillin (75 μg/mL) with constant shaking. Expression of the recombinant proteins is induced by addition of isopropyl-β-D-thiogalactopyranoside (IPTG, 0.5 m*M*) to the medium.

3. After 2 h, the bacteria are pelleted by centrifugation at 4°C and subsequently disrupted in lysis buffer A by three rounds of freeze–thawing. At this point, the lysate can be diluted in buffer I to reduce recombinant PKR protein concentration to levels appropriate for further use.

4. DNase I (25 mg/mL) is added (100 μL/500 mL culture) to the lysate and incubated on ice to break down *E. coli* genomic DNA, following which the lysate is clarified by centrifugation. A majority of recombinant PKR is soluble in *E. coli*, with little retained in inclusion bodies.

5. For immunopurification of PKR, supernatants generated as previously described are incubated with cyanogen bromide-activated beads coupled to a PKR-specific monoclonal antibody *(70)*.

6. After incubation at 4°C for 2 h, the beads are washed four times with buffer I and 3X with buffer II. The beads are then placed in disposable gel chromatography columns (Bio-Rad, Richmond, CA) and treated with ice-cold 1 *M* KSCN (pH 11.0). The eluant containing PKR is allowed to drip into neutralizing buffer and is dialyzed against 2 L of the same buffer.

7. PKR can subseqently be concentrated using a Centriprep column (Millipore) and examined for purity by 10% SDS-PAGE followed by Coomassie brilliant blue staining.

3.2. Immunoprecipitation and Immunoblot Analysis of PKR

1. To immunoprecipitate PKR from cultured human cells, the cells are first gently rinsed in ice cold phosphate-buffered saline solution (PBS) and disrupted in lysis buffer B. Lystaes are subsequently diluted in buffer I and incubated for 1 h at 4°C with anti-PKR monoclonal antibody *(70)* or polyclonal antisera *(38)*. Protein G-agarose (Invitrogen, Carlsbad, CA) is then added and the lysate incubated for a further hours at 4°C. The beads are subsequently washed four times in buffer I and three times in buffer II.

2. For immunoblot analysis of PKR, cells lysates or immunoprecipitates prepared as above are disrupted in SDS-PAGE loading buffer and denatured by boiling for 5 min at 95°C. Samples are then electrophoresed and transferred onto nitrocellulose membranes.

3. After soaking in blocking buffer for 1 h, the blot is subsequently incubated with PKR-specific monoclonal or polyclonal antibodies in blocking buffer for 12 h at 4°C.

4. Primary antibodies are detected by incubation with species-specific secondary antibody-horseradish peroxide (HRP) conjugates in blocking buffer for 3 h at room temperature followed by chemiluminiscence (SuperSignal, Pierce, Rockford, IL) **(Fig. 4)**.

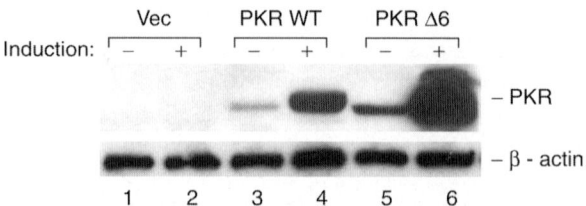

Fig. 4. Immunoblot analysis of PKR. Transformed murine fibroblasts were trans-fected with either empty pIND plasmid (Invitrogen), or with pIND encoding either wild-type PKR (PKR WT) or PKR Δ6, along with the the plasmid pVgRXR encoding a functional *Drosophila* ecdysone receptor. After 24 h posttransfection, expression of wild-type or mutant PKR was induced by stimulation of transfected cells with muris-terone A, and expression of PKR WT or PKR Δ6 assayed by immunoblot analysis using a human PKR-specific monoclonal antibody *(70)* (lanes 1 and 2) empty vector; (lanes 3 and 4) wild-type; (lanes 5 and 6) PKR Δ6 lanes (1, 3, and 5) without muristerone A; (lanes 2, 4, and 6) with muristerone A.

3.3. In Vitro Kinase Assay for PKR

Activity of wild-type or mutant PKR produced in *E. coli* or immunoprecipi-tated from lysates can be efficiently assayed using the following protocol:

1. Recombinant PKR purified from *E. coli* or beads containing bound PKR-antibody complexes are mixed with 30 μL buffer III. To this is added 4 μL buffer IV containing 2 μ*M* [γ-^{32}P] ATP (50 Ci/mmol) to each sample tube. At this point, various activators of PKR can be added to the samples as appropriate. Specifically, dsRNA, or 10 U of the polyanionic activator heparin can robustly activate PKR in vitro.
2. The tubes are gently vortexed and incubated at 30°C for 15 min. The kinase reac-tion is stopped by the addition of 30 μL SDS-PAGE loading buffer containing 100 m*M* EDTA. Samples are subsequently subjected to 10% SDS-PAGE.
3. After electrophoresis, the protein dye front containing unincorporated [γ-^{32}P] ATP is carefully cut away and the gel dried and exposed to X-ray film at −70°C. Occasionally, examination of the effect of PKR activity on its protein substrates (such as eIF2α or the NFARs) is required. In such cases, approx 0.1 μg of the respective proteins are added to the kinase reaction samples prior to the addition of the radiola-beled ATP, and the reactions incubated as previously described. Purification of eIF2α and the NFARs have been described previously *(26,71)* (**Fig. 5**).

3.4. In Vivo Kinase Assay for PKR

1. To examine PKR kinase activity in cultured mammalian cells in vivo, cells are usually seeded in 100-mm dishes (~2 × 10^6/dish) the day before the experiment.
2. On the day of the experiment, the cells are rinsed once in phosphate-free Dulbecco's modified Eagle's medium (DMEM) and incubated in DMEM contain-ing 5% (v/v) dialyzed FBS for 30 min.

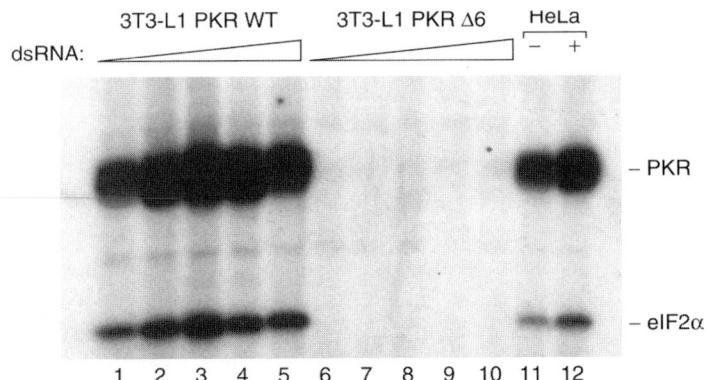

Fig. 5. In vitro kinase assay for PKR. Extracts from murine 3T3-L1 cells expressing wild-type human PKR (PKR WT) or human PKR Δ6 were immunoprecipitated using a human PKR-specific monoclonal antibody *(70)*. Immune-complexes were then subjected to in vitro kinase reactions in the presence of exogenously supplied eIF2α and increasing amounts (100 ng to 0.5 μg/reaction) of poly (I:C) or the anionic PKR activator heparin (10 U/reaction). Following electrophoresis, gels were subjected to autoradiography. Lanes 1–5 wild-type PKR; (lanes 6–10) PKR Δ6 (lanes 11–12), controls (immunoprecipitated PKR from IFN-treated HeLa cell extracts activated with 1 μg/mL dsRNA). Lanes 1 and 6, no dsRNA; (lanes 2 and 7) 0.1 μg/mL dsRNA; (lanes 3 and 8) 1 μg/mL dsRNA; (lanes 4 and 8) 5 μg/mL dsRNA; (lanes 5 and 10) heparin. Figure reproduced from **ref. *41***.

3. Each dish is then either left untreated, or is infected with virus at multiplicities of infection (m.o.i.s) of more than 100 in serum-free, phosphate-free DMEM, or transfected with poly (I):poly (C). After 30 min of virus infection, the medium on infected plates, as well as on untreated control plates, is replaced with 5 mL phosphate-free DMEM/2% (v/v) dialyzed FBS. 150 μCi/mL [^{32}P]-orthophosphate is then added to each plate.

4. After incubation for a further 3 h, the cells are lysed in 1 mL ice-cold buffer V, sonicated briefly to shear genomic DNA, and incubated with poly (I):poly (C) agarose beads for 2 h at 4°C. Alternatively, anti-PKR antibodies (1 μg/sample, [38, 70]) are added to each sample for 1 h at 4°C, followed by protein G-agarose approx 50 μL slurry/sample, washed well in buffer V for another 1 h at 4°C.

5. The beads are then washed five times with ice-cold buffer V, heated to 95°C for 5 min in 100 μL SDS-PAGE loading buffer, and resolved by 10% SDS-PAGE. After electrophoresis, the protein dye front is carefully cut away and the gel dried and exposed to X-ray film at −70°C (**Fig. 6**).

3.5. Detection of eIF2α Phosphorylation

1. Previous efforts at measuring levels of phosphorylated eIF2α relied on vertical slab isoelectric focusing electrophoresis to separate phorphrylated eIF2α from unphosphorylated eIF2α, followed by immunoblotting using a monoclonal

Agarose   Preimmune   Ag-poly (I:C)   Anti-PKR

dsRNA: `⌐ –  + ⌐` `⌐ –  + ⌐` `⌐ –  + ⌐` `⌐ –  + ⌐`

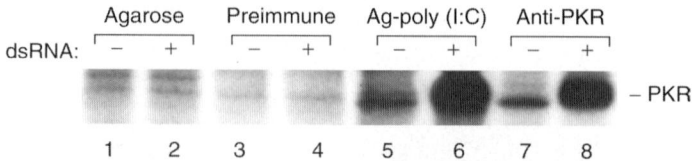

– PKR

1   2   3   4   5   6   7   8

Fig. 6. In vivo kinase assay for PKR. HeLa cells were treated with or without poly (I:C) (dsRNA) in the presence of $[^{32}P]$-orthophosphate for 3 h. PKR was precipitated from lysates prepared from these cells, and examined for autophosphorylation by SDS-PAGE and autoradiography. Lanes 1 and 2, HeLa lysate precipitated with agarose beads alone; (lanes 3 and 4) lysates precipitates with preimmune rabbit serum, followed by protein G coupled agarose; (lanes 5 and 6) lysates precipitates with agarose beads complexed with poly (I:C); (lanes 7 and 8) lysates precipitated with rabbit polyclonal antiserum specific for PKR, followed by protein G-coupled agarose. Lanes 1, 3, 5, and 7 untreated; (lanes 2, 4, 6, and 8) poly (I:C) treated.

antibody specific for eIF2α *(72)*. However, the recent availability of phosphoserine specific polyclonal antibodies that exclusively detect phosphorylated eIF2α has made the analysis of eIF2α phosphorylation significantly easier.

2. Subconfluent monolayers of cultured mammalian cells are treated with various eIF2α kinase activators (for example, high m.o.i.s of VSV, 3 µg/mL tunicamycin, or 6 µg poly (I):poly (C) + 8 µg Lipofectamine 2000/mL medium). After 3 h posttreatment, cells are lysed in buffer V, sonicated briefly to shear genoic DNA, and lysates resolved by 10% SDS-PAGE. Phosphorylated and total eIF2α are detected using antibodies that recognize either phosphrylated eIF2α (Biosource, Camarillo, CA) or total eIF2α *(73)*, respectively, following immunoblotting as described earlier (*see* **Subheading 3.2.**) for SDS-PAGE and immunoblotting (**Fig. 7**).

3.6. Guanine Nucleotide Exchange Assay for Determining eIF2B Activity in Mammalian Cells

1. Mammalian cells (~1×10^6/condition) in culture are either left untreated, or are treated with various eIF2α kinase activators (as described in the **Subheading 3.5.**). Following treatment, cells are washed once in ice-cold PBS and lysed in ice-cold homogenization buffer.

2. The lysates are clarified by centrifugation at 8000g for 10 min at 4°C. Protein concentrations in the resulting supernatants are quantitated by the Bradford assay (Pierce), and lysates normalized to contain equal amounts of protein (in a total volume 150 µL) are stored on ice until assayed for eIF2B activity. Meanwhile, the labeled eIF-2·$[^3H]$GDP binary complex is prepared by incubating rat liver eIF2 and $[^3H]$GDP in assay buffer at 30°C for 10 min. The Mg^{2+} concentration of the binary complex is then adjusted to 2 mM with magnesium acetate, and stored on ice before use.

3. To measure eIF2B activity, assay buffer (140 µL/sample) containing a 100-fold excess of cold GDP and 2 mM magnesium acetate is combined with 120 µL of cell lysate and warmed to 30°C for 1 min.

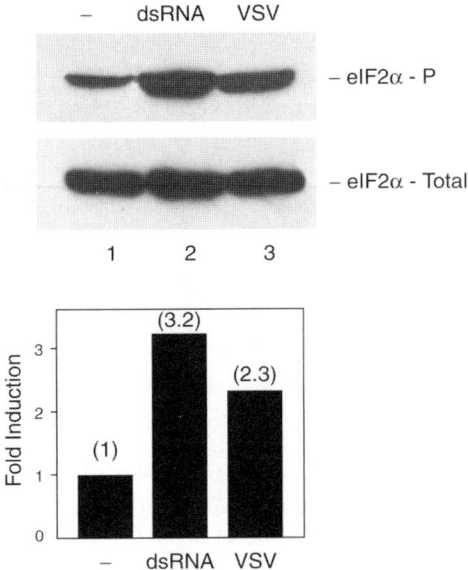

Fig. 7. Detection of eIF2α phosphorylation status by immunoblot analysis. HeLa cells were transfected with poly (I:C) or, infected with VSV for 3 h, after which lysates prepared from these cells were examined for eIF2α phosphorylation using antibodies specific for phosphorylated eIF2α (top panel). Subsequently, the blot was reprobed with an antibody that detects total cellular eIF2α. Lane 1, untreated; (lane 2) poly (I:C); (lane 3) VSV (middle panel). Relative eIF2α phosphorylation (represented as fold induction) was densitometrically determined from these blots (bottom panel).

4. The reaction is started by the addition of labeled binary complex (35 μL/sample) to this mixture. The samples are incubated at 30°C, and 60 μL aliquots of the mixture are taken immediately after addition of the labeled binary complex (time zero) and every min thereafter for a total of 3 min. The aliquots are immediately quenched in 1 mL wash buffer and filtered through moist cellulose nitrate filters previously positioned on a vacuum drum.
5. Filters are subsequently rinsed two times in wash buffer and bound [³H]GDP measured by liquid scintillation counting. The exchange reaction is measured as a decrease in eIF2-[³H]GDP bound to nitrocellulose filters with time. eIF2B activity is inversely proportional to the amount of [³H]GDP still bound to the membrane (**Fig. 8**).

3.7. Assay for Anchorage Independent Growth of Cells in Soft Agar

A hallmark of transformed mammalian fibroblasts, along with their faster doubling times and loss of contact inhibition, is their ability to proliferate even in the absence of a growth surface.

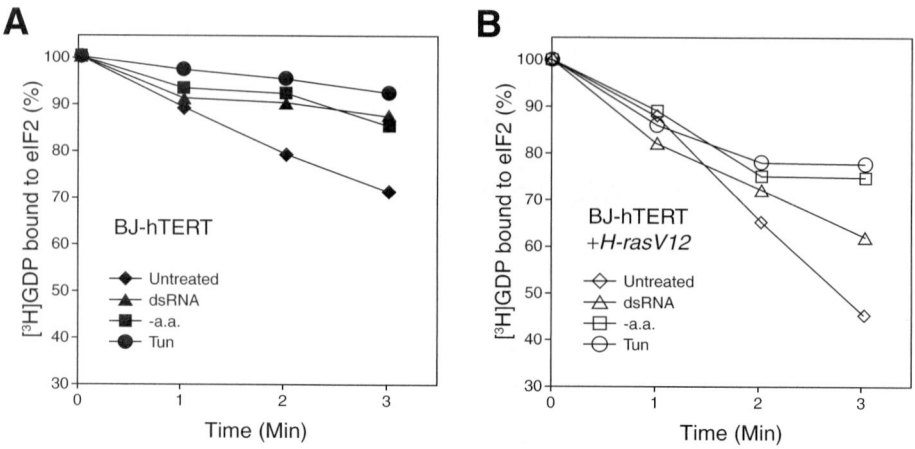

Fig. 8. Oncogenically-transformed human fibroblasts possess enhanced eIF2B guanine nucleotide exchange activity compared with normal cells. (**A**) BJ-hTERT cells
were treated with poly (I:C), tunicamycin, or were grown in medium devoid of glutamine (−aa.) for 3 h. Lysates prepared from these cells were subjected to eIF2B guanine
nucleotide exchange reactions as described in **Subheading 3.6.** (**B**) BJ-hTERT+H-
rasV12 cells were treated with as in **A**. Lysates prepared from these cells were subjected
to eIF2B guanine nucleotide exchange reactions. Figure reproduced from **ref. *58***.

The following protocol is routinely used by us to determine the ability of various human and rodent fibroblast cell lines to exhibit substrate-independent
growth.

1. Initially, a 2.1% (w/v) solution of tissue culture grade low melting temperature
   agarose is prepared, sterilized by autoclaving, and allowed to cool to 37°C.
2. Next, a mixture of complete medium containing 0.7% LMTA (w/v) is prepared by
   diluting the 2.1% (w/v) LMTA/water solution with appropriately concentrated
   complete medium. This solution is maintained at 37°C to prevent solidification.
   2.5 mL of this 0.7% (w/v) LMTA/1X complete medium mixture (hereafter
   referred to as 0.7% LMTA/medium) is then added to each well of a six-well plate
   and allowed to solidify.
3. Next, the cells to be assayed for anchorage-independent growth are counted and
   diluted at either 5×10^3 cells/mL or 5×10^2 cells/mL in complete medium.
4. One mL of each dilution is then mixed with 1 mL of 0.7% LMTA/medium and
   overlaid carefully on the solid 0.7% LMTA/medium currently at the bottom of
   each well. It is important to ensure that none of the overlay seeps under this bottom layer. Again, the mixture is allowed to solidify. It may be necessary to transfer the cells to 4°C for this process, since the concentration of LMTA in the
   overlay is now only 0.35% (w/v).
5. After this layer has solidified, 2.5 mL 0.7% LMTA/medium is again added to each
   well and allowed to solidify. The plates are then returned to a 37°C/5% CO_2

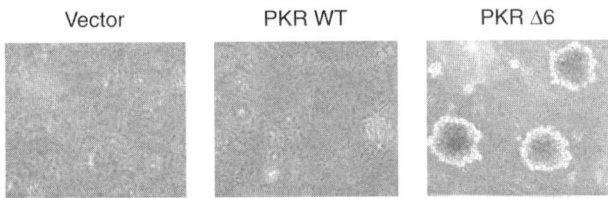

Fig. 9. Soft agar growth assay for anchorage-independent proliferation. 3T3-L1 cells expressing either control vector, wild-type PKR (PKR WT) or PKR Δ6 were allowed to grow in soft agar for 14 d, after which photomicrographs of foci (×200) were taken.

humidified incubator and the cells are examined daily. Cells capable of anchorage-independent growth will form clearly visible giant foci about 10–14 d after plating, whereas, nontransformed fibroblasts may at best form very small foci in this period. Using this protocol, it is not necessary to supplement the cells with medium (**Fig. 9**).

3.8. Generation of Recombinant VSV

1. Infectious VSV can be recovered from cells transfected with a plasmid vector encoding the entire 11161 nucleotide VSV antigenome, together with plasmids encoding VSV N, P, and L proteins *(74,75)* (**Fig. 10**).
2. To generate recombinant VSV expressing a transgene, the gene of interest is inserted between the *XhoI* and *NheI* sites of the plasmid pVSV-XN2 by standard recombinant DNA technology. 2–3 × 10⁶ baby hamster kidney (BHK) cells are plated on a 100-mm dish, such that they are approx 70% confluent after overnight culture.
3. On the day of the transfection, cell culture supernatant containing vaccinia virus encoding T7 polymerase is mixed with 0.25% (w/v) Trypsin in a 1:1 (v/v) ratio and incubated at 37°C for 30 min with occasional vortexing.
4. BHK cells are then infected with trypsinized vTF7-3 virus (m.o.i. = 10) in DMEM containing 2.5% FBS for 45 min at 37°C in a total volume of 3 mL. Meanwhile, 2.5 µg pVSV-XN2 (encoding the transgene of choice), along with 0.7 µg pB1-N, 1.25 µg pB1-P, and 0.5 µg pB1-L, plasmids encoding the N, P, and L proteins, respectively, are complexed with Lipofectamine Plus (Invitrogen) according to the manufacturer's instructions in serum-free DMEM for 30 min.
5. Following infection of BHK cells with the vTF7-3 virus, the medium is removed and replaced with 5 mL serum-free DMEM containing the Lipofectamine Plus–plasmid DNA complexes. After 3 h of posttransfection, the medium is replaced with DMEM containing 10% FBS, and the cells are incubated for a further 48 h. At this time, the cell culture supernatant is collected and centrifuged to remove cellular debris and filtered two times through 0.2-µm cellulose acetate filters (Corning, Corning, NY) to eliminate vaccinia virus.
6. The filtered medium is then layered on a fresh 100-mm dish of about 70% confluent BHK cells. Recombinant virus can be collected from the medium of these cells 24 h after infection.

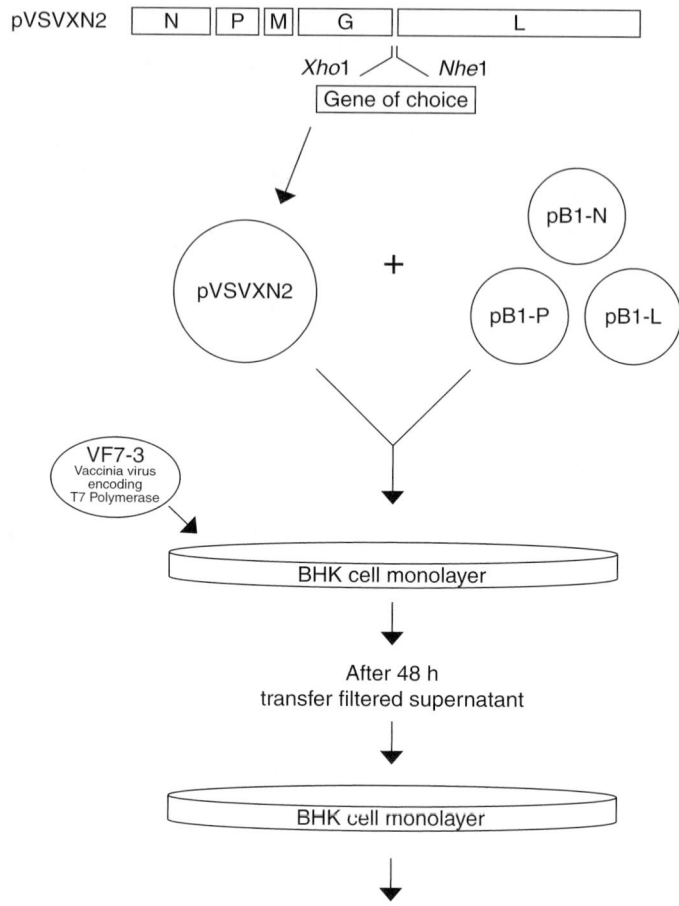

Fig. 10. Schematic of the procedure used to generate recombinant VSV. Plasmids encoding VSV L, P, and N proteins are cotransfected with the pVSVXN2 plasmid (encoding the VSV antigenome containing the transgene of choice) into BHK cells previously infected with vaccinia virus VTF7-3. Forty eight hours later, supernatants from these cells are filtered and used in a fresh round of infection to generate recombinant VSV (*see* **Subheading 3.8.** for details).

3.9. Infection of Cells With VSV and Titration of Progeny Virus

1. VSV can infect virtually any mammalian cell line, irrespective of tissue or species of origin. To infect adherent cells with wild-type or recombinant VSV, medium is removed from the cell monolayer, and replaced with serum free medium containing either wild-type or recombinant VSV at the desired m.o.i. in a volume that is just sufficient to completely cover the monolayer (typically 100 µL for one well

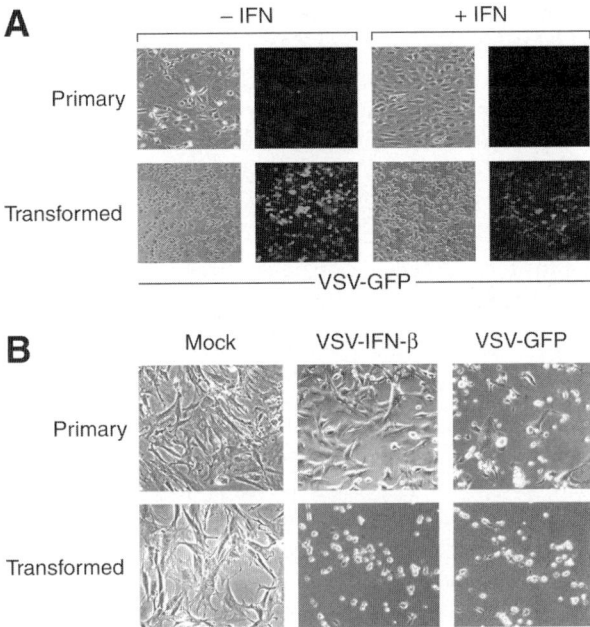

Fig. 11. Recombinant VSV as oncolytic agents. **(A)** Freshly explanted human microvascoular endothelial cells (Primary) or murine breast tumor-derived DA-3 cells (Transformed) were treated for 18 h with IFN-α and subsequently infected with VSV expressing GFP (VSV-GFP) at an m.o.i of 0.1. Identical fields were subsequently photographed by brightfield and fluorescence microscopy. Figure adapted from **ref. *62*. (B)** Primary or transformed embryonic fibroblasts derived from C57Bl/6 mice were infected with either recombinant VSV expressing GFP (VSV-GFP) or IFN-β (VSV-IFN-β) at an m.o.i. of 0.01 for 24 h and subsequently photographed. Note that VSV-IFN-β is cytopathic to transformed, but not primary cells. Figure adapted from **ref. *63*.**

    of a 12-well dish, or 1 mL for a 100-mm dish). The virus is allowed to infect the cells in a humidified incubator at 37°C for 1 h (with occasional rocking), following which, the cells are washed twice with PBS. Complete medium is then added to these cells **(Fig. 11)**.

2. Infections of suspension cells with VSV are carried out in small volumes of serum-free medium (typically ~50 μL/1 × 10^6 cells), usually in 96-well plates, essentially as described for adherent cells. The cell-virus mixture is mixed every 10 min by gentle pipetting. After 1 h, the cells are washed two times in PBS and returned to complete medium.

3. Determination of the amount of VSV in a sample can be performed on any permissive cell line that forms a uniform, confluent monolayer in culture. We typically use BHK cells to titrate VSV. BHK cells are seeded into six-well dishes at about 1 × 10^6 cells/well the day before the virus is to be titrated. On the day of virus titration, supernatants containing VSV are serially diluted into serum-free medium (typically

10-fold dilutions are made), and used to infect BHK monolayers, as described above. After infection, the medium is removed, cells are washed two times with PBS, and overlaid with DMEM/10% (v/v) FBS. The medium-LMTA mixture is maintained at 37°C after preparation and prior to use to prevent it from solidifying.
4. After the cells are overlaid with the medium-LMTA mixture, they are briefly shifted to 4°C to allow the overlay to solidify, following which they are returned to a humidified with 5% CO_2/37°C incubator for 24 h later, the overlay is removed and the monolayers stained with a crystal violet-based fixative. Plaques are the scored, and virus titer expressed in plaque forming units (PFU)/mL after accounting for the dilution factor.

4. Notes

1. When misfolded proteins accumulate in the ER, a specific and well-studied stress response, termed the UPR emanates from this organelle. The UPR results in the transcriptional upregulation of a defined group of genes that attenuate ER stress. In tandem, the UPR possesses a cytoplasmic component that attenuates protein translation rates and reduces input of newly synthesized polypeptides into the ER. The UPR thus allows the cell to adapt to and survive under conditions that cause misfolding of proteins in the ER (reviewed in **refs. *76–79***).

References

1. Medzhitov, R. and Janeway, C. A., Jr. (1997) Innate immunity: the virtues of a monclonal system of recognition. *Cell* **91,** 295–298.
2. Medzhitov, R. and Janeway, C. A., Jr. (1998) Innate immune recognition and control of adaptive immune responses. *Semin. Immunol.* **10,** 351–353.
3. Stark, G. R., Kerr, I. M., Williams, B. R., Silverman, R. H., and Schreiber, R. D. (1998) How cells respond to interferons. *Annu. Rev. Biochem.* **67,** 227–264.
4. Der, S. D., Zhou, A., Williams, B. R., and Silverman, R. H. (1998) Identification of genes differentially regulated by interferon α, β, or γ using oligonucleotide arrays. *Proc. Natl. Acad. Sci. USA* **95,** 15,623–15,628.
5. Meurs, E., Chong, K., Galabru, J., et al. (1990) Molecular cloning and characterization of the human double-stranded RNA-activated protein kinase induced by interferon. *Cell* **62,** 379–390.
6. Barber, G. N., Edelhoff, S., Katze, M. G., and Disteche, C. M. (1993) Chromosomal assignment of the interferon-inducible double-stranded RNA-dependent protein kinase (PRKR) to human chromosome 2p21–p22 and mouse chromosome 17 E2. *Genomics* **16,** 765–767.
7. Barber, G. N., Jagus, R., Meurs, E. F., Hovanessian, A. G., and Katze, M. G. (1995) Molecular mechanisms responsible for malignant transformation by regulatory and catalytic domain variants of the interferon-induced enzyme RNA-dependent protein kinase. *J. Biol. Chem.* **270,** 17,423–17,428.
8. Clemens, M. J. and Elia, A. (1997) The double-stranded RNA-dependent protein kinase PKR: structure and function. *J. Interferon Cytokine Res.* **17,** 503–524.

9. Meurs, E. F., Galabru, J., Barber, G. N., Katze, M. G., and Hovanessian, A. G. (1993) Tumor suppressor function of the interferon-induced double-stranded RNA-activated protein kinase. *Proc. Natl. Acad. Sci. USA* **90,** 232–236.

10. Koromilas, A. E., Roy, S., Barber, G. N., Katze, M. G., and Sonenberg, N. (1992) Malignant transformation by a mutant of the IFN-inducible dsRNA-dependent protein kinase. *Science* **257,** 1685–1689.

11. Wu, S. and Kaufman, R. J. (1996) Double-stranded (ds) RNA binding and not dimerization correlates with the activation of the dsRNA-dependent protein kinase (PKR). *J. Biol. Chem.* **271,** 1756–1763.

12. Webb, B. L. and Proud, C. G. (1997) *Eukaryotic* initiation factor 2B (eIF2B). *Intl. J. Biochem. Cell Biol.* **29,** 1127–1131.

13. Thomis, D. C. and Samuel, C. E. (1993) Mechanism of interferon action: evidence for intermolecular autophosphorylation and autoactivation of the interferon-induced, RNA-dependent protein kinase PKR. *J. Virol.* **67,** 7695–7700.

14. McMillan, N. A., Carpick, B. W., Hollis, B., Toone, W. M., Zamanian-Daryoush, M., and Williams, B. R. (1995) Mutational analysis of the double-stranded RNA (dsRNA) binding domain of the dsRNA-activated protein kinase, PKR. *J. Biol. Chem.* **270,** 2601–2606.

15. Jagus, R., Joshi, B., and Barber, G. N. (1999) PKR, apoptosis and cancer. *Int. J. Biochem. Cell Biol.* **31,** 123–138.

16. Williams, B. R., (1999) PKR; a sentinel kinase for cellular stress. *Oncogene* **18,** 6112–6120.

17. Panniers, R. and Henshaw, E. C. (1984) Mechanism of inhibition of polypeptide chain initiation in heat-shocked Ehrlich ascites tumour cells. *Eur. J. Biochem.* **140,** 209–214.

18. Cuddihy, A. R., Wong, A. H., Tam, N. W., Li, S., and Koromilas, A. E. (1999) The double-stranded RNA activated protein kinase PKR physically associates with the tumor suppressor p53 protein and phosphorylates human p53 on serine 392 in vitro. *Oncogene* **18,** 2690–2702.

19. Ito, T., Jagus, R., and May, W. S. (1994) Interleukin 3 stimulates protein synthesis by regulating double-stranded RNA-dependent protein kinase. *Proc. Natl. Acad. Sci. USA* **91,** 7455–7459.

20. Mundschau, L. J. and Faller, D. V. (1995) Platelet-derived growth factor signal transduction through the interferon-inducible kinase PKR. Immediate early gene induction. *J. Biol. Chem.* **270,** 3100–3106.

21. Yang, Y. L., Reis, L. F., Pavlovic, J., et al. (1995) Deficient signaling in mice devoid of double-stranded RNA-dependent protein kinase. *EMBO J.* **14,** 6095–6106.

22. Kumar, A., Yang, Y. L., Flati, V., et al. (1997) Deficient cytokine signaling in mouse embryo fibroblasts with a targeted deletion in the PKR gene: role of IRF-1 and NF-kappaB. *EMBO J.* **16,** 406–416.

23. Wong, A. H., Tam, N. W., Yang, Y. L., et al. (1997) Physical association between STAT1 and the interferon-inducible protein kinase PKR and implications for interferon and double-stranded RNA signaling pathways. *EMBO J.* **16,** 1291–1304.

24. Goh, K. C., deVeer, M. J., and Williams, B. R. (2000) The protein kinase PKR is required for p38 MAPK activation and the innate immune response to bacterial endotoxin. *EMBO J.* **19,** 4292–4297.

25. Barber, G. N., Thompson, S., Lee, T. G., et al. (1994) The 58-kilodalton inhibitor of the interferon-induced double-stranded RNA-activated protein kinase is a tetratricopeptide repeat protein with oncogenic properties. *Proc. Natl. Acad. Sci. USA* **91,** 4278–4282.

26. Saunders, L. R., Perkins, D. J., Balachandran, S., et al. (2001) Characterization of two evolutionarily conserved, alternatively spliced nuclear phosphoproteins, NFAR-1 and -2, that function in mRNA processing and interact with the double-stranded RNA-dependent protein kinase, PKR. *J. Biol. Chem.* **276,** 32,300–32,312.

27. Patel, R. C. and Sen, G. C. (1998) PACT, a protein activator of the interferon-induced protein kinase, PKR. *EMBO J.* **17,** 4379–4390.

28. Pataer, A., Vorburger, S. A., Barber, G. N., et al. (2002) Adenoviral transfer of the melanoma differentiation-associated gene 7 (*mda*7) induces apoptosis of lung cancer cells via up-regulation of the double-stranded RNA-dependent protein kinase (PKR). *Cancer Res.* **62,** 2239–2243.

29. Jagus, R. and Gray, M. M. (1994) Proteins that interact with PKR. *Biochimie* **76,** 779–791.

30. Williams, B. R. (2001) Signal integration via PKR. *Sci STKE* 2001 **89,** RE2.

31. Fisher, P. B. (2005) Is *mda*-7/IL-24 a magic bullet for cancer? *Cancer Res.* **65,** 10,128–10,138.

32. Katze, M. G. (1995) Regulation of the interferon-induced PKR: can viruses cope? *Trends Microbiol.* **3,** 75–78.

33. Katze, M. G., He, Y., and Gale, M. Jr. (2002) Viruses and interferon: a fight for supremacy. *Nat. Rev. Immunol.* **2(9),** 675–687.

34. Durbin, R. K., Mertz, S. E., Koromilas, A. E., and Durbin, J. E. (2002) PKR protection against intranasal vesicular stomatitis virus infection is mouse strain dependent. *Viral Immunol.* **15,** 41–51.

35. Balachandran, S., Roberts, P. C., Brown, L. E., et al. (2000) Essential role for the dsRNA-dependent protein kinase PKR in innate immunity to viral infection. *Immunity* **13,** 129–141.

36. Stojdl, D. F., Abraham, N., Knowles, S., et al. (2000) The murine double-stranded RNA-dependent protein kinase PKR is required for resistance to vesicular stomatitis virus. *J. Virol.* **74,** 9580–9585.

37. Barber, G. N., Wambach, M., Thompson, S., Jagus, R., and Katze, M. G. (1995) Mutants of the RNA-dependent protein kinase (PKR) lacking double-stranded RNA binding domain I can act as transdominant inhibitors and induce malignant transformation. *Mol. Cell Biol.* **15,** 3138–3146.

38. Barber, G. N., Tomita, J., Garfinkel, M. S., Meurs, E., Hovanessian, A., and Katze, M. G. (1992) Detection of protein kinase homologues and viral RNA-binding domains utilizing polyclonal antiserum prepared against a baculovirus-expressed ds RNA-activated 68,000-Da protein kinase. *Virology* **191,** 670–679.

39. Dever, T. E., Sripriya, R., McLachlin, J. R., et al. (1998) Disruption of cellular translational control by a viral truncated eukaryotic translation initiation factor-2α kinase homolog. *Proc. Natl. Acad. Sci. USA* **95,** 4164–4169.

40. Chong, K. L., Feng, L., Schappert, K., et al. (1992) Human p68 kinase exhibits growth suppression in yeast and homology to the translational regulator GCN2. *EMBO J.* **11,** 1553–1562.

41. Balachandran, S., Kim, C. N., Yeh, W. C., Mak, T. W., Bhalla, K., and Barber, G. N. (1998) Activation of the dsRNA-dependent protein kinase, PKR, induces apoptosis through FADD-mediated death signaling. *EMBO J.* **17,** 6888–6902.

42. Donze, O., Dostie, J., and Sonenberg, N. (1999) Regulatable expression of the interferon-induced double-stranded RNA dependent protein kinase PKR induces apoptosis and fas receptor expression. *Virology* **256,** 322–329.

43. Lee, S. B. and Esteban, M. (1994) The interferon-induced double-stranded RNA-activated protein kinase induces apoptosis. *Virology* **199,** 491–496.

44. Gil, J. and Esteban, M. (2000) Induction of apoptosis by the dsRNA-dependent protein kinase (PKR): mechanism of action. *Apoptosis* **5,** 107–114.

45. Donze, O., Deng, J., Curran, J., Sladek, R., Picard, D., and Sonenberg, N. (2004) The protein kinase PKR: a molecular clock that sequentially activates survival and death programs. *EMBO J.* **23,** 564–571.

46. Gil, J., Rullas, J., Garcia, M. A., Alcami, J., and Esteban, M. (2001) The catalytic activity of dsRNA-dependent protein kinase, PKR, is required for NF-κB activation. *Oncogene* **20,** 385–394.

47. Gil, J., Garcia, M. A., Gomez-Puertas, P., et al. (2004) TRAF family proteins link PKR with NF-kappa B activation. *Mol. Cell Biol.* **24,** 4502–4512.

48. Perkins, D. J. and Barber, G. N. (2004) Defects in translational regulation mediated by the alpha subunit of eukaryotic initiation factor 2 inhibit antiviral activity and facilitate the malignant transformation of human fibroblasts. *Mol. Cell Biol.* **24,** 2025–2040.

49. Hinnebusch, A. G. (1994) The eIF-2α kinases: regulators of protein synthesis in starvation and stress. *Semin. Cell Biol.* **5,** 417–426.

50. Han, A. P., Yu, C., Lu, L., et al. (2001) Heme-regulated eIF2α kinase (HRI) is required for translational regulation and survival of erythroid precursors in iron deficiency. *EMBO J.* **20,** 6909–6918.

51. Dever, T. E., Feng, L., Wek, R. C., Cigan, A. M., Donahue, T. F., and Hinnebusch, A. G. (1992) Phosphorylation of initiation factor-2α by protein kinase GCN2 mediates gene-specific translational control of GCN4 in yeast. *Cell* **68,** 585–596.

52. Harding, H. P., Zhang, Y., Zeng, H., et al. (2003) An integrated stress response regulates amino acid metabolism and resistance to oxidative stress. *Mol. Cell.* **11,** 619–633.

53. Harding, H. P., Zhang, Y., Bertolotti, A., Zeng, H., and Ron, D. (2000) Perk is essential for translational regulation and cell survival during the unfolded protein response. *Mol. Cell* **5,** 897–904.

54. Harding, H. P., Zhang, Y., and Ron, D. (1999) Protein translation and folding are coupled by an endoplasmic-reticulum-resident kinase (*see* comments). *Nature* **397,** 271–274 [erratum appears in *Nature* 1999;**398**:90].

55. Scheuner, D., Song, B., McEwen, E., et al. (2001) Translational control is required for the unfolded protein response and in vivo glucose homeostasis. *Mol. Cell* **7**, 1165–1176.

56. Harding, H. P. and Ron, D. (2002) Endoplasmic reticulum stress and the development of diabetes: a review. *Diabetes* **51**, S455–S461.

57. Donze, O., Jagus, R., Koromilas, A. E., Hershey, J. W., and Sonenberg, N. (1995) Abrogation of translation initiation factor eIF-2 phosphorylation causes malignant transformation of NIH 3T3 cells. *EMBO J.* **14**, 3828–3834.

58. Balachandran, S. and Barber, G. N. (2004) Defective translational control facilitates vesicular stomatitis virus oncolysis. *Cancer Cell* **5**, 51–65.

59. Wagner, R. R. and Rose, J. K. (1996) Rhabdoviridae: the Viruses and Their Replication, in *Fields Virology*, (Fields, B. N., Howley, P. M. et al., ed.), Lipincott-Raven Publishers: Philadelphia, PA, pp. 1121–1135.

60. Balachandran, S. and Barber, G. N. (2000) Vesicular stomatitis virus therapy of tumors. *IUBMB Life* **50**, 135–138.

61. Balachandran, S., Porosnicu, M., and Barber, G. N. (2001) Oncolytic activity of vesicular stomatitis virus is effective against tumors exhibiting aberrant p53, Ras, or Myc function and involves induction of apoptosis. *J. Virol.* **75**, 3474–3479.

62. Fernandez, M., Porosnicu, M., Markovic, D., and Barber, G. N. (2002) Genetically engineered vesicular stomatitis virus in gene therapy: application for treatment of malignant disease. *J. Virol.* **76**, 895–904.

63. Obuchi, M., Fernandez, M., and Barber, G. N. (2003) Development of recombinant vesicular stomatitis viruses that exploit defects in host defense to augment specific oncolytic activity. *J. Virol.* **77**, 8843–8856.

64. Stojdl, D. F., Lichty, B., Knowles, S., et al. (2000) Exploiting tumor-specific defects in the interferon pathway with a previously unknown oncolytic virus. *Nat. Med.* **6**, 821–825.

65. Stojdl, D. F., Lichty, B. D., tenOever, B. R., et al. (2003) VSV strains with defects in their ability to shutdown innate immunity are potent systemic anti-cancer agents. *Cancer Cell* **4**, 263–275.

66. Wagner, R. R. (1987) Rhabdovirus biology and infection, an overview, in *The Rhabdoviruses*, (Wagner, R. R., ed.), Plenum, New York, pp. 9–74.

67. Rose, R. R. and Joklik, W. (1996) Rhabdoviridae: the viruses and their replication, in *Fields Virology*, (D.M.K. B.N. Fields, P.M. Howley, eds.), Lippincott-Raven, Philadelphia, PA, pp. 1121–1136.

68. Porosnicu, M., Mian, A., and Barber, G. N. (2003) The oncolytic effect of recombinant vesicular stomatitis virus is enhanced by expression of the fusion cytosine deaminase/uracil phosphoribosyltransferase suicide gene. *Cancer Res.* **63**, 8366–8376.

69. Dever, T. E., Chen, J. J., Barber, G. N., et al. (1993) Mammalian eukaryotic initiation factor 2 α-kinases functionally substitute for GCN2 protein kinase in the GCN4 translational control mechanism of yeast. *Proc. Natl. Acad. Sci. USA* **90**, 4616–4620.

70. Laurent, A. G., Krust, B., Galabru, J., Svab, J., and Hovanessian, A. G. (1985) Monoclonal antibodies to an interferon-induced Mr 68,000 protein and their use for the detection of double-stranded RNA-dependent protein kinase in human cells. *Proc. Natl. Acad. Sci. USA* **82,** 4341–4345.

71. Konieczny, A. and Safer, B. (1983) Purification of the eukaryotic initiation factor 2-eukaryotic initiation factor 2B complex and characterization of its guanine nucleotide exchange activity during protein synthesis initiation. *J. Biol. Chem.* **258,** 3402–3408.

72. Savinova, O. and Jagus, R. (1997) Use of vertical slab isoelectric focusing and immunoblotting to evaluate steady-state phosphorylation of eIF2 alpha in cultured cells. *Methods* **11,** 419–425.

73. DeGracia, D. J., Sullivan, J. M., Neumar, R. W., et al. (1997) Effect of brain ischemia and reperfusion on the localization of phosphorylated eukaryotic initiation factor 2 alpha. *J. Cerebral Blood Flow Metab.* **17,** 1291–1302.

74. Lawson, N. D., Stillman, E. A., Whitt, M. A., and Rose, J. K. (1995) Recombinant vesicular stomatitis viruses from DNA [published erratum appears in Proc Natl Acad Sci USA 1995 Sep 12;92(19):9009]. *Proc. Natl. Acad. Sci. USA* **92,** 4477–4481.

75. Schnell, M. J., Buonocore, L., Kretzschmar, E., Johnson, E., and Rose, J. K. (1996) Foreign glycoproteins expressed from recombinant vesicular stomatitis viruses are incorporated efficiently into virus particles. *Proc. Natl. Acad. Sci. USA* **93,** 11,359–11,365.

76. Kaufman, R. J. (1999) Stress signaling from the lumen of the endoplasmic reticulum: coordination of gene transcriptional and translational controls. *Genes Dev.* **13,** 1211–1233.

77. Mori, K. (2000) Tripartite management of unfolded proteins in the endoplasmic reticulum. *Cell* **101,** 451–454.

78. Patil, C. and Walker, P. (2001) Intracellular signaling from the endoplasmic reticulum to the nucleus: the unfolded protein response in yeast and mammals. *Curr. Opin. Cell Biol.* **13,** 349–355.

79. Harding, H. P., Calfon, M., Urano, F., Novoa, I., and Ron, D. (2002) Transcriptional and translational control in the Mammalian unfolded protein response. *Ann. Rev. Cell Dev. Biol.* **18,** 575–599.

19

Analysis of Transformation and Tumorigenicity Using Mouse Embryonic Fibroblast Cells

Hong Sun and Reshma Taneja

Summary

An important step in cellular transformation and tumorigenesis is immortalization, in which cells gain the ability to grow indefinitely by bypassing cellular senescence that imposes a finite number of divisions in culture. Primary mouse embryonic fibroblast (MEF) cells have a limited growth capacity and on prolonged passaging spontaneously immortalize at a low frequency. In contrast to transformation of primary MEF cells that requires the presence of two cooperating oncogenes, immortalized MEF cells can be transformed by a single oncogene (*Ras*) resulting in a loss of contact inhibition, anchorage-independent growth, and tumor formation in nude mice. Studies of MEF cells have played an important role in the elucidation of the molecular mechanisms underlying cellular immortalization, transformation, and tumorigenesis. Additionally, utilization of MEF cells disrupted for specific genes has provided a powerful tool to analyze the genetic regulation of these cellular processes. In this chapter, methods for analysis of cellular immortalization using the 3T3 protocol, as well as transformation of MEF cells using oncogenic retroviruses are provided. This is followed by protocols for analysis of transformed cell characteristics such as foci formation, anchorage independent growth, and tumor formation in nude mice.

Key Words: Embryos; immortalization; immunodeficient; MEF; embryo; nude mice; oncogenes; retrovirus; transformation; tumors.

1. Introduction

Activation of oncogenes or inactivation of tumor suppressor pathways results in neoplastic transformation. Animal models with altered expression of oncogenes or tumor suppressors have been established to address the role of these genes in tumorigenesis and provide a molecular basis for cancer therapy (*1,2*). Primary mouse embryonic fibroblast (MEF) cells represent a well-defined cell type that has been instrumental in studying the consequences of gene-ablation in cellular growth and proliferation. Moreover, the ability to transform MEFs with oncogenes, and their response to growth arrest triggers has been well

From: *Methods in Molecular Biology, vol. 383: Cancer Genomics and Proteomics: Methods and Protocols*
Edited by: P. B. Fisher © Humana Press Inc., Totowa, NJ

characterized. Thus, MEF cells derived from such mutant mice have proven to be an invaluable tool to study immortalization, transformation, and tumorigenesis.

Like many other primary mammalian cells *(3)*, MEF cells have limited life span in cell culture and normally undergo cellular senescence resulting in irreversible growth arrest *(4,5)*. Senescence limits proliferation of cells and is thus considered to suppress cancers *(6)*. When cultured for several generations on a defined protocol (3T3), in which cells are split every 3 d with fixed seeding density of 3×10^5/6-cm cell culture dish, primary MEF cells can overcome cellular senescence and spontaneously immortalize at a low frequency *(7)*. Although the molecular mechanisms involved in immortalization are not fully understood, a majority of spontaneously immortalized MEF cells harbor mutations in the *p53* locus *(8,9)*, and MEF cells isolated from mice with inactivated *p53* or *Rb* genes do not undergo senescence, indicating that spontaneous immortalization of MEF cells is controlled primarily by these tumor suppressor pathways *(10–13)*. Thus, analysis of the ability of MEF cells to escape from senescence using the 3T3 protocol allows for an assessment of genes that regulate this pathway.

Transformation of primary MEF cells requires the presence of at least two cooperating oncogenes *(14,15)*. Expression of *Ras* alone in primary MEF cells induces senescence-like growth arrest *(16)*, but coexpression of *Ras* along with an immortalizing oncogene such as *c-myc (17)* or adenoviral *E1A (18)* induces a transformed phenotype that is characterized by loss of contact inhibition in monolayer culture, anchorage-independent growth in soft-agar and tumor formation in immunodeficient mice. In addition to growing indefinitely in vitro, immortalized MEF cells defective in the p53 or Rb tumor suppressor pathways can be transformed by *Ras* alone *(11,13)*. Thus, transformation assays with *Ras* alone, or a combination of *Ras* plus *E1A*, or *Ras* plus *c-myc*, provide a useful tool to dissect the genetic regulators of transformation and tumorigenesis.

In this chapter, protocols to study immortalization and transformation of MEF cells are provided. The 3T3 protocol can be used not only for testing the ability of cells to bypass cellular senescence, but also to establish immortalized MEF cell lines. For transformation assays, a single oncogene (*Ras*) or a combination of *Ras* plus *c-myc*, or *Ras* plus *E1A*, can be introduced into primary MEF cells by retroviral infection, followed by examining contact inhibition by focus formation assays, and anchorage-independent growth by soft-agar assays. The tumorigenic property of transformed MEF cells can be examined by injecting cells subcutaneously into nude mice and monitoring tumor formation in vivo.

2. Materials

1. Cell culture incubator with 5% CO_2 at 37°C.
2. Female athymic nude mice (nu/nu), 4 to 6 wk old.

3. Giemsa solution (Sigma, St. Louis, MO; cat. no. GS-500) or Giemsa blood stain (EM Diagnostic Systems, Gibbstown, NJ; cat. no. 619-71).

4. Glutaradehyde stock (25%, Sigma, cat. no. G-6257).

5. Growth medium: 10% heat-inactivated fetal bovine serum (FBS, Hyclone, heat inactivated at 56°C for 30 min) in Dulbecco's modified Eagle's medium (DMEM, Invitrogen [Carlsbad, CA], high glucose, with glutamine) plus 0.1 mM nonessential amino acids (Sigma, cat. no. M-7145), 55 μM 2-mercaptoethanol (Sigma, cat. no. M-7522), and 10 μg/mL gentamicin (Sigma, cat. no. G-1397).

6. Hygromycin (Clontech, Mountain View, CA).

7. Lipofectamine 2000 (Invitrogen).

8. Microscopes.

9. Opti-MEM medium (Invitrogen).

10. Retroviral vectors expressing *Ha-RasV12*, *c-Myc*, and *E1A12S* with either puromycin- or hygromycin-resistance markers.

11. Phosphate-buffered saline (PBS): dissolve 8 g NaCl, 0.2 g KCl, 1.44 g Na$_2$HPO$_4$, 0.24 g KH$_2$PO$_4$ in 800 mL distilled water, adjust pH to 7.4, then add water to 1000 mL, sterilize by autoclaving.

12. Plastic tissue culture dishes (10-, 6-, and 3.5-cm) and tissue culture plates (6-, 12-, 24-, and 96-well) from Nunc.

13. Phoenix-Eco cells (φNX-ε, an ecotropic retrovirus packaging line derived from transformed human embryonic kidney 293T cells).

14. 10 mg/mL Polybrene (Sigma, cat. no. H-9268) stock solution: dissolve in 0.9% NaCl, sterilize by filtration, store at 4°C.

15. Puromycin (Sigma, cat. no. P-8833): 50 mg/mL in distilled water, sterilize by filtration, aliquot and store at −20°C.

16. SeaPlaque low-melting temperature agarose (Biowhittaker Molecular Applications, Rockland, ME; cat. no. 50101).

3. Methods

3.1. 3T3 Protocol

1. Isolate MEF cells as described in Chapter 20. Plate passage (P) 1 MEF cells in a 6-cm cell culture dish at a density of 3×10^5 cells per dish.

2. Culture cells at 37°C. Change the medium after 2 d.

3. Three days after plating cells, trypsinize MEF cells, count the total cell number, and replate 3×10^5 cells into 6-cm cell culture dish. These MEF cells are considered P2 cells.

4. Repeat **steps 2** and **3**, count cell numbers at each passage, and replate 3×10^5 cells per dish in a 6-cm dish for 25–30 passages until immortalized cell lines appear (*see* **Note 1**).

5. For analysis of long-term growth ability of MEF cells, plot total cell numbers in each passage on a log scale, against passage numbers on a linear scale (*see* **Note 2**).

3.2. Transformation With Oncogenes

3.2.1. Preparing Retroviral Supernatants

1. Plate Phoenix cells into a 6-cm cell culture dish at a density of 2×10^6 cells/dish. Culture cells at 37°C in a 5% CO_2 incubator (*see* **Note 3**).
2. Next day, transfect Phoenix cells with a retroviral vector alone, or vectors containing *RasV12*, *c-myc* or *E1A* using Lipofectamine 2000 (*see* **Note 4**). For each 6-cm dish, mix 8 µg of retroviral DNA with 500 µL of Opti-MEM medium (without serum) in a 1.5-mL microfuge tube. Stand at room temperature for 5 min.
3. At the same time, dilute 20 µL of Lipofectamine 2000 with 500 µL of Opti-MEM medium (without serum) in another 1.5-mL microfuge tube.
4. After 5 min, combine DNA samples from **step 2** and Lipofectamine 2000 complexes from **step 3**. Mix well, and leave at room temperature for 20 min.
5. Remove the culture medium from cells, wash once with PBS and add 5 mL of fresh culture medium (without antibiotics) to each dish.
6. Add 1 mL of DNA-Lipofectamine 2000 mixture from **step 4** very slowly to the cells. Continue culturing cells at 37°C.
7. After 6–12 h, remove the medium containing transfection reagent and DNA complexes, wash once with PBS, add fresh medium to the cells, and culture cells at 37°C for 24 h.
8. Collect the culture medium from transfected Phoenix cells into 15-mL falcon tubes. If needed, add another 6 mL medium to the dish and harvest 12–24 h later (*see* **Note 5**).
9. Centrifuge the supernatant at 1000g for 10 min to remove cellular debris and filter the supernatant through a 0.45-µm filter. This supernatant contains retrovirus and can be used directly to infect MEF cells or frozen at –80°C for future use (*see* **Note 6**).

3.2.2. Infection of MEF Cells by Retroviruses

1. Plate low passage (P3-P5) MEF cells into 10-cm cell culture dishes at a density of 8×10^5 cells per dish and culture at 37°C (*see* **Note 7**).
2. Next day, remove the culture medium from cells and wash cells twice with PBS.
3. Add polybrene to the viral supernatant at a final concentration of 4 µg/mL. Apply 4–5 mL of this virus solution to each 10-cm dish and culture at 37°C for 6 h. Repeat one or two more cycles of infection by replacing fresh viral supernatant into cultured cells, and incubate for 6 h at 37°C for each round (total infection time will be 12–18 h).
4. Remove medium containing retrovirus, wash cells twice with PBS. Add fresh medium continue culture at 37°C.
5. Twenty four hours later, split cells in a 1:3 ratio into 10-cm dishes, and culture cells with medium containing 2 µg/mL puromycin or 75 µg/mL hygromycin B. The control plate with uninfected cells should show 100% killing about 3–5 d after infection depending on the drug being used for selection.

6. If a second retroviral infection is required, infect cells with the second retroviral supernatant 48 h after the first infection and then follow **steps 2–5**. Select with medium containing both puromycin and hygromycin.
7. The selected MEF cells can be used for following assays.

3.3. Transformation Assays

3.3.1. Focus Formation Assay

1. Mix 1×10^4 infected MEF cells with 3×10^5 uninfected MEF cells, and plate this mixture into a 10-cm cell culture dish (*see* **Note 8**).
2. Culture cells at 37°C in 5% CO_2 incubator, change the medium every 2 d.
3. After 14 d, remove culture medium from cells, wash cells once with PBS, and stain with Giemsa solution for about 1 min or until the cells become dark blue.
4. Wash the plates with a large amount of distilled water, air-dry.
5. Score visible foci.

3.3.2. Anchorage-Independent Growth by Soft Agar Assays

1. Melt 1% low-melting temperature agarose at 65°C and cool to 40°C in a water bath (*see* **Note 9**). Warm 2X DMEM medium (containing 20% FBS) to 37°C in a water bath. Allow at least 30 min for the temperature to equilibrate.
2. Mix equal volumes of the two solutions to get 0.5% agarose in 1X DMEM plus 10% FBS. Keep the solution warm at 37°C in a water bath.
3. To prepare the base agarose, add 2 mL of 0.5% agarose/DMEM/FBS mixture to each 35-mm dish and leave at room temperature for at least 30 min to allow agarose to solidify.
4. Trypsinize infected MEF cells, determine cell number, and prepare a cell suspension at 2×10^4 cell/mL in medium. Aliquot 2 mL of cell suspension into a 15-mL falcon tube.
5. From this step, process only one tube at a time. To prepare 0.3% top agarose, quickly transfer 3 mL of prewarmed 0.5% agarose/DMEM/FBS mixture from **step 2** into the tube with 2 mL of cell suspension, mix well by gently pipetting up and down. Avoid bubble formation at this step.
6. Quickly but gently apply 2 mL of cell/agarose mix from **step 5** to a 35-mm dish containing 2 mL 0.5% base agarose.
7. Finish all the cell samples, and leave the plates at room temperature to let the top agarose solidify for about 30–60 min before transferring into a cell culture incubator.
8. Incubate plates at 37°C for 21 d. Every 2 d, gently apply 0.5 mL growth medium to the surface of the top agarose to prevent it from drying out.
9. Count colonies (>50 cells) using a dissecting microscope (*see* **Note 10**).

3.4. Tumorigenicity in Nude Mice

1. Prepare infected and uninfected MEF cell suspensions in PBS at a concentration of 1×10^7 cells/mL.

2. Inject 100 µL of MEF cells (~10^6 cells) subcutaneously into each flank of 6-wk-old athymic nude mice.
3. Maintain mice in a pathogen-free facility. Monitor tumor growth every 3 d by measuring tumor size using a caliper.
4. The size of the tumor (V, volume) can be determined by the following equation, where L is length and W is the width of the xenograft: $V = (L \times W^2) \times 0.5$.
5. Around 2–3 wk later when the tumors are up to 1.5 cm in size, sacrifice the mice. Dissect each tumor, measure the weight, and process for histological analysis if required.

4. Notes

1. Cell numbers will drop dramatically between passages 10 and 20, and after that, cell numbers will slowly increase owing to spontaneous immortalization.
2. Alternatively, the increased cell numbers in each passage can be represented as the number of population doublings, which can be calculated according to the formula $\log(N_f/N_0)/\log_2$ (N_0 is the initial cell number and N_f is the final cell number).
3. Retroviral supernatants can be produced from either stable producer cell lines or transiently transfected producer cell lines. Retroviral supernatants are prepared by transiently transfecting retroviral vectors into a packaging cell line, Phoenix-eco cells, which are already transfected with two MoMLV packaging gene constructs: CMV-Env-PolyA, and RSV-Gag/Pol-Tyt2-PolyA (for further details, check the Nolan lab website at http://www.stanford.edu/group/nolan). In case Phoenix cells are not available, retroviral vectors can also be cotransfected with an ecotropic envelope virus vector (such as pCL-Eco) into 293T cells. For safety, the person handling retroviruses should follow proper institutional guidelines.
4. Phoenix cells are derived from 293T cells, and can be transfected with many standard transfection reagents. Similar to 293T cells, Phoenix cells need to be split at 1:3 or 1:4 ratio to prevent cell clumping which will decrease the transfection efficiency.
5. Because the virus is not very stable at 37°C (half-life is ~4 h), it is best to collect supernatant frequently at short durations. For instance, collect virus supernatant twice at 12 h intervals instead of collecting once after 24 h. Decreasing the temperature to 32°C during supernatant collection has been reported to result in a 5- to 15-fold increase in retroviral titer.
6. It is best to use freshly prepared retroviral supernatant, as freezing and thawing supernatants can dramatically decrease virus titers. For determining viral titer, plate NIH3T3 cells into six-well plates at a density of 2×10^4 cells/well. Prepare dilutions of the retroviral supernatant (1:100, 1:1000, 1:10^4, 1:10^5, 1:10^6) with culture medium. Add polybrene stock to 4 µg/mL, and infect NIH3T3 cells (plate in duplicate wells for each dilution) in the same way as described for MEF cells (**Subheading 3.2.2.**). After 4–5 d of selection with appropriate drug containing selection medium, count the number of colonies, which are obtained at the highest dilution. Viral titers can be determined by multiplying the colony number by the dilution factor.

7. Retrovirus can only infect proliferating cells. MEF cells grow very fast at early passages, and their growth rate drops around passages 8–10. Therefore, it is very important to use MEF cells at early passages (<P5) for retroviral infection.

8. Focus formation assay can be done with infected cells only. Because most transformed cells lose contact inhibition, they usually form a very thick multilayer culture instead of an individual focus. In this protocol, we mix infected cells and uninfected cells for better visualization of foci formation.

9. Instead of low-melting temperature agarose, Difco Bacto agar, or Noble agar, also give good results in soft-agar assays. The only advantage of using agarose is that it is easier to handle at such a low percentage.

10. Cell colonies can also be visualized by staining with 1 mL per well of 1% p-iodonitrotetrazolium violet (Sigma, cat. no. I-8377) in distilled water.

Acknowledgments

Supported in part by funds from NIH and a Scholar Award from the Leukemia and Lymphoma Society.

References

1. Pattengale, P. K., Stewart, T. A., Leder, A., et al. (1989) Animal models of human disease. Pathology and molecular biology of spontaneous neoplasms occurring in transgenic mice carrying and expressing activated cellular oncogenes. *Am. J. Pathol.* **135,** 39–61.

2. Hakem, R. and Mak, T. W. (2001) Animal models of tumor-suppressor genes. *Annu. Rev. Genet.* **35,** 209–241

3. Hayflick, L. and Moorhead, P. (1961) The serial cultivation of human diploid cell strains. *Exp. Cell Res.* **25,** 585–621.

4. Peacocke, M. and Campisi, J. (1991) Cellular senescence: a reflection of normal growth control, differentiation or aging? *J. Cell. Biochem.* **45,** 147–155.

5. Sherr, J. and DePinho, R. A. (2000) Cellular senescence: mitotic clock or culture shock? *Cell* **102,** 407–410.

6. Ishikawa, F. (2003) Cellular senescence, an unpopular yet trustworthy tumor suppressor mechanism. *Cancer Sci.* **94,** 944–947.

7. Todaro, G. J. and Green, H. (1963) Quantitative studies of the growth of mouse embryo cells in culture and their development into established cell lines. *J. Cell Biol.* **17,** 299–313.

8. Harvey, D. M. and Levine, A. J. (1991) p53 alteration is a common event in the spontaneous immortalization of primary BALB/c murine embryo fibroblasts. *Genes Dev.* **5,** 2375–2385.

9. Rittling, S. R. and Denhardt, D. T. (1992) p53 mutations in spontaneously immortalized 3T12 but not 3T3 mouse embryo cells. *Oncogene* **2,** 445–452.

10. Harvey, M., Sands, A. T., Weiss, R. S., et al. (1993) In vitro growth characteristics of embryo fibroblasts isolated from p53-deficient mice. *Oncogene* **8,** 2457–2467.

11. Kamijo, T., Zindy, F., Roussel, M. F., et al. (1997) Tumor suppression at the mouse INK4a locus mediated by the alternative reading frame product p19[ARF]. *Cell* **91,** 649–659.

12. Dannenberg, J. H., van Rossum, A., Schuijff, L., and te Riele, H. (2000) Ablation of the retinoblastoma gene family deregulates G_1 control causing immortalization and increased cell turnover under growth-restricting conditions. *Genes Dev.* **14,** 3051–3064.

13. Sage, J., Mulligan, G. J., Attardi, L. D., et al. (2000) Targeted disruption of the three Rb-related genes leads to loss of G_1 control and immortalization. *Gene Dev.* **14,** 3037–3050.

14. Land, H., Parada, L. F., and Weinberg, R. A. (1983) Tumorigenic conversion of primary embryo fibroblasts requires at least two cooperating oncogenes. *Nature* **304,** 596–602.

15. Ruley, H. E. (1983) Adenovirus early region 1A enables viral and cellular transforming genes to transform primary cells in culture. *Nature* **304,** 602–606.

16. Serrano, M., Lin, A. W., McCurrach, M. E., Beach, D., and Lowe, S. W. (1997) Oncogenic ras provokes premature cell senescence associated with accumulation of p53 and p16[INK4a]. *Cell* **88,** 593–602.

17. Latres, E., Malumbres, M., Sotillo, R., et al. (2000) Limited overlapping roles of p15[INK4b] and p18[INK4c] cell cycle inhibitors in proliferation and tumorigenesis. *EMBO J.* **19,** 3496–3506.

18. Lowe, S. W., Bodis, S., McClatchey, A., et al. (1994) p53 status and the efficacy of cancer therapy in vivo. *Science* **266,** 807–810.

20

Analysis of Growth Properties and Cell Cycle Regulation Using Mouse Embryonic Fibroblast Cells

Hong Sun, Neriman Tuba Gulbagci, and Reshma Taneja

Summary

A balance between proliferation and apoptosis is crucial for cellular homeostasis, and its disruption leading to enhanced cellular proliferation and uncontrolled growth are hallmarks of cancer. Genetic manipulation in the mouse offers a powerful approach to delineate the roles of genes in carcinogenesis and determine the molecular and cellular basis of their function. Mouse embryonic fibroblast cells derived from mice that are disrupted for tumor suppressors or oncogenes have served as an invaluable tool to study altered growth properties of cells and identify regulatory molecules involved in neoplastic transformation. In this chapter, protocols for isolation of mouse embryonic fibroblast cells from midgestation mouse embryos and their applications to study altered growth properties by growth curves and colony formation assays are provided. Methods to analyze cell cycle profiles by flow cytometry using bromodeoxyuridine and propidium iodide staining were also provided, entry of cells in S-phase by [³H] thymidine incorporation studies, and the analysis of cells in mitosis by staining with antiphospho-H3 antibodies are also provided.

Key Words: BrdU; cell cycle; checkpoint; embryos; FACS; growth; MEFs; mitosis; proliferation; S-phase.

1. Introduction

The history of isolation and use of primary embryonic fibroblasts can be traced back to the middle of the last century *(1,2)*. Primary fibroblasts are routinely isolated from midgestation chicken, rat, or mouse embryos. When cultured in vitro these cells grow and proliferate for a finite number of passages that varies between different species before entering a phase of irreversible growth arrest called cellular senescence *(3,4)*. For several years, studies on primary embryonic fibroblast cells were mainly focused on transformation studies with different viruses such as Sarcoma and Polyoma *(5,6)*. Subsequently mouse embryonic fibroblast (MEF) cells were used as feeder

From: *Methods in Molecular Biology, vol. 383: Cancer Genomics and Proteomics: Methods and Protocols*
Edited by: P. B. Fisher © Humana Press Inc., Totowa, NJ

layers to culture embryonic stem cells *(7)*. With the development of technologies for gene targeting by homologous recombination, MEF cells derived from mice mutant for oncogenes or tumor suppressors have provided an important tool to dissect the molecular regulators involved in growth control and neoplastic transformation *(8,9)*.

MEF cells provide several advantages as a model system for various studies compared to other cell types. First, is the relative ease with which they can be established and maintained. Additionally, MEF cells are isolated at early stages of embryonic development (E12.5–E14.5) thus enabling studies in cases, where gene disruption leads to embryonic lethality. MEF cells are facile to handle in cell culture and are readily transfected with plasmid DNAs or infected with retroviruses making them amenable for analysis of regulatory mechanisms ex vivo. Moreover, these cells can be treated with a number of triggers to induce apoptosis making them a suitable choice for analysis of genes involved in regulation of survival and apoptosis *(10)*. Unlike human cells, MEF cells upon prolonged culture spontaneously immortalize giving rise to a cell line *(11)*. Thus these cells can be used to study the regulation of cellular senescence and immortalization. In response to specific stimuli, MEF cells can also be differentiated into certain cell types making them a suitable choice for studying regulators of cellular differentiation *(12,13)*. Thus MEF cells can be used to study a number of biological properties such as cell cycle regulation, immortalization, transformation, senescence, apoptosis, and differentiation, which are often deregulated in cancer.

In this chapter, protocols for generating and maintenance of MEF cells as well as their uses in studying growth properties and cell cycle regulation are described. In the first part of this chapter, isolation of MEF cells from midgestation mouse embryos as well as procedures for handling of MEF cells during routine cell culture are described. After isolation of MEF cells, an initial assessment of altered growth properties of MEF cells can be obtained by determination of growth curves. In addition, colony formation assays can be used to determine the ability of single cells to form a visible colony, which is in part dependent on the proliferative potential of cells. Often, changes in growth are reflected in altered cell cycle profiles, which can be studied by flow cytometry using 5-Bromo-2′-deoxyuridine (BrdU) and propidium iodide (PI) staining. This method is widely used to estimate changes in various phases of the cell cycle, and can also be used to examine the G_1 checkpoint in response to specific stimuli. A more detailed characterization of the ability of cells to traverse the G_1 checkpoint can be determined by monitoring re-entry of cells into the S-phase by [^{3}H] thymidine incorporation assays. Additionally, the mitotic index can be measured by flow cytometric analysis of phosphorylated histone H3 levels, which is expressed exclusively in mitosis.

2. Materials

1. Becton Dickinson FACScan.
2. BrdU (Sigma, St. Louis, MO; cat. no. B-5002): make a 10 mM stock solution in PBS, sterilize by filtration, aliquot and store in the dark at –20°C (BrdU is light-sensitive).
3. Bovine serum albumin (BSA, Sigma, cat. no. A-2153).
4. 37°C, 5% CO_2 cell culture incubator.
5. 1% Crystal violet solution (Sigma, cat. no. C-3886): dissolve 1 g of crystal violet in 2 mL of ethanol, then add distilled water to a final volume of 100 mL.
6. Dissecting instruments: sterile fine scissors, fine forceps, sterile blades.
7. Ethanol, 200 Proof (Pharmco, Brookfield, CT).
8. FITC-labeled goat antimouse and antirabbit IgG antibody (Jackson Immunoresearch Labs, West Grove, PA).
9. Freezing medium: Dulbecco's modified Eagle's medium containing 20% heat-inactivated FBS and 10% DMSO (Sigma, cat. no. D-8779).
10. Glutaradehyde stock (25%, Sigma, cat. no. G-6257).
11. Growth medium: 10% heat-inactivated fetal bovine serum (FBS, Hyclone, [Logan, UT] heat inactivated at 56°C for 30 min) in Dulbecco's modified Eagle's medium (Invitrogen, Carlsbad, CA, high glucose with glutamine) plus 0.1 mM nonessential amino acids (Sigma, cat. no. M-7145), 55 μM 2-mercaptoethanol (Sigma, cat. no. M-7522), and 10 μg/mL gentamicin (Sigma, cat. no. G-1397).
12. Microplate reader (Bio-Rad, Hercules, CA; Model 550).
13. Microscopes.
14. Methanol and acetic acid (Fisher Scientific, Pittsburg, PA).
15. Mouse monoclonal anti-BrdU antibody (Sigma, cat. no. B-2531).
16. Phosphate-buffered saline (PBS): dissolve 8 g NaCl, 0.2 g KCl, 1.44 g Na_2HPO_4, 0.24 g KH_2PO_4 in 800 mL distilled water, adjust pH to 7.4, then add water to 1000 mL, sterilize by autoclaving.
17. 10X PI solution: 500 μg/mL of propidium iodide (PI, Sigma, cat. no. P-4107) in PBS.
18. Plastic tissue culture dishes (10, 6, and 3.5 cm) and tissue culture plates (6-, 12-, 24-, and 96-well) from Nunc (Rochester, NY).
19. Rabbit polyclonal antiphosphorylated histone H3 antibody (Upstate Biotechnology).
20. Beckman LS 6500 Multipurpose Scintillation Counter.
21. RNase A (Roche, Indianapolis, IN; cat. no. 109169) stock: 10 mg/mL in 10 mM Tris Cl, pH 7.5, and 15 mM NaCl. Boil at 100°C for 10 min to eliminate DNase activity, aliquot and store at –20°C.
22. Ecoscint A scintillation solution for aqueous or nonaqueous samples (National Diagnostics, Atlanta, GA).
23. [^{3}H] thymidine (NEN, cat. no. NET-027Z).
24. Trichloroacetic acid (TCA, Sigma, cat. no. 490-10).

3. Methods

3.1. Establishment and Maintenance of Primary MEF Cells

3.1.1. Isolation of Primary MEF Cells

1. Euthanize pregnant mice at embryonic day E12.5–E14.5 by CO_2 inhalation (*see* **Note 1**). Put mice on a solid ground and wipe the abdominal region with 70% ethanol. Open the abdomen, remove the whole uterus and transfer it to a 10-cm cell culture dish with 15 mL PBS.
2. Under a sterile hood, cut the wall of the uterus, and release embryos with their intact yolk sac. Place each embryo in a separate 6-cm dish with PBS (*see* **Note 2**).
3. Carefully remove the yolk sac surrounding the embryo and transfer it to a 1.5-mL microfuge tube to isolate DNA for genotyping if needed.
4. Cut the head of the embryo and carefully dissect out all the internal organs such as liver, lung, heart, kidney, and intestines, trying to remove them as completely as possible to prevent contamination from other cell types. Transfer the embryo body to a new 6-cm dish and wash twice with PBS.
5. Mince the embryo carcass into small pieces using sterile razor blades or scissors (*see* **Note 3**). Add 3 mL of 1X (0.05%) trypsin, and transfer all the pieces to a 15-mL falcon tube. Dissociate cells by pipetting the suspension vigorously several times with a 10-mL pipet. Incubate for 20 min in a 37°C water bath.
6. Inactivate trypsin by adding 10 mL of medium.
7. Spin down the cell suspension at $500g$ for 5 min. Discard the supernatant. Resuspend the cell pellet in 10 mL medium and transfer to a 10-cm cell culture dish. Incubate the cells at 37°C in a 5% CO_2 incubator.
8. Next day, change the medium to get rid of all the cellular debris and continue culturing cells until they reach almost 90% confluence (about 2 d). These cells are considered to be passage 1 MEF cells.

3.1.2. Passaging, Freezing, and Thawing of MEF Cells

1. When the cells are almost 90% confluent, aspirate the culture medium and wash cells twice with PBS.
2. Add 1.5 mL of trypsin solution to the dish and incubate at 37°C for about 2 min or at room temperature until all cells have detached. Add 8.5 mL of medium and mix well.
3. To passage MEF cells, plate at density of 1×10^6 cells in a 10-cm dish. Culture cells at 37°C in a CO_2 incubator.
4. To freeze MEF cells, spin them down at $500g$ for 5 min and resuspend cells in freezing medium at a concentration of 4×10^6 cells/mL. Aliquot into cryotubes (1 mL/vial), keep at –80°C for 24 h and then transfer to liquid nitrogen.
5. To thaw MEF cells, take one vial from liquid nitrogen. After thawing cells in a 37°C water bath, quickly transfer to a 15-mL falcon tube containing 10 mL

medium to dilute the DMSO. Spin down cells and resuspend them in 10 mL medium. Transfer to a 10-cm cell culture dish and culture at 37°C.

3.2. Growth Properties

For all the protocols described in this chapter, it is important to use at least two to three independent MEF cell isolates. For each MEF isolate, repeat the experiments twice.

3.2.1. Measurement of Cell Growth Rates by Growth Curves

1. At day 0, seed MEF cells at a density of 2.5×10^4 cells/well in a 12-well plate. For each time-point to be analyzed from day 0 to day 8, plate cells in triplicate. Incubate cells at 37°C in 5% CO_2 incubator, and change the medium every 48 h. About 6–7 h after seeding the cells, take one set of triplicate culture aliquots corresponding to day 0, and process as described in **steps 2** and **3** (*see* **Note 4**).
2. At each time-point (for example, every day or every other day, depending on the experimental design), process one set of culture aliquots. Carefully remove medium from the wells and wash cells three times with PBS.
3. Fix cells with 1 mL of 1% glutaraldehyde per well for 10 min and rinse the plates three times with PBS (*see* **Note 5**). The cells can be stored at 4°C in PBS until all the samples are ready to be stained.
4. Carefully remove PBS from the wells, add 1 mL of 0.1% crystal violet solution to each well and stain the cells for 30 min at room temperature. Rinse the plates three times with distilled water in a large container and air-dry at room temperature.
5. Add 2 mL of 30% methanol and 10% acetic acid to each well to extract the cell-associated dye (*see* **Note 6**). Put the plates on a shaker until the color is uniform.
6. Transfer 150 μL of the extract to a 96-well plate and read the absorbance at 590 nm. Normalize the absorbance with control cells (from **step 1**) to get the relative cell number.
7. Plot relative cell number (or absorbance) on a log scale, against time (day) on a linear scale.

3.2.2. Colony Formation Assay

1. Plate MEF cells at a very low density of 1×10^4 cells per 10-cm dish (*see* **Note 7**).
2. Culture the cells at 37°C. Change the medium for every 2 d.
3. After 2 wk, carefully remove the culture medium and wash twice with PBS.
4. Fix cells with 1% glutaradehyde in PBS for 15 min at room temperature. Rinse with PBS.
5. Stain with 0.1% crystal violet in distilled water for 30 min at room temperature.
6. Completely remove crystal violet and rinse with tap water in a large container.

7. Allow plates to air-dry and examine under microscope, count the number of colonies (*see* **Note 8**).

3.3. Cell Cycle Regulation

3.3.1. Analysis of Cell Cycle Profile by BrdU Incorporation

1. Plate the MEF cells in a 10-cm cell culture dish (1×10^6 cells per dish). After 24 or 48 h, add BrdU solution directly to the culture medium at a final concentration of 10 μ*M* (*see* **Note 9**).
2. Harvest the cells 2–4 h after BrdU addition, wash twice with ice-cold PBS, spin down and resuspend cells in 600 μL ice-cold PBS.
3. Slowly add 1.4 mL ice-cold ethanol, while gently vortexing (*see* **Note 10**), mix well, and leave cells on ice for at least 30 min. The fixed cells are stable and can be stored in the dark at –20°C.
4. Spin down cells, and carefully remove the alcohol. Wash cells once with PBS, and resuspend the cell pellet in 1 mL 2 *N* HCl solution (add slowly while vortexing). Incubate at room temperature for 30 min.
5. Add 5 mL PBS to each tube, mix well, and spin down cells.
6. Wash cells twice with blocking buffer (PBS containing 1% BSA, 1% Tween-20). Transfer cells to a 1.5-mL centrifuge tube and spin down.
7. Resuspend cells in 200 μL of blocking buffer, add monoclonal anti-BrdU antibody (1:200) to each tube, mix and incubate at room temperature for 1 h.
8. Wash cells twice as in **step 6**, resuspend cell pellet in 200 μL of blocking buffer.
9. Add 1 μL of FITC-labeled goat antimouse IgG, mix, and incubate for 30 min at room temperature.
10. Wash cells twice with blocking buffer, spin down and resuspend in PBS containing 25 μg/mL PI and 100 μg/mL RNase A (freshly made from stock, *see* **Note 11**), incubate at room temperature in the dark for 30 min.
11. Analyze by flow cytometry.

3.3.2. DNA Synthesis by [³H] Thymidine Incorporation

1. Seed MEF cells in 24-well plates at a density of 5×10^4 cells per well.
2. Next day, for synchronizing cells by serum deprivation, change to medium containing 0.1% FBS and continue to culture at 37°C 48 h later, remove low-serum containing medium, add normal growth medium (containing 10% FBS) and continue to culture for 8–12 h.
3. Add 0.5 μCi of [³H] thymidine to each well 4 h before harvesting the cells.
4. At the time of harvesting, aspirate the medium and wash cells at least three times with ice-cold PBS (2 mL per well).
5. Add 2 mL of ice-cold 5% TCA to each well and incubate on ice for 10 min. Discard TCA by aspiration, and incubate with another 2 mL of ice-cold 10% TCA for 10 min on ice.
6. Carefully remove TCA from the wells, add 500 μL of 0.5 *M* NaOH to each well, and incubate at 37°C for 2 h to lyse the cells.
7. Add 250 μL of 1 *N* HCl to neutralize the lysate, pipet up and down to mix well.

8. Transfer the lysate to scintillation vials containing 5 mL scintillation solution, and measure the radioactivity in a scintillation counter.

3.3.3. Mitotic Index by Histone H3 Phosphorylation

1. Plate MEF cells in 100-mm cell culture dishes at density of 1×10^6 cells per dish. Culture at 37°C.
2. Next day, irradiate cells with 10 gray of γ-ionizing irradiation (IR). Continue to culture cells at 37°C.
3. Collect cells at 0.5, 1, 1.5, and 2 h after irradiation (*see* **Note 12**). Wash cells twice with PBS, and fix them with 70% ethanol as earlier.
4. Spin down fixed cells and carefully remove alcohol. Wash cells twice with PBS, and resuspend in 1 mL of blocking buffer (1% BSA in PBS). Transfer the cell suspension to a 1.5-mL microfuge tube and incubate at room temperature for 30 min.
5. Spin down the cells and resuspend them in 100 µL of blocking buffer and antibody against phosphorylated form of histone H3 (1:200), incubate for 3 h at room temperature (vortex samples periodically to prevent them from settling at the bottom of the tubes).
6. Wash cells three times with blocking buffer. Resuspend cells in blocking buffer containing FITC-labeled goat anti-rabbit IgG antibody; incubate at room temperature for 30 min in the dark.
7. Wash cells three times with blocking buffer and resuspend in 1 mL of PBS containing 25 µg/mL PI and 100 µg/mL RNase A. Incubate at room temperature for 30 min in the dark before analysis by flow cytometry.

4. Notes

1. MEF cells can be made from E10.5–E17.5 embryos. Routinely use E12.5–E14.5 embryos (*14*) owing to the rather low yield of MEF cells from earlier embryonic stages and contamination of more differentiated cells in late stage embryos. To calculate the embryonic stage, set up crosses of male and female mice around 5 PM, and check for a plug early next morning. The noon of the day the plug is seen is considered to be E0.5.
2. To prevent contamination between different genotypes, it is important to put each embryo in separate dish that is clearly marked. It is also best to make MEF cells from individual embryos instead of a pool of embryos.
3. Alternatively, transfer the embryo carcass into a 15-mL Falcon tube with 5–10 mL PBS. Dissociate cells by passing through an 18-gauge needle three to five times. Spin down at 500*g* for 5 min, and resuspend cells in 3 mL trypsin. Incubate for 20 min at 37°C and proceed as described in **step 6**.
4. Carefully organize the wells according to the number of MEF cell isolates (or genotypes) and time-points that need to be tested. For instance, plate triplicates of wild-type and knockout MEFs that are to be harvested at the same time-point in a single plate, so that you can process the whole plate at each time-point.
5. Use a large container filled with PBS. Wash cells by submerging the whole plate under PBS. Remove PBS by inversing the plate and patting it on a thick layer of paper towels. Try to remove the remaining PBS as much as possible without detaching the cells.

6. One percent SDS solution can also be used for extracting the dyes.
7. The ability of primary MEFs to form colonies at an extremely low cell density is partly dependent on their proliferative capacity. Therefore, the proliferative potential of primary MEFs can be measured by the efficiency of colony-formation at a very low cell density ($<10^4$ cells per 10-cm dish).
8. Colonies can be counted by cell numbers in each colony (>30 cells) or by diameters (>2 mm).
9. Ten to twenty micromolars of BrdU is usually used for incorporation into cultured cells. Final concentrations of higher than 50 μM might affect cell proliferation and cell cycle profiles.
10. Keep the vortex at a very low speed, and add ice-cold ethanol very slowly drop wise (about 2 min for 1.4 mL ethanol) while vortexing cells. Alternatively, cells can be resuspended in a small amount of PBS, followed by addition of a large amount of 70% alcohol (5 mL, ice-cold) into cells with continued stirring on a low speed vortex.
11. RNA in cells can also bind to PI, which will interfere with the results. RNA should be removed with RNase A before performing flow cytometric analysis.
12. The number of cells entering M-phase start to drop within the first half to 1 h after γ-irradiation, but increases again after 12 h. Therefore, cell samples collected at early time-points (earlier than 2 h postirradiation) can provide information about activation of the G_2/M checkpoint. To test maintenance of the G_2 checkpoint, cells can be cultured with 0.2 µg/mL nocodazole (Sigma, cat. no. M1404) after γ-irradiation and assayed for histone H3 phosphorylation after 12 h.

Acknowledgments

Supported in part by funds from NIH and a Scholar Award from the Leukemia and Lymphoma Society (RT).

References

1. Dulbecco, R. and Freeman, G. (1959) Plaque production by the polyoma virus. *Virology* **8,** 396–397.
2. Pittman, D., St. John, R. C., and Shechmeister, I. L. (1965) Latent period of vesicular stomatitis virus in chick embryo cells. *Nature* **206,** 1228–1231.
3. Meek, R. L., Bowman, P. D., and Daniel, C. W. (1977) Establishment of mouse embryo cells in vitro. Relationship of DNA synthesis, senescence and malignant transformation. *Exp. Cell Res.* **107,** 277–284.
4. Lima, L. and Macieira-Coelho, A. (1972) Parameters of aging in chicken embryo fibroblasts cultivated in vitro. *Exp. Cell Res.* **70,** 279–284.
5. Boettiger, D. and Temin, H. M. (1970) Light inactivation of focus formation by chicken embryo fibroblasts infected with avian sarcoma virus in the presence of 5-bromodeoxyuridine. *Nature* **228,** 622–624.
6. Rassoulzadegan, M., Cowie, A., Carr, A., Glaichenhaus, N., Kamen, R., and Cuzin, F. (1982) The roles of individual polyoma virus early proteins in oncogenic transformation. *Nature* **300,** 713–718.

7. Doetschman, T. C., Eistetter, H., Katz, M., Schmidt, W., and Kemler, R. (1985) The in vitro development of blastocyst-derived embryonic stem cell lines: formation of visceral yolk sac, blood islands and myocardium. *J. Embryol. Exp. Morphol.* **87,** 27–45.

8. Livingstone, L. R., White, A., Sprouse, J., Livanos, E., Jacks, T., and Tlsty, T. D. (1992) Altered cell cycle arrest and gene amplification potential accompany loss of wild-type p53. *Cell* **70,** 923–935.

9. Almasan, A., Yin, Y., Kelly, R. E., et al. (1995) Deficiency of retinoblastoma protein leads to inappropriate S-phase entry, activation of E2F-responsive genes, and apoptosis. *Proc. Natl. Acad. Sci. USA* **92,** 5436–5440.

10. Stambolic, V., Suzuki, A., de la Pompa, J. L., et al. (1998) Negative regulation of PKB/Akt-dependent cell survival by the tumor suppressor PTEN. *Cell* **95,** 29–39.

11. Todaro, G. J. and Green, H. (1963) Quantitative studies of the growth of mouse embryo cells in culture and their development into established lines. *J. Cell Biol.* **17,** 299–313.

12. Landsberg, R. L., Sero, J. E., Danielian, P. S., Yuan, T. L., Lee, E. Y., and Lees, J. A. (2003) The role of E2F4 in adipogenesis is independent of its cell cycle regulatory activity. *Proc. Natl. Acad. Sci. USA* **100,** 2456–2461.

13. Lengner, C. J., Lepper, C., van Wijnen, A. J., Stein, J. L., Stein, G. S., and Lian, J. B. (2004) Primary mouse embryonic fibroblasts: a model of mesenchymal cartilage formation. *J. Cell. Physiol.* **200,** 327–333.

14. Sun, H., Lu, B., Li, R. Q., Flavell, R. A., and Taneja, R. (2001) Defective T cell activation and autoimmune disorder in Stra13-deficient mice. *Nat. Immunol.* **2,** 1040–1047.

21

Reverse Phase Protein Microarrays
for Monitoring Biological Responses

Virginia Espina, Julia D. Wulfkuhle, Valerie S. Calvert, Emanuel F. Petricoin III, and Lance A. Liotta

Summary

Cancer has a genomic and proteomic basis. Genomic information provides information about the somatic genetic changes existing in the tumor that provides a survival advantage driving neoplastic progression. On the other hand, proteomics aids in the identification of dysregulated cellular proteins, including known or novel drug targets, governing cellular survival, proliferation, invasion, and cell death. The clinical utility of reverse phase protein microarrays lies in their ability to generate a map of known cell signaling networks or pathways for an individual patient. This protein network map aids in identifying critical nodes or pathways that may serve as drug targets for individualized or combinatorial therapy. Reverse phase protein microarrays are one of the tools available for profiling the protein molecular pathways in a given cellular sample. This type of microarray can uniquely quantify phosphorylation states of proteins. An entire cellular proteome is immobilized on a substratum with subsequent immunodetection of total and activated forms of cell signaling proteins. The pattern of signal intensity generated by the protein spots can be correlated with biological and clinical information as diagnostic and prognostic indicators.

Key Words: Cancer; combinatorial therapy; laser capture microdissection; microarray; molecular profiling; protein; proteomics; tissue heterogeneity.

1. Introduction

Protein microarrays, in the simplest sense, are immobilized protein spots on a substratum (1–6). Individual protein spots may be heterogeneous or homogeneous, and may consist of cell or phage lysates, an antibody, a nucleic acid, drug, or recombinant protein (2,3,5–13). These immobilized bait molecules are detected by probing the microarray with a signal-generating molecule such as, a tagged antibody, ligand, serum, or cell lysate. The tagging molecule generates

From: *Methods in Molecular Biology, vol. 383: Cancer Genomics and Proteomics: Methods and Protocols*
Edited by: P. B. Fisher © Humana Press Inc., Totowa, NJ

a pattern of positive and negative spots. The signal intensity of each spot is proportional to the quantity of applied tagged molecules bound to the bait molecule. The spot pattern image is captured, analyzed, and correlated with biological information.

The focus of this protocol is on reverse phase protein microarrays, also referred to as protein lysate microarrays. This microarray format is made up of an immobilized cellular lysate that is probed with a primary antibody. Signal amplification is independent of the immobilized protein, permitting the coupling of detection strategies with highly sensitive tyramide amplification chemistries *(14–17)*.

Protein microarrays have broad applications for discovery and quantitative analysis, with applications in drug discovery, biomarker identification, and molecular profiling of cellular material. The clinical utility of protein microarrays lies in their ability to provide a map of the state of multiple in vivo kinase driven signal pathways and to provide crucial information about protein post-translational modifications, such as the phosphorylation states of these proteins *(18–20)*. These modifications reflect the activity state of signal pathways and networks, and this information cannot be measured by gene microarrays. Protein microarrays provide a view of the disrupted cellular machinery governing disease. Identification of critical nodes, or interactions, within these networks is a potential starting point for drug development and/or design of individual therapy regimens *(18,21–23)*. Characterization of molecular signatures of diseased/normal tissue serves two purposes (1) stratifying patients for therapy and (2) discovery of new therapeutic regimens for improving treatment outcomes.

The methods described herein encompass (1) preparation of a whole cell lysate from either cell culture or tissue samples, (2) protein lysate microarray printing, (3) immunostaining, and (4) microarray spot analysis.

2. Materials
2.1. Cellular Lysates

1. Cell culture, human, or animal cellular sample. Satisfactory tissue samples for protein analysis are (1) frozen tissue sections, (2) ethanol fixed, paraffin embedded tissue, or laser capture microdissected (LCM) cell populations from the aforementioned tissues (*see* **Note 1**).
2. Microdissected samples on LCM caps with Safe-Lock Eppendorf tubes, 0.5-mL (Brinkmann, Westbury, NY; cat. no. 22 36 361-1) or MicroAmp™ 500-mL thin-walled reaction tubes (Perkin-Elmer Applied Biosystems, Foster City, CA; cat. no. N801-0611).
3. Cell lysis buffer:
   a. T-PER™ (Tissue Protein Extraction Reagent, Pierce, Rockford, IL).
   b. 2X Tris-glycine SDS loading buffer (Invitrogen, Carlsbad, CA).

    c. 2-Mercaptoethanol (Sigma Aldrich, St. Louis, MO).

    d. 5 *M* NaCl (300 m*M* final conc.).

4. 200 m*M* PEFABLOC (4-[2-Aminoethyl]-benzenesulfonyl fluoride hydrochloride) (Roche Applied Science, Indianaoplis, IN; cat. no. 1 585 916).

5. 1 mg/mL Aprotinin (5 µg/mL final conc.) (Roche, cat. no. 1583794).

6. 5 mg/mL Pepstatin A (5 µg/mL final conc.) (Roche, cat. no. 1524488).

7. 5 mg/mL Leupeptin (5 µg/mL final conc.) (Roche, cat. no. 1529048).

8. Protease inhibitor cocktail (Sigma Aldrich, cat. no. P2714). Use 10 µL of protease inhibitor cocktail/mL of lysis buffer.

9. Phosphatase inhibitor: 1 m*M* orthovanadate (Sigma Aldrich).

10. Cell culture lysis buffer (per milliliter):

    a. 915 µL Tissue Protein Extraction Reagent (TPER™; Pierce).

    b. 60 µL 5 *M* NaCl (300 m*M* final conc.); final conc. of NaCl is 450 m*M* because TPER reagent contains 150 m*M* NaCl.

    c. 10 µL 100 m*M* Orthovanadate (1 m*M* final conc.) (boil vanadate 10 min in H_2O to solublize).

    d. 10 µL 200 m*M* PEFABLOC.

    e. 5 µL 1 mg/mL Aprotinin (5 µg/mL final conc.).

    f. 1 µL 5 mg/mL Pepstatin A (5 µg/mL final conc.).

    g. 1 µL 5 mg/mL Leupeptin (5 µg/mL final conc.).

11. Prepare tissue lysis buffer:

    a. Pipet 50 µL 2-mercaptoethanol, 950 µL 2X Tris-glycine SDS loading buffer and 1 mL TPER into a clean plastic tube. Mix well.

    b. Add 10 µL Sigma protease inhibitor cocktail for each 1 mL of lysis buffer.

    c. If desired, add 10 µL of 1 *M* orthovanadate to each 1 mL of lysis buffer (final concentration 10 m*M* orthovanadate).

    d. Prepare daily. Store at room temperature.

2.2. Printing Reverse Phase Protein Microarrays

1. Nitrocellulose coated glass slides (FAST slides, Florham Park, NJ and Aushon 2470, Aushon Biosystems, Burlington, MA).

2. Microarray printing device (GMS 417 Arrayer, MWG Biotech, High Point, NC).

3. Cool mist humidifer (Ultrasonic Humidifier, Kaz, Inc., Hudson, NY).

4. Cellular lysate sample (minimum volume 30 µL).

5. 96- or 384-Well polypropylene microtiter plate.

6. Purified water (type I reagent grade water).

7. 70% Ethanol.

8. Desiccant (Drierite anhydrous calcium sulfate, W. A. Hammond Drierite Co. Xenia, OH).

9. Ziplock-style plastic bags.

10. Lysis buffer for diluting the samples in a microtiter plate:

    a. 50 µL 2-mercaptoethanol.

    b. 950 µL 2X Tris-glycine SDS loading buffer.

    c. 1 mL TPER.

2.3. Immunostaining Microarrays

2.3.1. Microarray Pretreatment and Blocking

1. Reblot™ Mild Antigen Stripping solution 10X (Chemicon, Temecula, CA) (*see* **Note 2**).
2. Phosphate buffered saline (PBS) without calcium or magnesium.
3. I-Block™ Protein Blocking Solution (Applied Biosystems, Foster City, CA).
4. Tween-20 (Dako, Carpinteria, CA).
5. Preparation of I-block™:
   a. Pour 500 mL of 1X PBS without calcium and magnesium into a 1000 mL beaker and add a magnetic stir bar.
   b. Add 1 g of I-Block™ powder. Place beaker on hot plate with magnetic stirrer. Heat gently with stirring until solution becomes clearer (~5–15 min). Do not boil.
   c. Allow I-Block™ solution to cool to room temperature.
   d. Add 500 μL of Tween-20.
   e. Store at 2–4°C for up to 7 d.

2.3.2. Dako Autostainer Immunostaining

Immunostaining may be performed manually if an Autostainer is not available.

1. Validated primary antibody of choice (*see* **Note 3**).
2. Biotinylated, species-specific secondary antibody.
3. Dako Autostainer (Dako).
4. Catalyzed Signal Amplification (CSA) kit (Dako).
5. Biotin blocking system (Dako).
6. Antibody diluent with background reducing components (Dako).
7. Tris Buffered Saline with Tween (Dako).
8. Liquid DAB+ (Diaminobenzidene) (Dako).

2.3.3. Total Protein Blot Stain for Microarrays

1. Sypro Ruby protein blot stain (Invitrogen/Molecular Probes, Carlsbad, CA; cat. no. S-11791).
2. Fixative solution—7% acetic acid, 10% methanol in dH_2O. (Acetic acid, glacial, Mallinckrodt, cat. no. V193; methanol, absolute, Sigma Diagnostics, cat. no. 17-5).
3. Standard UV transilluminator or a laser scanner (FluorChem 8800, Alpha Innotech Imager, San Leandro, CA) (*see* **Note 4**).
4. Sypro Red/Texas Red filter (490 nm longpass filter) or a 600 nm bandpass filter.
5. Fixative solution:
   a. 5 mL Methanol.
   b. 3.5 mL Acetic acid.
   c. 41.5 mL dH_2O.

2.4. Microarray Spot Analysis

1. Spot analysis software of choice: examples are ImageQuant® (Amersham Biosciences, Piscataway, NJ), MicroVigene™ (Vigene Tech, Billerica, MA) or P-SCAN (Peak quantitation with Statistical Comparative Analysis [http://abs.cit.nih. gov]).
2. High resolution flatbed scanner (Epson® Perfection Scanner series 1640, Long Beach, CA or UMAX PowerLook 1120, Dallas, TX).
3. Adobe® Photoshop software.

3. Methods

The protocols described below illustrate (1) protein lysate preparation, (2) protein (tissue) lysate microarray printing, (3) immunostaining, and (4) spot analysis. These protocols are uniquely designed for quantitative analysis of protein phosphorylation events in cellular lysates, with the concomitant analysis of the corresponding total (phosphorylated and nonphosphorylated) protein.

3.1. Protein Lysate Preparation

1. Samples may be cell culture lysates, microdissected tissue lysates or other cellular lysates. These protocols are compatible with LCM material.
2. Prepare lysates within 1 wk of microarray printing. Long-term storage (>6 mo) of protein lysates may result in protein degradation or diminished protein yield. If needed, store protein lysates at –80°C or in the vapor phase of liquid nitrogen.

3.1.1. Preparation of Cell Culture Lysate

1. Prepare cell culture lysis buffer (**Subheading 2.1.**, **items 10a–g**).
2. Thaw cell pellets on ice for 20 min to 1 h depending on size of pellet.
3. Spin at approx 900*g* for 10 min in refrigerated (4°C) centrifuge and remove any excess liquid.
4. If cells were stored in DMSO, wash cell pellet twice with PBS without calcium or magnesium. Wash in approx 15 mL of PBS without calcium or magnesium, spin at approx 900*g* for 10 min. Decant supernatant and repeat wash step with an additional 15 mL of PBS without calcium or magnesium. Thoroughly decant supernatant.
5. Suspend cell pellet in appropriate volume of lysis buffer, vortex 15 s, spin briefly, and incubate on ice for 20 min. Volume of lysis buffer: 1×10^6 cells/100 µL of extraction buffer or, for lymphocytes, 1×10^6 cells/µL of extraction buffer.
6. Spin at approx 3000*g*, for 5 min and transfer supernatant to clean tube.
7. Perform protein assay of choice to determine protein concentration.
8. Dilute sample with 2X Tris-glycine SDS Sample Buffer (Invitrogen) + 2.5% 2-mercaptoethanol to 2 mg/mL based on the results of the protein assay (*see* **Note 5**).
9. Store lysate at –80°C if needed.

3.1.2. Preparation of Tissue Lysates

The limited volume of cellular material from biopsy samples makes it necessary to use a slight modification of the lysis buffer recipe as compared to the cell culture lysis buffer. The following protocol describes a method for preparing whole cell lysates from LCM cells. The limited biopsy sample generally precludes the use of total protein assays prior to microarray printing. To compensate for this, one microarray slide is stained for total protein using a Sypro Ruby Protein blot stain (*see* **Subheading 3.3.3.**).

1. Prepare tissue lysis buffer.
2. As a rule of thumb, use 15 μL of lysis buffer to solubilze 15,000 cells for microdissected tissue. More than one LCM cap may be used with a given volume of lysis buffer to concentrate the amount of protein in a given volume of lysis buffer (*see* **Note 6**).
3. Pipet the desired volume of lysis buffer in the bottom of the 0.5 mL microcentrifuge tube. Take care to avoid leaving droplets of lysis buffer near the lip of the tube. Place the LCM cap snugly on the tube. Invert the tube and mix well. Do not vortex.
4. Place the tube, cap side down, in an oven or heat block at 70°C ± 2°C for 15 min.
5. Mix well after 15 min. Place back in oven for an additional 15 min. Mix well again.
6. Spin tube at approx 3000*g* for 1–2 min. Remove and discard LCM cap (*see* **Note 7**).
7. Transfer extracted protein sample to a clean, labeled screw cap tube.
8. Boil samples for 5–10 min before storing/printing microarrays. Store at –80°C if necessary.

3.2. Protein Tissue Lysate Microarray Printing

Each microarray is made up of multiple samples, printed in dilution curves, on a single slide. This format allows the comparison of multiple samples across an array for a given antibody (**Fig. 1**). Additionally, the dilution curve allows each antibody affinity to be matched with its optimal sample concentration. Nitrocellulose cannot be stripped and reprobed, therefore, each slide is probed with a different antibody, generating a set of microarray slides for each set of antibody probes.

Quality control and comparison of microarrays across platforms and time requires the use of control lysates that are printed on each array. These samples may be test samples, commercial cell lysates, peptides, or peptide mixtures. Ideally, the control lysate should contain the protein(s) of interest that are being investigated in the test samples. Control samples should be diluted in the same manner as the test samples and the same control lysate should be printed on each array.

1. A minimum volume of 30 μL of lysate is required for printing microarrays in a dilution format from a 384-well microtiter plate. 15 μL of neat sample is used to prepare serial 1:2 dilutions of the lysate. An additional 15 μL of lysate is required for the neat sample.

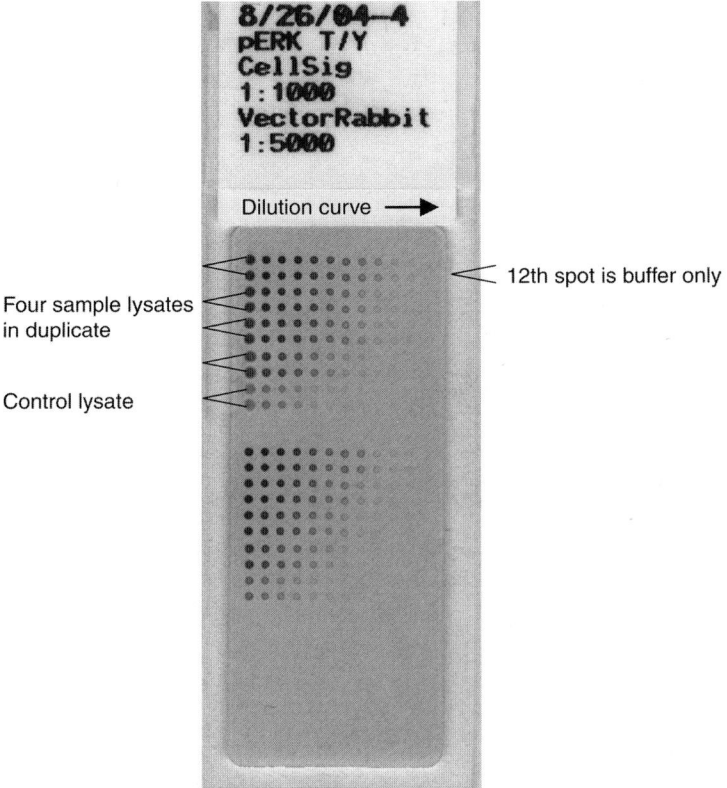

Fig. 1. Reverse phase protein microarray layout printed with a GMS 417 arrayer. The array consists of 10 samples printed in two areas on the array. The first 10 rows represent four samples and one control printed in duplicate, in a 12-point dilution curve. The 12th point is buffer only and is not visible. The second block of 10 rows represents another set of samples and controls printed in the same format as described earlier.

2. A minimum volume of 50 µL of lysate is required for printing microarrays in a dilution format from a 96-well plate (25 µL of lysate to prepare the dilutions and 25 µL for the neat lysate).

3. Prepare additional lysis buffer for diluting the samples in a microtiter plate. Mix well. Prepare daily. Store at room temperature. It is not necessary to add additional protease or phosphatase inhibitors.

4. Boil whole cell lysate: place screw cap tube containing the thawed lysate sample in a boiling water bath or heat block at 100°C for 5 min. Remove tubes from heat and allow tubes to cool to room temperature. Spin briefly in a microcentrifuge.

5. Dilute samples to be printed on the array in appropriate dilutions in a microtiter plate. Refer to **Table 1** for a typical 96-well plate format. Refer to **Table 2** for a typical 384-well plate format (*see* **Note 8**). Typical dilutions are 1:2 serial dilutions

Table 1
A 96-Well Microtiter Plate Format Compatible With Printing by a GMS 417 Arrayer

Sample 1					
Neat	1:2	1:4	1:8	1:16	Buffer
A1	A3	A5	A7	A9	A11
Sample 2					
Neat	1:2	1:4	1:8	1:16	Buffer
A2	A4	A6	A8	A10	A12

Samples are prepared in 1:2 serial dilution curves in the microtiter plate. Sample no. 1 occupies row A in every other well starting with the *odd* wells. Sample no. 2 occupies row A in every other well starting with the *even* wells. This format may be repeated for rows B–H.

Table 2
A 384-Well Microtiter Plate Format Compatible With Printing by a GMS 417 Arrayer

PIN 1						PIN 2					
Sample 1						*Sample 2*					
Neat	1:2	1:4	1:8	1:16	Buffer	Neat	1:2	1:4	1:8	1:16	Buffer
A1	A9	A17	A2	A10	A18	A3	A11	A19	A4	A12	A20
Sample 3						*Sample 4*					
B1	B9	B17	B2	B10	B18	B3	B11	B19	B4	B12	B20
PIN 3						PIN 4					
Sample 5						*Sample 6*					
C1	C9	C17	C2	C10	C18	C3	C11	C19	C4	C12	C20
Sample 7						*Sample 8*					
D1	D9	D17	D2	D10	D18	D3	D11	D19	D4	D12	D20

The sample lysate is placed in wells in a manner permitting six spots to be printed on each side of the microarray slide, with 12 spots horizontally in a row on the array. This 384 well format allows a greater *x*-distance between the printed spots on the microarray, facilitating spot analysis. Sample no. 1 undiluted is placed in well A1. Serial dilutions of the sample are made in wells A9, A17, A2, and A10. Well A18 is buffer only. Sample no 2 undiluted is placed in well A3. Serial dilutions of the sample are made in wells A11, A19, A4, and A12. Well A20 is buffer only. The pattern can be repeated in rows B–P.

with the final dilution lysis buffer only (neat, 1:2, 1:4, 1:8, 1:16, buffer only). The buffer only serves as a negative control spot.

6. Set up the GMS 417 arrayer.
7. Select microtiter plate—96- or 384-well.
8. Select number of microtiter plates.
9. Select number of hits per spot. Five hits/spot is recommended with a 384-well plate. Ten hits/spot is recommended with a 96-well plate (*see* **Note 9**).

10. Calibrate the pins with a microarray slide of the same lot number as the arrays to be printed. The pins should just touch the nitrocellulose. This can be determined by hearing an audible click as the pins touch the slide. Take care not to compress the nitrocellulose by lowering the pins further onto the nitrocellulose.
11. Calibrate the rings in the same format (96- or 384-well) microtiter plate to be used for printing the microarrays. The rings should be positioned just below the surface of the fluid in the well, ensuring the ring is completely submerged, yet not touching the bottom of the well.
12. Set wash preferences: enable second wash bath, dry time 15 s, wash time 5 s.
13. Fill appropriate wash container with water and a second wash container with 70% ethanol.
14. Position FAST™ slides on printing platen. Load microtiter plate. Turn on vacuum pump. Turn on humdifier if desired to achieve 30–50% humidity.
15. Program appropriate dot spacing based on number of samples, replicate spots and size of nitrocellulose pad (*see* **Note 10**).
16. Print microarrays. Store printed slides in a slide box, inside a Ziplock-style plastic bag. Add desiccant to the plastic bag prior to freezing. Store printed microarrays at –20°C.

3.3. Immunostaining

3.3.1. Microarray Slide Pretreatment and Blocking

Comparison of multiple phosphoproteomic end points for a group of samples on the array permits evaluation of the interconnected cell signaling proteins in the sample populations.

Each microarray is probed with a single primary antibody directed against the protein of interest. Microarray slides used for immunostaining should be blocked prior to staining. Microarray slides used for Sypro Ruby protein staining do not require blocking.

1. Remove microarray slide(s) from freezer and leave at room temperature for approx 5–10 min.
2. Prepare a 1X solution of Reblot mild solution in deionized water. Wash microarray slides with gentle rocking in 1X Reblot mild solution for 15 min (*see* **Note 11**).
3. Discard Reblot solution. Wash slides in PBS without calcium or magnesium twice, for 5 min each.
4. Block slides in I-block™ solution for a minimum of 1 h at room temperature (*see* **Note 12**).

3.3.2. Microarray Immunostaining

Immunostaining requires the use of control slides for determining background staining resulting from secondary antibody alone. Each species-specific secondary antibody should be used to stain an individual microarray. Therefore, for any immunostaining run, one slide must be used as a negative (secondary antibody alone) control slide. The total number of slides to be stained is determined by the number and species of primary antibodies. For example, if three antirabbit primary antibodies and one mouse primary antibody are selected, a total of six microarray slides will be needed (four antibodies + two controls).

1. Select primary antibodies of choice.
2. Select species specific, biotinylated secondary antibodies.
3. Prepare CSA reagents following manufacturer's directions. Program Dako Autostainer as described in **Table 3** (*see* **Note 13**).
4. Prepare 1X TBST buffer per manufacturer's directions. Fill water reservoir with deionized water. Empty waste container if appropriate.
5. Load reagents and microarray slides on the Autostainer (*see* **Note 14**).
6. Prime water. Prime buffer. Start run.
7. Remove slides promptly at end of Autostainer run. Do not allow the Autostainer to add TBST to the slides after staining is complete (*see* **Note 15**). Allow slides to air-dry away from direct light. Covering the slides lightly with a paper towel is adequate to reduce exposure to direct light.
8. Label slides as to which antibody was used for immunostaining.
9. Scan slides and store the stained slides in the dark at room temparature.

3.3.3. Sypro Ruby Total Protein Staining of Microarray

The limited sample material from microdissected biopsy lysates precludes the use of spectrophotometeric analysis of total protein prior to microarray printing. A microarray slide stained for total protein serves as a tool for normalizing spot intensity between samples (*see* **Note 16**).

1. Remove microarray slide(s) from freezer, and leave at room temperature for approx 5–10 min (*see* **Note 17**).
2. Prepare fixative solution, final concentrations 10% methanol, 7% acetic acid. Mix well and store tightly sealed.
3. Wash slide in dH_2O twice for 5 min each with continuous agitation or rocking.
4. Fix slides by immersing in fixative solution (volume sufficient to cover slides) for 15 min. Place on orbital shaker or rocker.
5. Wash in dH_2O four times for 5 min each with continuous agitation or rocking.
6. Immerse slides in Ruby Red stain (volume sufficient to cover slides) for 30 min at room temperature, in the dark. Cover container with aluminum foil to prevent photobleaching of the stain. Place on orbital shaker or rocker.
7. Wash in dH_2O four times for 1 min each, in the dark, with continuous agitation or rocking.
8. Air-dry slides at room temperature in the dark.
9. Scan slides with a standard UV transilluminator or a laser scanner. Refer to http://probes.invitrogen.com/media/pis/x11791.pdf for a list of compatible scanning platforms.

3.4. Scanning and Data Analysis

The chromogenic detection system described allows the microarrays to be scanned on any high-resolution scanner (1200 dpi) with software capable of a

Table 3
Dako Autostainer Program Compatible With Reverse Phase Protein Microarrays

Reagent	Time (min)	Reagent category
Buffer 1X TBST		Rinse
Hydrogen peroxide	5	Endogenous enzyme block
Buffer 1X TBST		Rinse
Avidin block	10	Auxiliary
Buffer 1X TBST		Rinse
Biotin block	10	Auxiliary
Buffer 1X TBST		Rinse
Protein block		Protein block
Blow air		Rinse
Primary antibody	30	Primary antibody
Buffer 1X TBST		Rinse
Buffer 1X TBST	3	Auxiliary
Buffer 1X TBST		Rinse
Biotinylated secondary antibody	15	Secondary reagent
Buffer 1X TBST		Rinse
Buffer 1X TBST	3	Auxiliary
Buffer 1X TBST		Rinse
Streptavidin–biotin complex	15	Secondary reagent
Buffer 1X TBST		Rinse
Buffer 1X TBST	3	Auxiliary
Buffer 1X TBST		Rinse
Amplification reagent	15	Auxiliary
Buffer 1X TBST		Rinse
Buffer 1X TBST	3	Auxiliary
Buffer 1X TBST		Rinse
Streptavidin–HRP	15	Tertiary reagent
Buffer 1X TBST		Rinse
Buffer 1X TBST	3	Auxiliary
Buffer 1X TBST		Rinse
Switch to toxic waste		–
Diaminobenzidine	5	Chromogen
Water		Rinse
Overnight water[a]	840	Auxiliary

[a]Optional step.
This program is a modification of the manufacturer's CSA protocol and is compatible with protein lysate microarrays printed on nitrocellulose-coated slides.

14- or 16-bit scanning option for grayscale. Electronic images may be imported into a variety of spot analysis software programs, such as PSCAN (http://abs.cit.nih.gov), MicroVigene (Vigene Tech), or ImageQuant (Amersham

Biosciences). Each array is scanned, the spot intensity analyzed, data is normalized to total protein, and a standardized, single data value is generated for each microarray sample. The data may be used to generate histograms, dendograms based on hierarchical clustering algorithms or Bayesian clustering analysis for generation of cell signaling network profiles.

3.4.1. Scanner Settings

1. Place the microarray slides on the scanner.
2. Scan at 600- to 1200-dpi. Convert the image to grayscale. Save image as a 14- or 16-bit grayscale image.
3. Save the image as a tiff file, compatible with the spot analysis software.

4. Notes

1. Methods are under development for analysis of protein from formalin fixed tissues, although protein yield may be less than that from frozen tissue. Extraction of high yield, quality protein from formalin fixed tissue is generally difficult owing to the extensive cross-links formed in formalin fixed samples. Currently, the optimal tissue specimen for use with protein lysate microarrays is fresh tissue, immediately embedded in a cryoprotectant solution and frozen or snap frozen tissue.
2. Reblot solution is manufactured in two strengths: mild and strong. It is imperative that the mild solution be used with nitrocellulose-coated slides. The Reblot strong solution may cause the nitrocellulose to detach from the glass slide.
3. The primary antibody should be validated by Western blot prior to use on a microarray. A validated antibody should show a single band at the specified molecular weight. A list of validated antibodies is available on the NCI-FDA Clinical Proteomics Program website: http://home.ccr.cancer.gov/ncifdaproteomics/contact. asp. Primary antibodies should be unconjugated and may be any species, with the only caveat being the biotinylated secondary antibody should be species matched with the primary antibody. Primary and secondary antibody concentration should be optimized for use on a microarray prior to use with patient samples. Typically primary antibody dilutions of 1:250, 1:500, and 1:1000 are used to immunostain a set of microarrays. The secondary antibody concentration is held constant. Comparison of signal to noise ratio is used to determine the optimal concentration to use on the microarray. Secondary antibodies are optimized in a similar manner. A set of microarrays is immunostained with secondary antibody alone at a variety of concentrations and antibody diluent is used as the primary antibody. In general, the secondary antibody dilution that shows minimal staining is the optimal dilution. Verification of the optimal secondary antibody concentration may be performed by staining a microarray with an optimized primary antibody and the secondary antibody dilution of choice and accessing the signal to noise ratio.
4. Sypro Ruby blot stain is a ruthenium complex, exhibiting luminescence on excitation with either UV-B or blue light. The luminescence may be visualized with UV epi-illumination sources, UV or blue-light transilluminators, or laser scanning

instruments with excitation light at 450, 473, 488, or 532 nm. The emission peak of the ruthenium complex is approx 618 nm *(24)*.

5. The binding capacity of the nitrocellulose is 9 μg/mm^3 with a 0.2 μm pore membrane *(25)*. Samples with protein concentrations >2 mg/mL may saturate the nitrocellulose with concomitant loss of the dilution series in the printed diluted spots.

6. A common issue with microdissected tissue is minimal cell number on any given LCM cap. To effectively increase the cell number and protein concentration for a volume of lysis buffer, it may be necessary to solubilize two to three LCM caps in one volume of lysis buffer. This is done by sequentially solubilizing the cells on LCM caps in the same volume of lysis buffer. To illustrate the process, one LCM cap is solubilized in lysis buffer. This first cap is discarded and a fresh LCM cap containing microdissected material is placed on a tube containing the lysis buffer used previously to solubilize the first LCM cap. The cells on this second cap are solubilized in the lysis buffer, effectively increasing the protein concentration in a given volume of buffer.

7. After solubilizing cells from an LCM cap, the efficiency of extraction may be determined by examining the cap surface with the aid of a microscope. Place the cap, polymer side down, on a clean glass microscope slide. Observe the surface of the polymer with a standard light microscope. Cellular material should not be visible on the polymer surface. If cells are still present, place the cap on the tube containing lysis buffer and continue heating the cap at 70°C for an additional 15 min. Repeat the above process until the cells are solubilized.

8. Dot spacing on the microarray is a critical factor for ensuring successful spot intensity analysis. Careful planning of the array configuration prior to printing is essential. Spots placed too closely prevent discrimination of background areas from spot area with most spot analysis software programs. The GMS 417 arrayer is equipped with a pin and ring™ print head assembly with four pins in a 9-mm square format. A 96-well plate format allows more flexibility in printing with the GMS 417 arrayer because the wells are 9-mm apart. A 384-well plate formats are restricted to printing in rows of four as a result of the 4.5-mm well spacing. **Table 2** lists a typical 384-well plate configuration that is compatible with the GMS 417 arrayer. This format results in 12 horizontal spots, in two dilution curves for two different samples. Alternatively, as shown in **Fig. 1**, one sample could be diluted in a longer dilution series and all the dilutions could be printed horizontally across the slide for each sample.

9. The pin and ring™ format permits multiple aliquots of a sample to be printed in the same spot, effectively increasing the protein concentration/spot. Printing 5–10 hits/spot maximizes the protein binding capacity of nitrocellulose without saturating the nitrocellulose. The recommendation to print 10 hits/spot from a 96-well plate and five hits from a 384-well plate is based on dot spacing restrictions in the *x*- and *y*-axis from the different microtiter plate formats.

10. Typical microarrays with human samples are printed in triplicate for quality control and statistical analysis. Dot spacing is a function of pin diameter—dot spacing must be a minimum of 1.5X the pin diameter. FAST™ slides are manufactured in a variety of formats, with a 20 × 50-mm nitrocellulose pad being compatible with reverse phase microarray format and the DakoCytomation Autostainer.

11. Wash time is critical with the Reblot mild solution. Extended wash times may cause the nitrocellulose to detach from the glass.
12. Microarrays may be blocked overnight at 4°C. The minimum blocking time is 1 h.
13. The Autostainer grid layout in **Table 3** is a modification of the CSA grid provided by the manufacturer. Additional TBST rinse steps have been added to ensure adequate rinsing of the microarray slides. Four hundred fifty microliters µL of reagent is added to the slide in each of the three drop zones.
14. Do not let the microarray slides dry while loading and prior to starting the Autostainer run. If necessary, pour I-block or 1X TBST on the slides. If the ambient humidity is low, a shallow tray of water may be placed inside the Autostainer during the staining run. Operate the Autostainer with the lid closed.
15. The Dako Autostainer is designed primarily for immunohistochemistry. In these procedures, the slides are kept moist with periodic rinses of TBST. The high salt content causes crystal formation on the nitrocellulose if the TBST is allowed to dry on the microarray slide. If crystal formation inadvertently happens, it may be possible to dissolve the crystals by washing the microarrays with PBS + 0.2% Tween-20 for 1 h and then washing with PBS only. A final rinse in water is usually sufficient to remove the crystals. Additionally, the Autostainer may be programmed to include a final water step. Adding an "auxiliary" reagent step to the end of the program allows the Autostainer to run in an "overnight" mode. This auxiliary reagent is deionized water and the time is 840 min (14 h). The microarray slides will be rinsed with water following the DAB detection step. The Autostainer will remain in idle, waiting to dispense water in 840 min, allowing the operator to return in 14 h.
16. Syrpo Ruby staining is a permanent protein stain, detected by fluorescence with an excitation wavelength of 300 or 480 nm and an emission wavelength of 618 nm. The stain is photostable, allowing long emission lifetime and the ability to measure fluorescence over a longer time frame, minimizing background fluorescence.
17. Do not block (with I-block) the microarray slide used for total protein staining. Because of slight changes in the protein concentration per spot from the first slide printed to the last slide printed, an average protein concentration per spot for all slides printed within a run may be determined from one slide. Typically a slide representing the median slide of the printing run is selected for total protein staining. For example, if 25 slides were printed in a run, slide 12 would be selected for total protein staining.

Acknowledgments

The authors wish to thank Dan Harness (Affymetrix) for constructive technical advice.

References

1. Liotta, L. A., Espina, V., Mehta, A. I., et al. (2003) Protein microarrays: meeting analytical challenges for clinical applications. *Cancer Cell* **3,** 317–325.

2. Haab, B. B., Dunham M. J., and Brown P. O. (2001) Protein microarrays for highly parallel detection and quantitation of specific proteins and antibodies in complex solutions. *Genome Biol.* **2,** RESEARCH0004. Epub 2001 Jan 22.
3. Macbeath, G. and Schreiber, S. L. (2000) Printing proteins as microarrays for high-throughput function determination. *Science* **289,** 1760–1763.
4. Macbeath, G. (2002) Protein microarrays and proteomics. *Nat Genet.* **32 Suppl,** 526–532.
5. Paweletz, C. P., Charboneau, L., Bichsel, V. E., et al. (2001) Reverse phase protein microarrays which capture disease progression show activation of pro-survival pathways at the cancer invasion front. *Oncogene* **20,** 1981–1989.
6. Zhu, H. and Snyder, M. (2003) Protein chip technology. *Curr. Opin. Chem. Biol.* **7,** 55–63.
7. Wilson, D. S. and Nock, S. (2003) Recent developments in protein microarray technology. *Angew Chem. Int. Ed. Engl.* **42,** 494–500.
8. Templin, M. F., Stoll, D., Schrenk, M., Traub, P. C., Vohringer, C. F., and Joos, T. O. (2002) Protein microarray technology. *Trends Biotechnol.* **20,** 160–166.
9. Schaeferling, M., Schiller, S., Paul, H., et al. (2002) Application of self-assembly techniques in the design of biocompatible protein microarray surfaces. *Electrophoresis* **23,** 3097–3105.
10. Weng, S., Gu, K., Hammond, P. W., et al. (2002) Generating addressable protein microarrays with PROfusion covalent mRNA-protein fusion technology. *Proteomics* **2,** 48–57.
11. Petach, H. and Gold, L. (2002) Dimensionality is the issue: use of photoaptamers in protein microarrays. *Cur. Opin. Biotechnol.* **13,** 309–314.
12. Lal, S. P., Christopherson, R. I., and Dos Remedios, C. G. (2002) Antibody arrays: an embryonic but rapidly growing technology. *Drug Discov. Today* **7(18 Suppl),** S143–S149.
13. Humphery-Smith, I., Wischerhoff, E., and Hashimoto, R. (2002) Protein arrays for assessment of target selectivity. *Drug Discov. World* **4,** 17–27.
14. Bobrow, M. N., Harris, T. D., Shaughnessy, K. J., and Litt, G. J. (1989) Catalyzed reporter deposition, a novel method of signal amplification. Application to immunoassays. *J. Immunol Methods* **125(1–2),** 279–285.
15. Bobrow, M. N., Shaughnessy, K. J., and Litt, G. J. (1991) Catalyzed reporter deposition, a novel method of signal amplification. II. Application to membrane immunoassays. *J. Immunol. Methods* **137,** 103–112.
16. Hunyady, B., Krempels, K., Harta, G., and Mezey, E. (1996) Immunohistochemical signal amplification by catalyzed reporter deposition and its application in double immunostaining. *J. Histochem. Cytochem.* **44,** 1353–1362.
17. King, G., Payne, S., Walker, F., and Murray, G. I. (1997) A highly sensitive detection method for immunohistochemistry using biotinylated tyramine. *J. Pathol.* **183,** 237–241.
18. Petricoin, E., Wulfkuhle, J., Espina, V., and Liotta, L. A. (2004) Clinical proteomics: revolutionizing disease detection and patient tailoring therapy. *J. Proteome Res.* **3,** 209–217.

19. Grubb, R. L., Calvert, V. S., Wulkuhle, J. D., et al. (2003) Signal pathway profiling of prostate cancer using reverse phase protein arrays. *Proteomics* **3,** 2142–2146.

20. Wulfkuhle, J. D., Aquino, J. A., Calvert, V. S., et al. (2003) Signal pathway profiling of ovarian cancer from human tissue specimens using reverse-phase protein microarrays. *Proteomics* **3,** 2085–2090.

21. Liotta, L. A., Kohn, E. C., and Petricoin, E. F. (2001) Clinical proteomics: personalized molecular medicine. *JAMA* **286,** 2211–2214.

22. Petricoin, E. F., Zoon, K. C., Kohn, E. C., Barrett, J. C., and Liotta, L. A. (2002) Clinical proteomics: translating benchside promise into bedside reality. *Nat. Rev. Drug Discov.* **1,** 683–695.

23. Petricoin, E. F. and Liotta, L. A. (2004) Clinical proteomics: application at the bedside. *Contrib. Nephrol.* **141,** 93–103.

24. Berggren, K., Steinberg, T. H., Lauber, W. M., et al. (1999) A luminescent ruthenium complex for ultrasensitive detection of proteins immobilized on membrane supports. *Anal. Biochem.* **276,** 129–143.

25. Tonkinson, J. L. and Stillman, B. A. (2002) Nitrocellulose: a tried and true polymer finds utility as a postgenomic substrate. *Front. Biosci.* **7,** C1–C12.

22

Protein Crystallization

Champion Deivanayagam, William J. Cook, and Mark R. Walter

Summary

X-ray crystallography is a powerful method for obtaining the three-dimensional structures of biological macromolecules and macromolecular complexes. Improvements in protein production, crystallization, data collection, as well as structure solution and refinement methods have brought the field to the verge of rapid high-throughput genomic scale structure determination. The major bottle neck to this process remains protein production and crystallization. This chapter describes essential information on standard protein production and crystallization methods and ongoing efforts to perform this work using high-throughput robotics.

Key Words: Robotics; protein crystallization; X-ray diffraction; protein structure.

1. Introduction

Three-dimensional protein structures play a critical role in structure-aided drug design, protein engineering, and numerous other studies that further the understanding of the molecular basis of life. Sequencing of the human genome *(1,2)*, the genomes of several model organisms (e.g., yeast *[3]*, fly *[4]*, rat *[5]*, mouse *[6]*), and numerous pathogens *(7,8)*, has resulted in an avalanche of protein sequences. Comparisons of the amino sequences with known protein architectures suggests a very limited understanding of protein folds that control the biology of a simple bacterium, let alone a complex eukaryotic cell *(9–13)*. To overcome this deficiency requires an exponential increase in the speed of protein structure determination.

X-ray crystallography is the most powerful method for determining the three-dimensional structures of complicated biological molecules. However, the technique is fairly slow and often requires large amounts (2–20 mg) of protein. On average, the time required to determine the crystal structure of a soluble globular protein starting from the gene requires about 1–2 yr. A number of structural

From: *Methods in Molecular Biology, vol. 383: Cancer Genomics and Proteomics: Methods and Protocols*
Edited by: P. B. Fisher © Humana Press Inc., Totowa, NJ

genomics efforts have been initiated to try to reduce this time window and some progress has been made. However, the rate limiting steps in structure determination remain the production of the macromolecule of interest and the growth of crystals of sufficient size and quality to permit X-ray analysis. In this chapter, the basic theory and methods required to crystallize a protein of interest using cheap and easily obtainable materials are described. Recent developments in robotics and miniaturization of the experiment are also discussed.

1.1. Protein Crystallization 101

Crystallization is the process of finding a set of conditions, where the crystalline form of a protein is more favored than the soluble form. The experiment is easy to perform, although interpretation of the results and deciding on the next steps are more difficult. The most common crystallization experiment is the vapor diffusion equilibration technique using the hanging drop arrangement. In this experiment, a drop consisting of protein and precipitating agent are placed in a sealed chamber with a liquid reservoir of precipitating agent. Normally, the drop contains a lower precipitating agent concentration than the reservoir. The concentration gradient causes the aqueous solution in the protein drop to be transferred through the vapor phase to the more concentrated well solution until the two are in equilibrium. Crystal growth occurs as the protein is concentrated in the drop.

Crystallization consists of a nucleation and a growth phase that must be carefully controlled. The ideal experiment provides a situation, where 5–10 nucleation events occur in a crystallization experiment that subsequently grow into crystals of suitable size for X-ray diffraction analysis. If nucleation dominates the experiment, one observes hundreds to thousands of small crystals or aggregates that are too small to be analyzed. However, nucleation, the initial protein aggregate of protein molecules that lead to crystals, must occur or the protein drops will remain clear. Changing a number of variables in the crystallization experiment can control nucleation and growth. In some instances, crystal-seeding methods also provide ways to separate the crystallization experiment into separate nucleation and growth experiments.

Because of the large number of potential variables in crystallization (e.g., buffer, pH, precipitating agent, temperature, protein concentration, ions or detergents, substrates, or inhibitors) an exhaustive search of all possible crystallization conditions is not possible. However, analysis of previously successful crystallization conditions, first tabulated in the biological macromolecule crystallization database *(14)*, has resulted in a series of commercially available sparse matrix crystallization screens (*see* **Note 6**). Each screen contains between approx 50 and 100 different crystallization conditions, which were predominantly selected because they have previously been successful in growing X-ray diffraction quality protein crystals. The sparse-matrix strategy has been remarkably successful for obtaining initial crystallization conditions for thousands of proteins *(15,16)*. Once initial crystallization conditions are obtained, further optimization of the variables is sometimes necessary to increase the size of the crystals to approx 50–100 μms on edge, which is generally required for X-ray diffraction experiments.

Although obtaining protein crystals is a great achievement, success can only be claimed once the crystals are shown to diffract X-rays to at least a moderate resolution of about 3 Å. Ultimately, the resolution where significant new insights into the biology of a system can be gained determines whether the crystals grown are useful for structure determination. In the case of membrane proteins and large complex assemblies, 4 Å resolution data may be more than acceptable, while deciphering enzyme mechanisms may require crystals that diffract to 1.5–2 Å resolution.

2. Materials

1. Purified protein of interest (*see* **Notes 1–4**).
2. Protein concentrators such as Amicon Centripreps, Centricons, or stirred cells (*see* **Note 5**) (Amicon, Beverly, MA).
3. Crystal screening reagents (*see* **Note 6**).
4. Three constant temperature incubators are set at 4, 12, and 25°C, respectively.
5. 24-Well Linbro plates (Linbro MP Biomedicals, Irvine, CA; cat. no. 76-0033-05).
6. 22-mm Round cover slips.
7. Aqua sil or Rain-X siliconizing agents.
8. Vacuum grease.
9. Stereo microscope.

3. Methods

3.1. Hanging Drop Technique

The hanging drop method is quite straightforward. In a typical experiment, 2–10 μL protein drops are suspended over a 1 mL reservoir solution. The reservoir solution is made up of a precipitating agent that is buffered at a particular pH. The protein drop contains 1–5 μL of protein solution and usually an equal volume of the reservoir solution (*see* **Note 11**). The two solutions are usually dispensed with a 10-μL Hamilton syringe, mixed on a siliconized cover slip, and then inverted over the well. An airtight chamber is formed by placing a small bead of silicone grease around the circumference of the cover slip before to inversion over the reservoir.

1. Concentrate the protein solution to 5–10 mg/mL.
2. Prepare and dispense 1 mL of the reservoir solutions into the wells of a Linbro plate (*see* **Note 7**).
3. Apply a thin layer of grease to the rim of each well in the Linbro plate (*see* **Notes 8** and **9**).

4. Using a Hamilton syringe, dispense 1–5 µL of protein solution onto a siliconized cover slip (*see* **Notes 10** and **11**).
5. Rinse the syringe chamber one or two times using distilled water.
6. Dispense 1–5 µL of reservoir solution onto the cover slip and gently mix with the tip of the syringe.
7. Invert the cover slip and place it over the well containing precipitating agent. Be sure that the seal is complete so that the chamber is airtight (*see* **Note 12**).
8. Place the tray in a constant temperature incubator (*see* **Note 13**).
9. Evaluate the crystallization trays at least every other day with a stereoscopic microscope (*see* **Note 14**).
10. If seeding is required, the following technique is suggested (*see* **Note 15**). To obtain seeds, crystalline aggregates or small crystals may be crushed with a needle and stirred.
11. Streak a fresh drop with the needle. Small perfect crystals will often grow along the needle trail through the drop.
12. Wash these single crystals in a stabilizing solution with a slighter lower concentration of precipitant than is present in the well.
13. Transfer the seed with as little liquid as possible to 2–4 µL drops containing the complex and the precipitant at a slightly lower concentration than was used to obtain the original seed crystals.
14. Equilibrate these drops against 1 mL of the precipitant.

4. Notes

1. Protein structure determination generally requires approx 2–20 mg of protein. As a result, robust protein expression systems are required. To date, the system of choice is the Gram-negative bacterium, *Escherichia coli*. As described by Mancia et al. (*17*), of the 14,011 protein structures deposited in the Protein Data Bank that contain expression system entry records, more than 90% were expressed using *E. coli*. For the remaining entries, 3.5% were produced in yeast, 2.5% in insect cells, 1.5% in mammalian cells, and a very small fraction were expressed in natural sources or in cell-free systems. The success and popularity of *E. coli* for producing proteins for structural studies is owing to many factors including a clear understanding of the transcription and translation machinery, numerous expression vectors, ease of use, rapid growth characteristics, and low cost. In addition, defined media is available that permits the large scale expression of proteins labeled with seleno-methionine, rather than methionine, which greatly facilitates their structure determination (*18*).

   Despite these advantages, prokaryotic expression systems do not always produce properly folded and soluble proteins. Overexpressed protein is often deposited into insoluble inclusion bodies from which the protein must be refolded. This is a very common result because approx 50% of the proteins from the *Thermotoga maritima* genome are insoluble when overexpressed in *E. coli* (*19*). Similar results have been observed by other groups in other systems (*20*). Although protein refolding is sometimes possible, it is slow and often results in poor yields and mixtures of folding variants. Thus, it is not a useful method for high-throughput protein expression required for structural genomics studies.

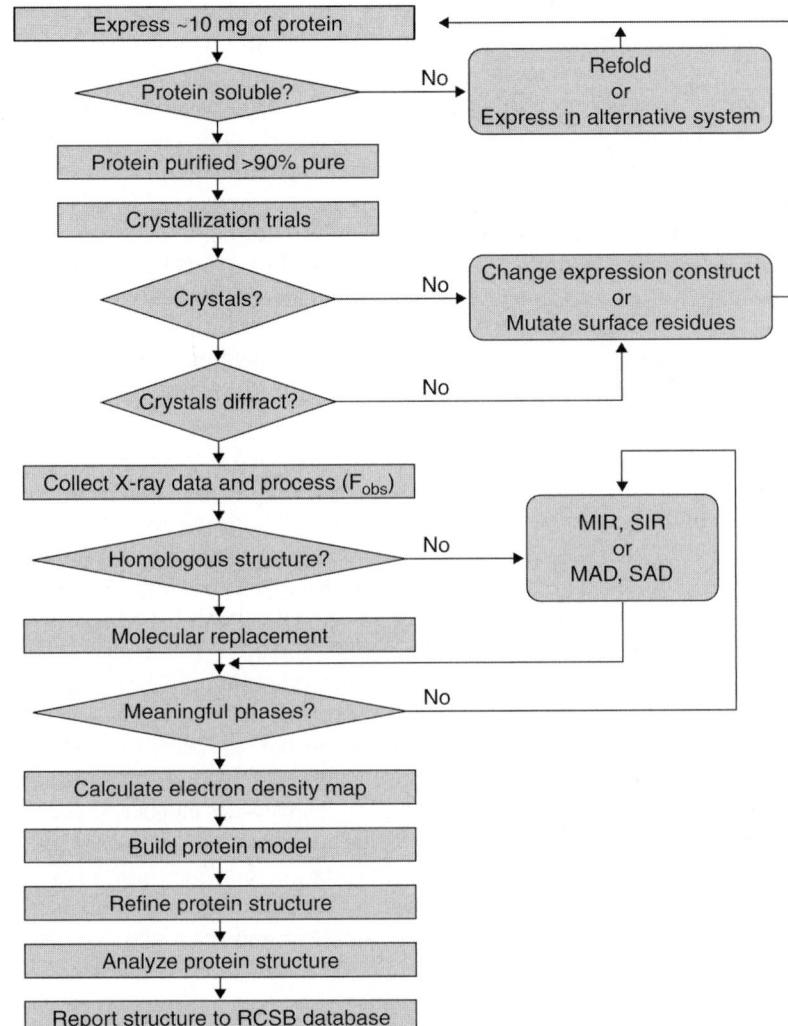

Although yeast, insect cell, and mammalian expression systems are more labor intensive, they have many benefits including a complement of expected chaperones to promote protein folding, and the ability to carry out N- and *O*-linked glycosylation, sulfonation, and other posttranslational modifications *(17,21–23)*. A current drawback of eukaryotic cell expression systems is the inability to homogeneously label proteins with seleno-methionine (Se-Met) to aide structure determination.

2. *Protein purification:* many problems associated with the purification of recombinant proteins have been overcome by fusing the gene of interest to an affinity tag *(24)*. Common affinity tags include short peptides such as the poly histidine tag (HHH-HHH) and Strep tag (WSHPQFEK) as well as several proteins including glutathione s-transferase (GST), maltose binding protein, and thioredoxin. Using these tags,

protein purification is normally reduced, three or four purification steps. As a result, in most instances, protein purification is not an obstacle to protein crystallization. The impact of an affinity tag on protein crystallization cannot be predicted but must be determined by crystallization studies. In some instances, crystallization has been dependent on the tag, while in other instances it has hindered crystallization *(19,25,26)*. If the affinity-tag is a problem, specific protease sites can be inserted between the affinity tag and the protein allowing it to be removed *(27)*. In addition to aiding purification, the maltose-binding protein affinity tag has been shown to dramatically increase the solubility of a large number of its fusion partners *(28,29)*.

3. Robotic systems for automated, parallel, and miniaturized protein expression and purification have been developed. In many cases, the small scale expressions (normally 96-well plate formats) are used to test for protein solubility, followed by "traditional" 1–2 L expressions of proteins that give positive solubility and expression tests in the small-scale experiments *(30–32)*. This strategy has been problematic; because the small-scale results have not accurately predicted large-scale expression behavior. However, recent work suggests this problem can be overcome by optimizing the aeration of the small-scale cultures *(33)*. Another strategy is to perform intermediate scale (~100 mL) expressions that provide enough protein for testing and crystallization studies *(34)*. This has been accomplished with a specially designed fermentor system, which simultaneously grows 96 bacterial cultures (~65 mL each) to very high cell densities (optical density values at 600 nm of 40), which has resulted in impressive protein production levels (~6.5 mg/65 mL culture). The proteins are subsequently purified in an automated and parallel manner by a second specially designed robot. Although this particular system is not commercially available, other robotics systems have been adapted for small-scale protein expression *(33,34)*. The benefits of the automated methods are they greatly increase the number of proteins, which can be studied at one time. However, they can also be used in more targeted studies such as generating a series of mutants for structure-function studies *(35)*.

4. Despite the many advances in crystallization technology, many recombinant proteins do not crystallize. This fact emphasizes the most important variable in protein crystallization experiments is the protein itself! A large body of literature, and results from the lab, show the most important characteristics of protein samples, which readily crystallize are:
   a. High purity (>95%).
   b. Soluble at high concentrations (~5–10 mg/mL).
   c. Conformational and chemical homogeneity.
   d. Monodisperse molecular aggregate.

Chemical heterogeneity, beyond what can be observed on a sodium dodecyl sulfate-polyacrylamide gel electrophoresis, must be addressed if crystallization fails. This is most easily monitored by MALDI-TOF mass spectrometry *(36)*. Common examples of this problem have been observed during the overexpression of protein kinases and proteases, which partially phosphorylate or cleave themselves, respectively. In addition, extracellular receptor proteins invariably are heterogeneously glycosylated leading to unsatisfactory protein for crystallization

(37,38). Molecular strategies for removing these sources of heterogeneity, such as site-directed mutagenesis, have been identified *(39)*.

Conformational heterogeneity is often exhibited by large multidomain proteins. To avoid this problem, crystallographers have used the "divide and conquer" approach by cloning and expressing the individual domains of these large molecules for structural studies *(35,40–42)*. This is not to say that multidomain proteins cannot be crystallized. For example, multidomain antibodies (Ab) were thought to be impossible to crystallize. This problem was initially overcome by protease digestions of the Abs into Fab and Fc fragments, which has led to numerous highly successful structure determinations. Through additional effort, using a number of different Abs, crystallization, and structure determination of intact Abs has now been successful *(43,44)*.

Protein conformational heterogeneity is also caused by large surface loops, which adopt multiple conformational states. Often, these loops form part of an active site and can be "locked" into a single conformation by the addition of small molecule substrates or inhibitors. In other cases, heroic molecular biology efforts have been undertaken to remove surface loops and allow crystallization *(45)*. In other studies, structural information has been obtained by performing crystallization experiments on a series of orthologs of the protein of interest *(46)*. The general philosophy of these methods is if it does not crystallize; change the protein *(35)*. Another method of changing the protein is to form complexes between the protein of interest and a monoclonal Fab *(47–50)*. Complex formation effectively changes the molecular species to be crystallized and has often been found to be a successful strategy *(51)*.

Monodispersity is equivalent to conformational heterogeneity at the quaternary structure level. Thus, a protein that exists as a mixture of monomers, dimers, and tetramers in approximately equal proportions is unlikely to crystallize. It should be emphasized that the goal is not to find conditions, where the protein is monomeric, but conditions where the molecule exists in a single quaternary state which may be monomeric or a large multisubunit complex. In many cases, the aggregation state of protein may be observed by size exclusion chromatography. However, dynamic light scattering is the preferred method because of its low protein requirements, higher sensitivity, better quantification, and the sample is not diluted during the experiment *(52)*. All of the previous described biochemical strategies to overcome conformational heterogeneity may be used to achieve a mono disperse protein sample.

5. Although crystals have been grown from solutions containing 1–20 mg/mL, the most common concentration range is 5 to 10 mg/mL. A variety of concentration devices are available depending on the volume of the sample. For smaller volumes, centrifugal filtration units from Amicon (Centricons, Centripreps) can concentrate volumes as large as 15 mL. A variety of molecular weight cutoffs are available. Larger volumes can be concentrated with stirred cells (Amicon) that use nitrogen gas to force solute through a membrane effectively concentrating the sample. The major trouble associated with these concentration devices is loss of protein on the porous membranes. This can be checked by cutting the membranes out of the concentrators followed by boiling in sodium dodecyl sulfate-polyacrylamide gel electrophoresis sample buffer and running the samples on a gel.

6. Preliminary screening kits: Hampton Research (Laguna Nigel, CA, www.hampton research.com), Emerald Biosystems (Bainbridge, WA, www.emeraldbiostructures. com), Molecular Dimensions (Apopka, FL, www.moleculardimensions.com), Jena Bioscience (Jena, Germany, www.jenabioscience.com) all sell a variety of screening kits for rapidly setting up protein crystallization experiments. Many of these screens are sparse-matrix screens described in **Subheading 1.1.**

7. Some type of sparse matrix screen, such as those suggested in **Note 6**, is always performed first. Although these broad screens are not always successful in yielding usable crystals, they usually provide a starting place for further screening. For optimization after initial conditions are identified, the primary reagents required are ammonium sulfate and polyethylene glycol of various molecular weights. The crystallization medium is almost always buffered; typical buffers include acetate, citrate, MES, HEPES, and Tris. Occasionally other ions may also be added; the most common additives are various salts of chloride and sulfate.

8. One of the most attractive features of a hanging drop experiment in Linbro plates is the opportunity to do a two-dimensional grid search. Typically, the two parameters that are varied are pH and the primary precipitating agent. For example, one might screen six pH values at intervals of 0.2 and four concentrations of ammonium sulfate at intervals of 0.2 M.

9. Although almost any buffer may be used, found that phosphate buffer is usually a poor choice, because many phosphate salts are insoluble and may crystallize, giving the crystallographer a transient feeling of success. Furthermore, many heavy metals used in preparing derivatives will precipitate in the presence of phosphate.

10. Although micropipet tips may be used to dispense the solutions, Hamilton syringe is easier to use and deliver accurate amount of desired volumes are found. Furthermore, the fine tip of the needle allows gentle mixing of the drop.

11. The traditional protein crystallization experiment described earlier has been automated and miniaturized *(53,54)*. Several robotic crystallization systems are now commercially available. To advance protein crystallization research, the Hauptman-Woodward Institute has set up a robotic crystallization service that is free to the structural biology community *(55)* (http://www.hwi.buffalo.edu/ProductsServices/ highthroughput/HighThroug.htm). Robotic handling of the crystallization experiment eliminates possible human error and increases the number of experiments that can be performed. At the same time, the protein drop size has been reduced, by at least a factor of 20, from 2 µL to 100 nL. This simple change greatly reduces the amount of protein required for a given number of experiments. For example, 500 experiments performed with 2 µL or 100 nL drops requires 5 mg or 250 µg of protein, respectively. This has allowed additional sampling of crystallization space without increasing the amount of protein required. Because diffusion rates are faster in the smaller drops, crystals are generally observed in shorter times, which may be beneficial for unstable protein preparations. It should be emphasized that scaling up the small-scale experiments to obtain crystals large enough for data collection is not always straightforward.

12. In a properly sealed vapor diffusion experiment, the protein drops will remain hydrated for 1–2 yr. However, crystals of suitable size for X-ray diffraction should

be removed from the drops as soon as possible. Old drops often acquire a thick film, which can attach itself to the crystals making them hard to recover.

13. In a typical laboratory the temperature in the room may vary by ±2°C. Whereas, this may not generally be a problem, some proteins cannot tolerate this temperature variation. This can be manifested by crystals appearing and then dissolving when viewed at a later time. In general, all experiments should be performed at 4°C, 12°C, and room temperature, because the processes of nucleation and crystal growth may differ radically at different temperatures.

14. Even if no crystals are obtained, the presence and amount of precipitated protein can generally be used to guide further experiments. Experiments should not be discarded as long as the drops are still liquid, because crystals have been observed only after several months in some cases. Automated imaging of the crystallization experiments at defined time windows is now possible. Image analysis software that can accurately score the results of each experiment still needs to be developed. Nonetheless, the storage of millions of high quality images of successful crystallization experiments, as well as the negative ones, provides a wealth of data for understanding the basic principles of crystallization and possibly new characteristics of proteins themselves. Preliminary studies using crystal scoring algorithms and neural net technology have been used to predict improved crystallization conditions from initial crystallization experiments *(56)*.

15. Unfortunately, sometimes crystals only grow as aggregated plates and needles. In order to grow large single crystals suitable for X-ray diffraction, a macro-seeding technique may be used. One such technique used to grow large perfect crystals has been described in detail in the crystallizations of an ubiquitin conjugating enzyme (2) and interleukin (IL)-10 (3). For example, ubiquitin conjugating enzyme crystals were originally grown from solutions containing 35% saturated ammonium sulfate in 0.05 *M* MES buffer (pH 6.7). After 2–3 d at room temperature, aggregates of orthorhombic crystals with dimensions up to $0.5 \times 0.2 \times 0.1$ mm^3 were obtained. To obtain seeds, crystalline aggregates were crushed with a needle and stirred; a fresh drop was then streaked with the needle. A number of small perfect crystals generally grew along the needle trail through the drop. These single crystals ($0.05 \times 0.02 \times 0.03$ mm^3) were washed in a stabilizing solution of 30% saturated ammonium sulfate in 0.05 *M* MES buffer, pH 5.5, and then transferred with as little liquid as possible to 2 µL drops containing 18 mg of protein/mL and 25% saturated ammonium sulfate in 0.05 *M* MES, pH 5.5. Introduction of the seed crystals into the drops was accomplished by using a 0.5 mm capillary and a micromanipulator. These drops were equilibrated against 1 mL of 25% saturated ammonium sulfate in 0.05 *M* MES, pH 5.5. After 2–4 d, large plate-like crystals with dimensions up to $0.6 \times 0.4 \times 0.2$ mm^3 were obtained. With careful seeding technique, only one crystal grew in each drop. This technique has been used a number of times in our laboratory with various protein crystals.

References

1. Venter, J. C., Adams, M. D., Myers, E. W., et al. (2001) The sequence of the human genome. *Science* **291,** 1304–1351.

2. Lander, E. S., Linton, L. M., Birren, B. C., et al. (2001) International Human Genome Sequencing Consortium: initial sequencing and analysis of the human genome. *Nature* **409,** 860–921.
3. Goffeau, A., Barrell, B. G., Bussey, H., et al. (1996) Life with 6000 genes. *Science* **274,** 546, 563–567.
4. Adams, M. D., Celniker, S. E., Holt, R. A., et al. (2000) The genome sequence of *Drosophila melanogaster*. *Science* **287,** 2185–2195.
5. Gibbs, R. A., Weinstock, G. M., Metzker, M. L., et al. (2004) Rat Genome Sequencing Project Consortium: genome sequence of the Brown Norway rat yields insights into mammalian evolution. *Nature* **428,** 493–521.
6. Waterston, R. H., Lindblad-Toh, K., Birney, E., et al. (2002) Mouse Genome Sequencing Consortium: initial sequencing and comparative analysis of the mouse genome. *Nature* **420,** 520–562.
7. Holt, R. A., Subramanian, G. M., Halpern, A., et al. (2002) The genome sequence of the malaria mosquito *Anopheles gambiae*. *Science* **298,** 129–149.
8. Cole, S. T., Brosch, R., Parkhill, J., et al. (1998) Deciphering the biology of *Mycobacterium tuberculosis* from the complete genome sequence. *Nature* **393,** 537–544.
9. Liu, J. and Rost, B. (2002) Target space for structural genomics revisited. *Bioinformatics* **18,** 922–933.
10. Liu, J. and Rost, B. (2004) CHOP proteins into structural domain-like fragments. *Proteins* **55,** 678–688.
11. Liu, J., Hegyi, H., Acton, T. B., Montelione, G. T., and Rost, B. (2004) Automatic target selection for structural genomics on eukaryotes. *Proteins* **56,** 188–200.
12. Portugaly, E., Kifer, I., and Linial, M. (2002) Selecting targets for structural determination by navigating in a graph of protein families. *Bioinformatics* **18,** 899–907.
13. Portugaly, E. and Linial, M. (2000) Estimating the probability for a protein to have a new fold: a statistical computational model. *Proc. Natl. Acad. Sci. USA* **97,** 5161–5166.
14. Gilliland, G. L., Tung, M., and Ladner, J. (1996) The Biological Macromolecule Crystallization Database and NASA Protein Crystal Growth Archive. *J. Res. Natl. Inst. Stand Technol.* **101,** 309–320.
15. Cudney, R. (1994) Screening and optimization strategies for macromolecular crystal growth. *Acta Crystallogr. D. Biol. Crystallogr.* **50,** 414–423.
16. Jancarik, J. A. K. S. (1991) Sparse Matrix Sampling: a screening method for crystallization of proteins. *J. Appl. Cryst.* **24,** 409.
17. Mancia, F., Patel, S. D., Rajala, M. W., et al. (2004) Optimization of protein production in mammalian cells with a coexpressed fluorescent marker. *Structure* (Camb) **12,** 1355–1360.
18. Hendrickson, W. A., Horton, J. R., and LeMaster, D. M. (1990) Selenomethionyl proteins produced for analysis by multiwavelength anomalous diffraction (MAD): a vehicle for direct determination of three-dimensional structure. *EMBO J.* **9,** 1665–1672.

19. Lesley, S. A., Kuhn, P., Godzik, A., et al. (2002) Structural genomics of the *Thermotoga maritima* proteome implemented in a high-throughput structure determination pipeline. *Proc. Natl. Acad. Sci. USA* **99,** 11,664–11,669.
20. Yee, A., Chang, X., Pineda-Lucena, A., et al. (2002) An NMR approach to structural proteomics. *Proc. Natl. Acad. Sci. USA* **99,** 1825–1830.
21. Boettner, M., Prinz, B., Holz, C., Stahl, U., and Lang, C. (2002) High-throughput screening for expression of heterologous proteins in the yeast *Pichia pastoris.* *J. Biotechnol.* **99,** 51–62.
22. Bailey, C. G., Tait, A. S., and Sunstrom, N. A. (2002) High-throughput clonal selection of recombinant CHO cells using a dominant selectable and amplifiable metallothionein-GFP fusion protein. *Biotechnol. Bioeng.* **80,** 670–676.
23. Berger, I., Fitzgerald, D. J., and Richmond, T. J. (2004) Baculovirus expression system for heterologous multiprotein complexes. *Nat. Biotechnol.* **22,** 1583–1587.
24. Terpe, K. (2003) Overview of tag protein fusions: from molecular and biochemical fundamentals to commercial systems. *Appl. Microbiol. Biotechnol.* **60,** 523–533.
25. Song, J. J., Liu, J., Tolia, N. H., et al. (2003) The crystal structure of the Argonaute2 PAZ domain reveals an RNA binding motif in RNAi effector complexes. *Nat. Struct. Biol.* **10,** 1026–1032.
26. Smyth, D. R., Mrozkiewicz, M. K., McGrath, W. J., Listwan, P., and Kobe, B. (2003) Crystal structures of fusion proteins with large-affinity tags. *Protein Sci.* **12,** 1313–1322.
27. Jenny, R. J., Mann, K. G., and Lundblad, R. L. (2003) A critical review of the methods for cleavage of fusion proteins with thrombin and factor Xa. *Protein Expr. Purif.* **31,** 1–11.
28. Kapust, R. B. and Waugh, D. S. (1999) *Escherichia coli* maltose-binding protein is uncommonly effective at promoting the solubility of polypeptides to which it is fused. *Protein Sci.* **8,** 1668–1674.
29. Routzahn, K. M. and Waugh, D. S. (2002) Differential effects of supplementary affinity tags on the solubility of MBP fusion proteins. *J. Struct. Funct. Genomics* **2,** 83–92.
30. Knaust, R. K. and Nordlund, P. (2001) Screening for soluble expression of recombinant proteins in a 96-well format. *Anal. Biochem.* **297,** 79–85.
31. Braun, P., Hu, Y., Shen, B., et al. (2002) Proteome-scale purification of human proteins from bacteria. *Proc. Natl. Acad. Sci. USA* **99,** 2654–2659.
32. Chance, M. R., Bresnick, A. R., Burley, S. K., et al. (2002) Structural genomics: a pipeline for providing structures for the biologist. *Protein Sci.* **11,** 723–738.
33. Page, R., Moy, K., Sims, E. C., et al. (2004) Scalable high-throughput microexpression device for recombinant proteins. *Biotechniques* **37,** 364, 366, 368 passim.
34. Lesley, S. A. (2001) High-throughput proteomics: protein expression and purification in the postgenomic world. *Protein Expr. Purif.* **22,** 153–160.
35. Choi, K. H., Groarke, J. M., Young, D. C., et al. (2004) Design, expression, and purification of a Flaviviridae polymerase using a high-throughput approach to facilitate crystal structure determination. *Protein Sci.* **13,** 2685–2692.

36. Cohen, S. L. and Chait, B. T. (2001) Mass spectrometry as a tool for protein crystallography. *Annu. Rev. Biophys. Biomol. Struct.* **30,** 67–85.
37. Stura, E. A., Nermerow, G. R., and Wilson, I. A. (1992) Strategies in the crystallization of glycoproteins and protein complexes. *J. Cryst. Growth.* **122,** 273–285.
38. Xu, T., Logsdon, N. J., and Walter, M. R. (2004) Crystallization and X-ray diffraction analysis of insect cell derived IL-22. *Acta Crystallogr. D. Biol. Crystallogr.* **D60,** 1295–1298.
39. Josephson, K., McPherson, D. T., and Walter, M. R. (2001) Purification, crystallization and preliminary X-ray diffraction of a complex between IL-10 and soluble IL-10R1. *Acta Crystallogr. D. Biol. Crystallogr.* **57,** 1908–1911.
40. Strelkov, S. V., Herrmann, H., and Geisler, N. (2001) Divide-and-conquer crystallographic approach towards an atomic structure of intermediate filaments. *J. Mol. Biol.* **306,** 773–781.
41. Barwell, J. A., Bochkarev, A., Pfuetzner, R. A., et al. (1995) Overexpression, purification, and crystallization of the DNA binding and dimerization domains of the Epstein-Barr virus nuclear antigen 1. *J. Biol. Chem.* **270,** 20,556–20,559.
42. Pantazatos, D., Kim, J. S., Klock, H. E., et al. (2004) Rapid refinement of crystallographic protein construct definition employing enhanced hydrogen/deuterium exchange MS. *Proc. Natl. Acad. Sci. USA* **101,** 751–756.
43. Harris, L. J., Larson, S. B., Hasel, K. W., Day, J., Greenwood, A., and McPherson, A. (1992) The three-dimensional structure of an intact monoclonal antibody for canine lymphoma. *Nature* **360,** 369–372.
44. Harris, L. J., Skaletsky, E., and McPherson, A. (1998) Crystallographic structure of an intact IgG1 monoclonal antibody. *J. Mol. Biol.* **275,** 861–872.
45. Kwong, P. D., Wyatt, R., Desjardins, E., et al. (1999) Probability analysis of variational crystallization and its application to gp120, the exterior envelope glycoprotein of type 1 human immunodeficiency virus (HIV-1). *J. Biol. Chem.* **274,** 4115–4123.
46. Grimm, C., Klebe, G., Ficner, R., and Reuter, K. (2000) Screening orthologs as an important variable in crystallization: preliminary X-ray diffraction studies of the tRNA-modifying enzyme S-adenosyl-methionine:tRNA ribosyl transferase/isomerase. *Acta Crystallogr. D. Biol. Crystallogr.* **56,** 484–488.
47. Ruf, W., Stura, E. A., LaPolla, R. J., Syed, T., Edgington, T. S., and Wilson I. A. (1992) Purification, sequence and crystallization of an anti-tissue factor Fab and its use for the crystallization of tissue factor. *J. Cryst. Growth* **122,** 253–264.
48. Stura, E. A., Graille, M. J., and Charbonnier, J. B. (2001) Crystallization of macromolecular complexes: combinatorial complex crystallization. *J. Cryst. Growth* **232,** 573–579.
49. Josephson, K., Jones, B. C., Walter, L. J., DiGiacomo, R., Indelicato, S. R., and Walter, M. R. (2002) Non-competitive antibody neutralization of IL-10 revealed by protein engineering and X-ray crystallography. *Structure* **10,** 981–987.
50. Kwong, P. D., Wyatt, R., Robinson, J., Sweet, R. W., Sodroski, J., and Hendrickson, W. A. (1998) Structure of an HIV gp120 envelope glycoprotein in complex with the CD4 receptor and a neutralizing human antibody. *Nature* **393,** 648–659.

51. Xiang, S. H., Kwong, P. D., Gupta, R., et al. (2002) Mutagenic stabilization and/or disruption of a CD4-bound state reveals distinct conformations of the human immunodeficiency virus type 1 gp120 envelope glycoprotein. *J. Virol.* **76,** 9699–9888.

52. Ferre-D'Amare, A. R. and Burley, S. K. (1994) Use of dynamic light scattering to assess crystallizability of macromolecules and macromolecular assemblies. *Structure* **2,** 357–359.

53. Bard, J., Ercolani, K., Svenson, K., Olland, A., and Somers, W. (2004) Automated systems for protein crystallization. *Methods* **34,** 329–347.

54. Weselak, M., Patch, M. G., Selby, T. L., Knebel, G., and Stevens, R. C. (2003) Robotics for automated crystal formation and analysis. *Methods Enzymol.* **368,** 45–76.

55. Cumbaa, C. A., Lauricella, A., Fehrman, N., et al. (2003) Automatic classification of sub-microlitre protein-crystallization trials in 1536-well plates. *Acta Crystallogr. D. Biol. Crystallogr.* **59,** 1619–1627.

56. DeLucas, L. J., Bray, T. L., Nagy, L., et al. (2003) Efficient protein crystallization. *J. Struct. Biol.* **142,** 188–206.

Index

A

3t3 protocol, 304–5
Adaptors, 18, 22, 49, 125
Agonists, 104, 207
Antagonists, 207
Antibodies, 74, 114, 136, 141, 143, 203,
    205, 209, 233, 245–6, 248–9, 251–2,
    254–6, 260–1, 263–6, 287–290, 313,
    316–7, 330, 343
  monoclonal antibodies (MAb), 203,
      205, 233, 245–9, 254–6, 287,
      289–290, 313, 316–7
    human anti-murine immunoglobulin
        antibodies, 246
  phospho-specific antibodies, 260–261,
      263, 265
  polyclonal antibodies, 233, 246–7,
      252, 287, 290, 313
  single chain antibodies, 203
Apoptosis, 24, 36, 82, 84, 87, 102–5,
    255, 280, 282, 312
  cell death, 87, 154–5, 215
Assays, 24, 91, 110, 113–4,135–6, 140–6,
    210, 216, 221–2, 234, 237, 254–5,
    260–4, 266–8, 288, 290–293, 304,
    307, 312, 315–6. *See also kinases*
  [³H] thymidine incorporation assay, 312
  apoptosis assays, 254–5
  assay for anchorage independent
      growth of cells in soft agar, 291–3,
      304, 307
  colony formation assay, 312, 315–6
  colony-PCR assay, 24
  ELISA, 250–1, 253–4
  FACS analysis, 254
  guanine nucleotide exchange
      assay, 290–1

hanging drop technique, 339–340
immune complex kinase assays,
    263–264, 266–268, 288. *See also
    Kinases*
AKT, 261–262, 266–267
internalization assays, 254
microarray assays. *See microarrays*
RT-PCR assays, 113. *See also PCR*
transformation assays, 304, 307
transfection assays, 110
Atomic force microscopy, 153, 155
ATP, 8–9, 18, 24, 259–262, 267–268,
    278, 288
  ³²P ATP, 8–9, 18, 24, 259–262,
      267–268, 288

B

Bacterial artificial chromosomes (BACs),
    153–161, 181
  BAC transfection, 153–161
    generation of stable lines by BAC
        transfection, 157, 160
  BAC DNA preparation, 156, 158
  BAC cotransfection, 157, 160
Bioinformatics, 81–2, 85
Biological display system, 203
Biopanning, 205–7, 209–210

C

Cancer, 16, 34–6, 67, 69–71, 76, 81–2,
    87, 102, 107, 109–111, 125, 165–6,
    176, 178, 183, 187, 229, 237, 245–9,
    253, 264, 303, 321. *See also cell
    cycle; cell lines; cellular transforma-
    tion; chemoprevention; chemoresis-
    tance; chemotherapy; genes; genome;
    kinase; immunotherapy; microarray;*